Analog Circuit Design

Analog Circuit Design

Operational Amplifiers, Analog to Digital Convertors, Analog Computer Aided Design

Edited by

Johan H. Huijsing
T.U. Delft,
Delft, The Netherlands

Rudy J. van der Plassche
Radio and Data Transmission Systems,
Philips Reasearch,
Eindhoven, The Netherlands

and

Willy Sansen
K.U. Leuven,
Heverlee, Belgium

KLUWER ACADEMIC PUBLISHERS
DORDRECHT / BOSTON / LONDON

ISBN 978-1-4419-5131-1

Published by Kluwer Academic Publishers,
P.O. Box 17, 3300 AA Dordrecht, The Netherlands.

Kluwer Academic Publishers incorporates
the publishing programmes of
D. Reidel, Martinus Nijhoff, Dr W. Junk and MTP Press.

Sold and distributed in the U.S.A. and Canada
by Kluwer Academic Publishers,
101 Philip Drive, Norwell, MA 02061, U.S.A.

In all other countries, sold and distributed
by Kluwer Academic Publishers Group,
P.O. Box 322, 3300 AH Dordrecht, The Netherlands.

Printed on acid-free paper

Contents

ADVANCES IN ANALOG CIRCUIT DESIGN

Operational Amplifiers
Analog to Digital Convertors
Analog Computer Aided Design

Preface

The fragmented coverage of analog circuits and the restriction to a few sessions at international circuit conferences such as ISSCC and ESSCIRC dissatisfied me for a long time. I would rather have seen a conference fully dedicated to analog circuits.

When it occurred to me that possibly the densest concentration of analog designers in Europe can be found around Philips in Eindhoven, The Netherlands, the idea arose to organize an international workshop on Advances in Analog Circuit Design in Europe. I was happy to find Rudy J. van der Plassche of Philips and Willy Sansen of the Katholieke Universiteit Leuven supporting this idea. Together we succesfully organized this workshop.

The main intention of the workshop was to brainstorm with a group of about 50 to 100 analog-design experts on new possibilities and future developments for a few selected topics, and to change this selection year by year. For this first year the three topics that were chosen are:
- Operational Amplifiers
- Analog-to-Digital conversion
- Analog Computer Aided Design.

On each topic six top experts presented tutorials on one day. In this way, a coherent coverage of the field of each topic has been achieved. The tutorial papers on the three topics are now presented in this book.

Kluwer Publishers has decided to yearly publish a volume dedicated to this workshop, containing six tutorial papers on the three topics of the workshop. In this way a valuable series will be built up with a high quality coverage of all important analog areas.

I hope that this first volume will contribute to helping designers of analog circuits.

Johan H. Huijsing
Delft University of Technology.

Operational Amplifiers

Introduction

Many interesting design trends are shown by the six papers on operational amplifiers (Op Amps).

Firstly, there is the line of stand-alone Op Amps using a bipolar IC technology which combines high-frequency and high voltage. This line is represented in papers by Bill Gross and Derek Bowers. Bill Gross shows an improved high-frequency compensation technique of a high quality three stage Op Amp. Derek Bowers improves the gain and frequency behaviour of the stages of a two-stage Op Amp. Both papers also present trends in current-mode feedback Op Amps.

Low-voltage bipolar Op Amp design is presented by Jeroen Fonderie. He shows how multipath nested Miller compensation can be applied to turn rail-to-rail input and output stages into high quality low-voltage Op Amps.

Two papers on CMOS Op Amps by Michael Steyaert and Klaas Bult show how high speed and high gain VLSI building blocks can be realised. Without departing from a single-stage OTA structure with a folded cascode output, a thorough high frequency design technique and a gain-boosting technique contributed to the high-speed and the high-gain achieved with these Op Amps.

Finally, Rinaldo Castello shows us how to provide output power with CMOS buffer amplifiers. The combination of class A and AB stages in a multipath nested Miller structure provides the required linearity and bandwidth.

Johan Huijsing

New High Speed Amplifier Designs, Design Techniques and Layout Problems

William H. Gross

Linear Technology Corporation
Milpitas, California
USA

Abstract

This paper presents a short review of high speed circuit topologies and several new improvements. The described circuits include both voltage and current feedback amplifiers. Complete amplifier circuits for both complementary and standard bipolar processes are summarized along with techniques for individual amplifier stage implementations. The presentation utilizes intuition and experience more than rigorous mathematics. All of the circuits described have been implemented in silicon.

1. Introduction

Making op amps faster has always been one of the goals of op amp designers. The first op amps were basically DC amplifiers; even voice quality audio was impossible because of the low slew rates. The first monolithic op amps were faster and able to pass voice quality audio without a problem. The introduction of feed-forward and integrated P-Channel JFETs improved the speed of monolithic amplifiers by about an order of magnitude. These amplifiers were more than capable of music quality audio, but still far short of being able to pass video signals. In the last decade, complementary bipolar processes have made possible true video op amps. Non-traditional topologies, such as current feedback, have improved the speed of op amps even more. Today there are many different processes that are used to make high speed op amps. The circuits described in this paper apply to any bipolar process; however the ultimate speed of the integrated amplifier will be determined by the cutoff frequency of the transistors and the size of the parasitic capacitors and resistors.

2. Speed versus DC Accuracy

The design of all op amps involves many trade-offs. Some of these are well known, such as amplifier bandwidth and slew rate versus supply current, amplifier bandwidth versus transistor cut-off frequency, and slew rate versus minimum closed loop gain. Experienced op amp designers often trade DC accuracy for speed. It is very important when comparing high speed op amps to look at the DC specs. Degenerating the input stage of an amplifier increases the slew rate, but it decreases the DC gain by the same factor. Wafer sort trimming improves the room temperature specifications; but looking closely, what is improved over temperature? The input bias and offset currents often cause larger errors than the offset voltage in trimmed op amps. Input degeneration increases the amplifier's noise voltage.

The input stages described in this paper do not show any resistive degeneration. The circuits will all slew faster with higher tail currents and input degeneration; in fact, most of these circuits were built with resistive degeneration. Often the amplifiers are optimized for closed loop gains greater than one by reducing the amount of resistive degeneration.

3. Folded Cascode Amplifiers

There are many ways to make an op amp, but all have one thing in common: the input stage must have its common mode input voltage independent of the output voltage. The way the signal is "level shifted" from the input to the output is often used to describe and categorize op amps. The fastest way to get the signal from the input stage to the output is with a "folded cascode". This type of level shift feeds the input current into the emitter of a transistor whose collector goes to the input of the output buffer. There are at least two basic ways of implementing a folded cascode: using two input stages in parallel, coming off each single-ended and summing the signals at the output (Fig.1), or using a balanced circuit with a current mirror to convert to single-ended (Fig.2). Both circuits are very fast because there is only one high impedance node and there is no Miller capacitance associated with the inputs. The two-input circuit is faster because there is no second pole due to the mirror, however in practice the differential circuit is more common because the DC accuracy is much better. High speed op amps with these topologies are usually implemented in complementary processes because the signal is passed by both NPNs and PNPs.

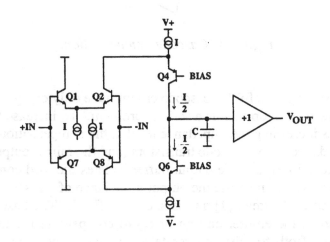

Figure 1: Two input folded cascode.

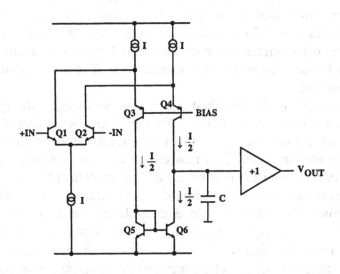

Figure 2: Balanced folded cascode.

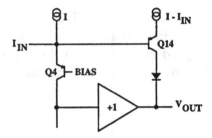

Figure 3: Cascode compensation.

Both versions of the folded cascode circuit have only one gain stage and therefore low open loop gain is a serious problem. To increase the voltage gain requires increasing the impedance level at the compensation node. The output impedance of the cascode transistors (and the mirror output in Fig.2) will limit the gain even if the output buffer isolates the load completely. A very effective way to increase the output impedance of a cascode transistor was described by Cotreau [1] and is shown in Fig.3. The base current of Q_{14} is added to the emitter current of Q_4 to compensate the loss of Q_4's base current. Both transistors operate at the same collector current and have the same base-collector voltage; therefore the two base currents are equal. The circuit has a theoretical mismatch of one divide by beta, but in reality the operating current and transistor beta matching will limit the performance. The circuit will increase the output impedance by ten to twenty times; note that the collector-base capacitance is also cancelled which may enhance AC performance. The disadvantage of this circuit is that it causes offset errors that must be compensated with a similar circuit and it must be done twice for maximum effect. The circuit also increases power dissipation.

Nelson and Feliz [2] described a new way to increase the gain of the balanced folded cascade circuit. Rather than increasing the output impedance of just one transistor, this new circuit increases the impedance at the compensation node. This new circuit does not introduce offset and requires very little power. It does not add to, or cancel the capacitance at the compensation node and therefore does not affect the AC performance.

Fig.4 shows the balanced folded cascade circuit with the new gain enhancement circuit (Q_{Z1}, Q_{Z2} and I_Z). R_Z represents the equivalent resistance at the high impedance node. The circuit operation is based on sensing the collector voltage of a common base transistor by measuring the changes in the voltage at the emitter. The circuit differentially senses the voltage change in the base-emitter voltage of the cascode transistors Q_3 and Q_4. The current that must be inserted into the mirror to compensate for R_Z so that Q_4's collector current does not change is

$$\Delta I = \frac{\Delta V_Z}{R_Z} \tag{1}$$

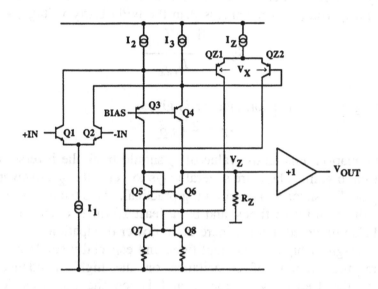

Figure 4: New improved gain enhancement.

The change in voltage V_X, due to the change in Q_4's collector voltage for a constant collector current in Q_3 and Q_4 is

$$\Delta V_X = V_t \frac{\Delta V_Z}{V_{a,Q_4}} \tag{2}$$

where V_{a,Q_4} is the transistor's Early voltage and ΔV_Z is small compared to it. The Q_{Z1}, Q_{Z2} differential current is

$$I_{C,Q_{Z1}} - I_{C,Q_{Z2}} = \frac{\frac{1}{2}I_Z}{V_t} \cdot \Delta V_X = \frac{\frac{1}{2}I_Z}{V_t} \cdot \left(\frac{V_t \cdot \Delta V_Z}{V_{a,Q_4}}\right) = \frac{I_Z}{2V_{a,Q_4}} \cdot \Delta V_Z \tag{3}$$

We have substituted for ΔV_X based on constant collector current in Q_4. Setting

$$\Delta I = I_{C,Q_{ZI}} - I_{C,Q_{Z2}} \tag{4}$$

results in

$$I_Z = \frac{2 \cdot V_{a,Q_4}}{R_Z} \tag{5}$$

For the typical case where the output buffer is not contributing to R_Z and the PNP Early voltage is much less than the NPN Early voltage, we get

$$R_Z = \frac{\beta_{Q_4} \cdot V_{a,Q_4}}{I_{c,Q_4}} \tag{6}$$

Substituting this into the formula for I_Z gives

$$I_Z = 2 \cdot I_{B,Q_4} \tag{7}$$

This last equation tells us the value of I_Z should track the inverse of PNP β for optimum gain. There are several ways to generate I_Z, however using a PNP "pinch" resistor makes trimming practical. This technique typically increases the gain by 25 times, and by at least 10 times over temperature. It should also be noted that the increase in power dissipation is very small.

In the design of high speed amplifiers, and especially single gain stage amplifiers, the amount of capacitance on the high impedance node determines the frequency response, and hence the compensation and stability of the amplifier. Therefore extra care and attention in mask design are required.

High speed amplifier design requires using a circuit simulator such as SPICE. SPICE models the parasitic capacitors associated with transistors very well; when a transistor is stretched or modified in mask design, it is a simple matter to evaluate the effects by modifying the models. Unfortunately the parasitic capacitance associated with the metal traces can be larger than that of the transistors. For this reason the parasitic capacitance due to the metal must be added as a separate component in the simulation. It is usually only necessary to add these capacitors to the high impedance nodes. It is important when characterizing the IC process to use actual measured data of the metal capacitance. The best way is to have a test pattern that allows the area and perimeter contributions to be evaluated as independently as possible. Due to fringing effects, the perimeter of a narrow trace dominates the total capacitance of that trace. It is also important to evaluate fully passivated parts rather that die that have only been processed through metal. The passivation can increase the capacitance of a trace by 20% or more. When debugging a circuit on the probe station, it is useful to estimate the effects of passivation and still have the benefits of exposed metal. Coat the die with vegetable oil to estimate the additional capacitance due to passivation.

4. Three Stage Amplifiers with Feed Forward

The three stage amplifier is widely used because it results in excellent DC performance and very high gain. Another advantage of the three stage amplifier topology is that the slow PNP level shift transistors found in conventional non-complementary processes are all in one stage and it is easy to use feed-forward to feed the signal around them. Fig.5 shows the traditional topology.

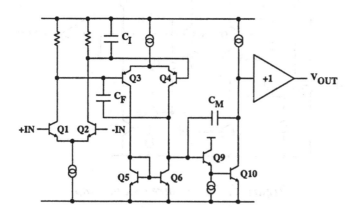

Figure 5: Traditional three stage amp.

Much has been written about this topology and its difficulties, but in a nutshell it has three AC problems. The first is the pole caused by C_I that is required to make the drive to the second stage single ended; this pole causes a doublet that slows down settling. The second problem is that the signal goes through the mirror of the second stage and therefore it is not as broad-band as it could be; this requires that the feed-forward pole be at a lower frequency and that slows down settling. The last problem is that the feed-forward doublet causes settling problems of its own.

The circuit of Fig.6, first described by Wright [3], eliminates the above three problems. The output of the first stage is single ended to the second stage. There is no connection to the collector of Q_2 and therefore there is no need for an input capacitor. In order to preserve the balanced second stage and it's inherently low offset, the reference input to the second stage is generated with Q_{11} and its load resistor. This voltage is identical with the voltage at the collector of Q_1, except there is no signal at the collector of Q_{11}. Operating the PNPs in a common base configuration eliminates the second problem. There is no signal in Q_7, Q_3, Q_5 or Q_6. The speed of the second stage is limited only by the PNP Q_4 (assuming NPN Q_8 is much faster). A doublet is formed by the feed-forward capacitor C_F and the cutoff frequency of Q_4. The feed-forward doublet is inside the Miller integrator feedback loop formed by C_M and the gain stage Q_9 and Q_{10}. This solves the third problem by reducing the effect of the doublet by the loop gain of the Miller integrator.

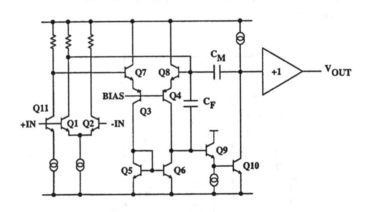

Figure 6: Faster three stage amp.

This circuit has been implemented in an 800 MHz, 15 V process with 15 MHz lateral PNPs. The unity gain stable amplifier has a 90 MHz gain bandwidth product and 45 degrees of phase margin. The voltage gain is 45,000 and the settling time to 1 mV is 160 ns.

5. Two Stage Miller Integrator Amplifiers with Feed-Forward

The most common op amp topology uses two stages where the second is a Miller integrator. Fig.7 shows how feed-forward can even improve this topology. The input PNP differential amp (Q_1 and Q_2) does not limit the bandwidth of this amplifier when implemented in a complementary process. The amplifier bandwidth is limited by the speed of the Miller integrator formed with C_{M2} and Q_5, Q_6 and Q_7. The two emitter followers Q_5 and Q_6

increase the gain of the circuit to over a million, but they also add interstage poles. The addition of R_F and C_F fixes the problem. Now the integrator is stable and has no "second stage bump" as described by Solomon [4]. This technique is especially useful when the impedance levels of the emitter followers are high because they are running at low currents.

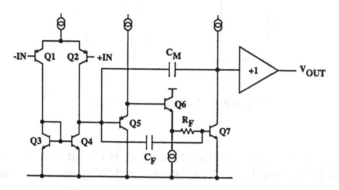

Figure 7: Two stage with feed forward.

6. Current Feedback Amplifiers

Current feedback amplifiers are popular because of their inherent AC advantages: excellent large signal performance and bandwidth that is not a function of closed loop gain. Current feedback amplifiers have a low impedance inverting input that follows the non-inverting input. The feedback is the current that flows in the inverting input. Because the current flows both into and out of the inverting input, fast NPNs and PNPs make the job a lot easier. The most critical blocks in the current feedback topology are the current mirrors that take the feedback current and turn it around into the high impedance node.

The speed of the mirrors determines the speed of the amplifier. There are two AC criteria to use when evaluating current mirrors for current feedback amplifiers. The small signal response determines the bandwidth of the amplifier and the large signal response determines the output slew rate and settling time. One way to evaluate the small signal performance is to note the frequency where a given amount of phase shift occurs, usually 30 degrees or so. The large signal performance is more difficult to quantify, but the maximum current that the mirror can handle before any transistors saturate is a good starting point.

12

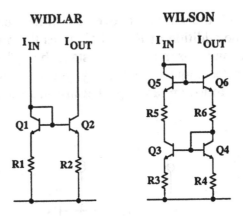

Figure 8: Standard mirrors.

Comparing the response of several mirrors is the first step in designing a current feedback amplifier. The Widlar and Wilson mirrors are shown in Fig.8. The R_b, C_{in} and C_{ob} of the transistors determine the small signal response of the Widlar mirror. The amount of current the diode connected transistor can handle before it saturates determines the large signal response. The most popular current mirror in current feedback amplifiers is the Wilson mirror because of its excellent DC performance. Q_5 and R_5 are optional to improve DC accuracy. R_6 reduces the amount of feedback and makes the Wilson more stable. Without R_6 or capacitance on the collector of Q_3 there usually is peaking in the small signal response. The peaking will cause gain margin problems in the overall amplifier. The emitter resistors in both mirrors improve the large signal behavior and the DC matching.

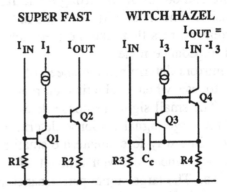

Figure 9: Non-saturating mirrors.

Two simple mirrors that do not have saturation problems are shown in Fig.9; these mirrors are not as DC accurate as the first two. The first circuit is the fastest mirror. Only the speed of Q_2 (R_b, C_{in} and C_{ob}) limits the small signal response since the emitter follower presents a low impedance drive to the base. The "Witch Hazel"[1] mirror has higher output impedance and has long been used to improve high current lateral PNP current sources. The small capacitor eliminates peaking. The large signal response of these mirrors is excellent since it is not limited by the saturation of any diode connected transistors. The turn off response time is enhanced because the emitter follower actively pulls base current out of the output transistor.

Conditions	Widlar	Wilson	Super Fast	Witch Hazel
0.1mA, 30° phase shift	31MHz	23MHz	57MHz	29MHz
0.25mA, 30° phase shift	45MHz	30MHz	67MHz	41MHz
0.5mA, 30° phase shift	53MHz	34MHz	74MHz	59MHz
Maximum Current	1.5mA	1.5mA	N/A	N/A

Table 1: Comparison of various current mirror circuits.

Table 1 shows the AC data for these mirrors when implemented in a 36 V, 600 MHz complementary process. The emitter degeneration resistors were fixed at 400 Ω; the bias current sources in the Super Fast and Witch Hazel mirrors were set to 100 μA. All the mirrors were compensated for flat amplitude response while driving a capacitive load of 5 pF.

The availability of complementary bipolar processes made monolithic current feedback amplifiers practical. The first monolithic current feedback amplifiers were basically implementations of discrete designs; they did not take advantage of the inherent matching of the NPNs and PNPs of a complementary process. The circuit of Fig.10 shows a very simple and very fast current feedback amplifier.

The circuit uses the simplest input stage possible, just four transistors, and the super fast current mirrors. What sets this circuit apart from others

---

[1]The origins of the name "Witch Hazel" are lost. It was common usage in the early 1970s at Motorola.

is the unusual biasing scheme. A floating current source feeds a diode connected NPN (Q_1) whose base-emitter voltage is impressed upon a PNP base-emitter (Q_2). The collector current of the PNP is therefore the current source times the ratio of the PNP beta divided by the NPN beta. The input stage current will vary plus or minus 50% with processing.

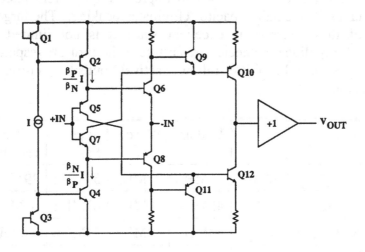

Figure 10: Very fast current feedback amplifier.

The advantage of this biasing scheme becomes apparent when we notice that the V_{be} of PNP Q_5 is now the same as the V_{be} of NPN Q_1. This means that the V_{be} of PNP Q_5 is going to match the V_{be} of NPN Q_6 and the circuit will have low input offset voltage. Notice that the same is true of Q_3, Q_4, Q_7 and Q_8. In practice this improves the input offset by a factor of three over the same circuit with constant currents feeding Q_5 and Q_7. The collector currents of Q_5 and Q_7 are then used in the current mirrors. Again the V_{be} of Q_{11} is now set by an NPN (Q_1) and it therefore matches the V_{be} of Q_{12}. This improves the DC accuracy of the mirror by six times compared with feeding Q_9 and Q_{11} with constant currents. Significantly improved accuracy can be obtained by cascoding Q_2 and Q_4 at the expense of input common mode range. This circuit, without the cascodes, has been implemented in a 36 V, 600 MHz complementary process. The resulting amplifier operates on supplies from ±2 V to ±15 V and has a bandwidth of 140 MHz with an output slew rate of 1100 V/µs. The circuit is fast because it is simple. It is accurate because of the unusual biasing.

7. Output Buffers

Almost all op amps have output buffers that isolate the load from a high impedance node. High speed op amps are no exception and they usually have to drive larger output currents than slower op amps. This is because high speed op amps are often used to drive cables and other low impedance loads as well as lower values of feedback resistors. All the high speed output buffers are based on the simple emitter follower. The fastest and most interesting output buffers use complementary NPNs and PNPs.

Fig.11 shows the basic complementary emitter follower output stage and the widely used cascaded complementary emitter follower stage. The basic stage isolates the load from the high impedance node with the current gain of the NPN *or* the PNP, depending on which is conducting the most current. At high frequencies, the gain is determined by the cutoff frequency of the transistors. To minimize second harmonic distortion requires the gain (and therefore the cutoff frequency) of the NPN and the PNP to match reasonably well. Because the basic stage has only one beta of isolation, it is very fast and well behaved, even into capacitive loads.

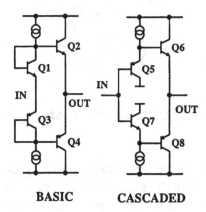

BASIC   CASCADED

Figure 11: The basic buffers.

The cascaded stage is very popular because it isolates the load with an NPN times a PNP current gain when the output is both sinking and sourcing current. This improves the second harmonic distortion and of course the additional isolation increases the amplifier's gain. This output stage is quite well behaved and usually does not require any additional components to stabilize, even with capacitive loads.

Fig.12 shows a modification to the cascaded stage[2]. This buffer has the same isolation as the cascaded stage, but requires less quiescent current thanks to Q_6 and Q_{10}. There is an important improvement in this circuit compared to the previous circuit where the outputs (Q_2 and Q_4) were made into darlingtons. Q_1 and Q_3 pull base current out of the output transistors to prevent them from destroying themselves during high frequency, large signal conditions. This output stage is well behaved, but it will peak more than the previous two buffers. The resistors in series with the output transistor bases slow the output stage a little, but improve the ability to drive capacitive loads a lot.

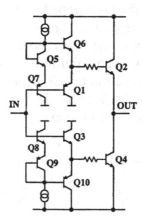

Figure 12: Modified cascade buffer.

There are times when more load isolation is required. In order to drive lower impedance loads with the same accuracy, or to improve accuracy into standard loads, requires raising the impedance at the input of the buffer. Fig.13 shows a buffer with the current gain of three transistors isolating the load. Again the resistors improve the circuit's ability to drive capacitive loads. The diode connected transistors Q_{14} and Q_{19} must be large enough to remove the output transistors' base current during large signal high frequency operation without saturating or excessive power dissipation will result.

---

[2]First used in the EL2002 IC.

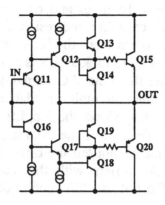

Figure 13: Triple buffer.

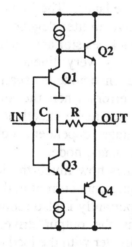

Figure 14: The fix for capacitive loads.

Recently several high speed amplifiers have been introduced that are stable with any capacitive load. Fig.14 shows the way most of these amplifiers accomplish this amazing feat. The capacitor and resistor are boot-strapped when there is a light load on the output, so with a light load, they have no effect on the amplifier's frequency response. When a capacitive load is applied, current flows in the output stage to drive it. This causes a difference between the input and output of the buffer and that voltage is forced across the capacitor and resistor. The current flowing in the capacitor and resistor comes from the high impedance node and it is therefore loaded. Another way to look at this is to note that a large enough output capacitance will result in no AC signal at the output; then the capacitor and resistor will load the high impedance load directly adding a pole and a zero. This loading slows the amplifier, reduces its gain and broad-bands the high impedance node making the amplifier stable. Unfortunately this technique cannot tell the difference between a resistive and a capacitive load. Amplifiers with this circuit do not drive low impedance loads well. Do not use this technique in video cable driver circuits.

Composite video amplifiers must keep their frequency response as constant as possible when the output DC level changes. The parameters that measure this performance are called differential gain and differential phase. The main causes of differential gain and phase errors are: non-linear input stage transconductance, the voltage coefficient of junction capacitors, and changes in the output impedance. The input stage transconductance is usually linearized with resistive emitter degeneration. In current feedback amplifiers the external feedback resistor acts as the transconductance stage and therefore they are usually very linear. The change in junction capacitance on the high impedance node is usually the major contributor to differential gain and phase errors when the amplifier is lightly loaded. Dielectrically isolated processes make excellent video amplifiers because they do not have the voltage dependent collector-substrate junction capacitance on the high impedance node.

The output stage determines how much the load affects the frequency response. Running a lot of quiescent current in the output stage reduces the output impedance; more importantly it also reduces the amount of change in the output impedance as the output drives the load. The output impedance forms a voltage divider with the load resistor and therefore must be constant for the gain to remain constant. This change in output impedance is a major cause of differential gain and phase errors in video cable driver amplifiers.

8. Current Sources and Cascodes

There probably are an infinite number of ways to make current sources for op amps. In high speed op amps, the current sources and cascode transistors slew quickly and should not generate any problems. When the collector of a transistor slews, the current that charges and discharges the collector-base capacitance must flow out of the base.

This base current is often as large as the collector current and the normal guidelines used to design for DC performance will not necessarily deliver good AC performance.

The AC performance of the simple Widlar mirror (Fig.15) with a constant input current is quite good. When the collector of Q_2 slews, the additional base current is mirrored and shows up as collector current. The amount of additional current Q_1 can take before it saturates limits the positive slew rate. I_{IN} limits the negative slew rate; when the slew induced base current is equal to it, the circuit turns off. Exceeding the slew rate the circuit can handle causes recovery delays and degrades the settling time of the amplifier. The input current and the current gain of the mirror should be optimized for the slew rate required at the collector. For symmetrical slew rate limits, Q_1 should not saturate at twice I_{IN}. A disadvantage of this simple circuit is its output capacitance. The collector-base capacitance of Q_2 is multiplied by the current gain of the mirror. Replacing Q_1 and R_1 with three series connected Schottky diodes makes for an interesting variation of this current source. The Schottky diodes reduce the gain of the mirror and therefore lower the output capacitance. There is no limit on the positive slew rate because Schottky diodes do not saturate. The negative slew rate limit is unchanged.

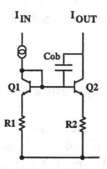

Figure 15: Current source dynamics.

Biasing the current source transistor base with a very low impedance that can sink and source a lot of current results in the highest slew rates and the lowest power dissipation. The low impedance on the base eliminates the multiplication of the collector-base capacitance and therefore minimizes the capacitance that has to be slewed. Fig.16 shows an excellent implementation of this concept. Q_5 and Q_6 form a class AB output stage that drives Q_7. They can be operated at a lower quiescent current than the slew induced base current of Q_7 and still the base voltage will not move much.

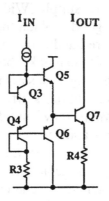

Figure 16: Biasing current source.

Q_6 must be large enough not to saturate during positive slews. Additional capacitance on the base of Q_5 (and Q_6) reduces the amount of input current required; the collector-substrate capacitance of Q_3 (and Q_4) is an excellent source. Of course cascode transistors benefit from this class AB bias technique as well.

9. Power Dissipation and Temperature Coefficients

The biggest influence on the slew rate of high speed amplifiers is the amount of current available. If unlimited supply current were available, amplifiers would go faster. In most ICs, package dissipation limits the maximum supply current a designer can use without exceeding the maximum reliable junction temperature. This is especially true when the package of choice is the SO-8! Designing for a lower supply voltage is an obvious way to get more current without exceeding the package limitations. The junction capacitors get larger (and more non-linear) as the supply voltages drop; however the increase in current available usually will more than make up for it. Lower supplies also reduce the dynamic range of the amplifier; to make up for this, the circuit design becomes more difficult, especially when cascades are required.

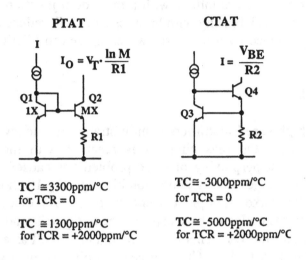

Figure 17: Typical bias current generator.

When considering power dissipation, it is important to pay attention to the temperature coefficient of the currents in the amplifier. Traditional op amp designs use "ΔV_{be} over R" or PTAT[3] biasing. This minimizes the offset drift of the amplifier. Unfortunately it also increases the power dissipation of the amplifier at high temperature. In most linear IC processes, the collector resistance of the transistors determines the maximum current they can carry. With PTAT biasing, the transistors are at a double disadvantage; the current is maximized when the transistor's R_C is maximized. In order for the amplifier to perform well hot, the transistors must be sized large enough not to saturate. The larger the transistor, the more capacitance it has and the slower the circuit.

The obvious way to minimize the transistor's size is to operate the transistor with a current that keeps it from saturating. This requires that the current reduce with temperature since the collector resistance increases with temperature. The popular "V_{be} over R" or CTAT[4] biasing scheme is a good way to minimize transistor size. Fig.17 shows simple PTAT and

---

[3]PTAT is an acronym for Proportional To Absolute Temperature.

[4]CTAT is an acronym for Complementary To Absolute Temperature.

CTAT biasing circuits and gives the approximate temperature coefficients for each with both thin film (TCR = 0 ppm/°C) and base (TCR = 2000 ppm/°C) resistors. In an amplifier with a maximum junction temperature of 150 °C, using CTAT biasing with base resistors will reduce the required transistor's capacitance by a factor of two compared to PTAT biasing.

10. Cross Talk

Very often high speed amplifiers exhibit strange frequency response or oscillations because the bias circuitry is reacting with the AC signal circuitry. The most perplexing of these problems is caused by coupling between amplifier stages through common biasing. Fig.18 shows the most common way to make multiple current sources. As discussed in the previous section, the collector-base capacitance of Q_2 will send a signal into Q_1 that will mirror into Q_3! This is a small signal as well as a large signal problem; V_1 will talk to V_2. This can cause problems even when a very low impedance is driving the bases of Q_2 and Q_3.

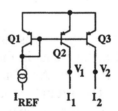

Figure 18: Common bias.

In the beginning phases of a high speed amplifier design, use completely separate bias generators for each stage of the amplifier. Keep the bias generators separate if the total power dissipation and die size are not severely impacted by this "additional" overhead. When combining bias circuitry, be sure to keep the output stage separate from the input. It is very hard to "debug" slow settling time when it is caused by the output slew affecting the input bias circuitry.

A last note of caution: dual and quad amplifiers that share common bias circuitry will almost certainly talk to each other. A pair of 100 MHz amplifiers in one package is enough trouble; do not compound it with common bias problems.

One of the most unexpected cross talk problems is caused by the mutual inductance between wire bonds in the package. When current flows in a wire bond, it magnetically couples to the other wire bonds; the result is a voltage source in series with each of the bonds. The wire bonds that carry the output current are the sources of this problem. The amount of voltage generated in the other bonds is proportional to the current in the output. Fig.19 shows the crosstalk from the output of one amplifier to the input of the other amplifier for a dual 120 MHz current feedback amplifier IC. The data is for the die in a 8 pin DIP package with the standard dual op amp pin-out. The data was taken with both 100 ohms and 1 kohm loads to show the output current dependence.

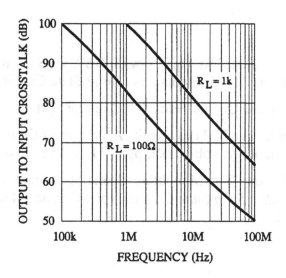

Figure 19: Output to input cross talk.

In application specific circuits where the pin-out is not defined by history, the differential inputs should be next to each other with parallel wire bonds. The output and supplies should be on the other side of the die with wire bonds parallel to the inputs. This will minimize the crosstalk.

11. Summary

High speed amplifier development is an exciting area of monolithic circuit design because the successful circuit techniques are the ones that are totally interwoven with the new process capabilities. The next big improvement in high speed amplifiers will involve not just new circuits, not just new processes, but the synergy that comes from both. The general purpose and precision amplifier design techniques already learned are the necessary foundation to build on for high speed amplifier design. The knowledge of good DC circuit design is important in high speed amplifier design because doing more with less (transistors) is the way to faster amplifiers. The 1970s saw enormous growth and development of general purpose op amps, the 1980s were the decade of precision op amps and the 1990s will be the decade of the high speed op amps. The fun is just beginning.

References

[1]   Gerald M. Cotreau, "An Opamp with 375 V/μs Slew Rate, ±100 mA Output Current" 1985 International Solid State Circuits Conference.

[2]   Carl Nelson and George Feliz, "LTC internal patent application".

[3]   John Wright, "The LT1190 Family, A Product of Design Innovation," Linear Technology, vol.1, no.2, pp.3, 11, 12, 15 October 1991.

[4]   James Solomon, "The Monolithic Operational Amplifier: A Tutorial Study," *IEEE J. Solid-State Circuits*, vol. SC-9, no.6, December 1974, pp.314-332.

Biography

Bill Gross is a Design Manager for Linear Technology Corporation, heading a team of design engineers developing high speed amplifier products. Mr. Gross has been designing integrated circuits for the semiconductor industry for 20 years, first at National Semiconductor, including three years of living and working in Japan, and later at Elantec. He has a B.S.E.E. from California State Polytechnic University at Pomona and an M.S.E.E. from the University of Arizona at Tucson. Mr. Gross is married and the father of two teen-age sons, whose sports activities keep him quite busy.

The Impact of New Architectures on the Ubiquitous Operational Amplifier

D. F. Bowers

Analog Devices Inc. (PMI Division)
Santa Clara, California
USA

Abstract

The operational amplifier has steadily evolved since its inception almost fifty years ago. Sometimes though, radically new architectures appear to fill market requirements. Also, the availability of new I.C. processes such as BiCMOS and Complementary Bipolar yield components which have not previously been available to the op-amp designer. This paper is a review of some recent and not so recent developments.

1. Introduction

The term "operational amplifier" was first used in a paper by Ragazzini et al. in May 1947 [1], though op-amps were (at least) in some use during the Second World War years; notably by Lovell and Parkinson in the development of the Western Electric (Mk.9) electronic anti-aircraft gun director. Monolithic op-amps emerged in the mid 1960's, notably those designed by Bob Widlar at Fairchild. By 1968 he had designed the LM101 [2] (at National Semiconductor), and this product set a standard topology for many future monolithic op-amps.

Much of the progress in monolithic op-amp performance has since then been evolutionary in nature. Process improvements, such as complementary bipolar processing and the bipolar-compatible JFET have definitely led to better op-amps. Design tricks such as feedforward compensation, slew-enhancement and all-NPN output stages have also been great contributors. This paper looks at some of the more radical departures from mainstream op-amp design.

2. The Precision Op-Amp

One of the first departures from the LM101 type "two-stage" topology was the µA725 designed by George Erdi at Fairchild in 1968 [3]. Several new concepts were added to this design to improve the precision performance. First, three gain stages were used to improve the overall open-loop gain. This also enabled a relatively low gain resistively-loaded input stage to be used, which can be shown to give the best offset and drift performance. Second, it was realised that gain can be limited by thermal feedback from the output stage to preceding stages. This was overcome by using a layout which was "thermally symmetric" with respect to critical transistor pairs. Third, a "cross-coupled quad" input stage (Fig.1) was used to improve the offset voltage.

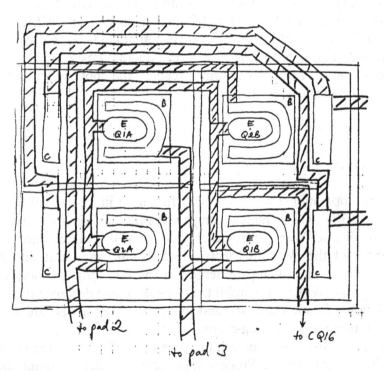

Figure 1: Cross-coupled quad technique (this is George Erdi's original hand sketch for the OP-27 input stage).

The theory here is that this arrangement substantially cancels process and thermal gradients across the chip by providing a common-centroid for the input stage pair. The real breakthrough in improving offset, however, came with the development of offset voltage trimming. In the late 1960's it was discovered (at Fairchild) that if the base-emitter junction of a bipolar transistor was avalanched at high current, then a permanent short would result. It was realised at Intersil and Precision Monolithics in the mid 1970's that such an effect could provide on-chip adjustment for integrated circuits. The natural first use of this technique was to trim the offset voltage of precision op-amps, and it became known as "zener-zapping".

"Zener-zapping" was applied to the PMI OP-07 in 1975 [4], where small zeners were used to short out portions of the input stage load resistors. This resulted in an offset voltage of less than 100 μV. Meanwhile, Analog Devices had been working along different lines.

Since 1973 they had been trimming thin film resistors by means of a laser, but these trims had been performed after devices were packaged (but not capped). In 1975 they adapted a Teradyne W311 laser trim system to perform trims at wafer level. This enabled them to compete in the OP-07 market place, and they came out with the AD-510 op-amp in 1976 with an offset specification of 25 μV maximum.

A few years ago our marketing department decided that the marketplace needed a quad precision op-amp with the same specifications as the OP-07, except that they wanted ten times the gain and the same power consumption for all four amplifiers as a single OP-07. This obviously demands a different approach.

First, in the interests of power consumption and die area, I decided that one of the gain stages had to go, which left two gain stages to provide the target open-loop gain of five million. Fig.2 shows the simplified schematic of what eventually became the OP-400.

Superbeta transistors Q_3 and Q_4 are used in the input stage to achieve a low input bias current. These are cascoded by a voltage limiting network which is merged with the input protection devices to reduce the die size. An active load Q_1 and Q_2 provides a very high stage gain ($\approx$ 10,000), and is regulated via a feedback loop consisting of Q_5 and the second stage transistors. Normally, to avoid degrading the first stage gain, emitter followers would be used to buffer the second stage. These, of course increase power consumption and die size. If a folded-cascode is used as the second stage, transistors Q_6 and Q_7 see very little collector-base voltage, and can be made superbeta types also, removing the need for the followers. Finally, a triple-buffered output stage is used to maintain the gain with resistive loads on the amplifier. This architecture meets all the specifications of an OP-07, but with a gain of 10,000,000 and a supply current of 500 µA per amplifier.

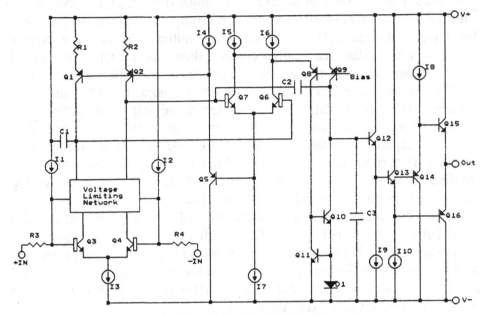

Figure 2: Simplified OP-400 schematic.

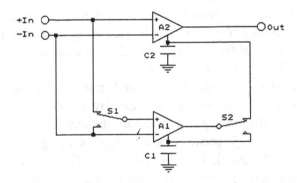

Figure 3: Chopper stabilised op-amp scheme.

The modern precision op-amp has errors almost as low as basic transistor physics will allow, and there does not seem to be room for any real break-throughs in this area. One area where the precision bipolar op-amp has seen some competition is from chopper stabilised amplifiers. The most used configuration is shown in Fig.3, and this evolved from the Harris HA2900 introduced in 1973. Amplifier A_1 spends half the time nulling its own offset, caused by shorting the inputs together with an MOS switch (S_1) and applying feedback to an internal amplifier point. When S_2 is opened, C_1 stores the amount of feedback necessary to produce the null, and the amplifier is theoretically offset free. For the other half of its time, A_1 nulls the main amplifier, A_2. Capacitor C_2 stores the voltage needed to null the main amplifier when A_1 is nulling itself. A_2 is thus continuously available for use in normal op-amp configurations. It is normal to integrate such amplifiers on a CMOS process, since the switches are available and MOS inputs can be used to obtain low input currents. Early versions used external storage capacitors and ran at frequencies of a hundred Hertz or so, but the modern trend is to provide internal capacitors with much higher chopping frequencies; 10 kHz being not uncommon.

Bipolar op-amp designers (such as myself) love to point out all the problems associated with chopper stabilisation; input current glitches, intermodulation distortion, overload recovery problems et cetera, but nevertheless these amplifiers do serve a large portion of the market.

3. Improving Speed

High speed amplifiers are the biggest current growth area of the op-amp marketplace. However as Bob Widlar once pointed out [5], "No one can dispute that zero offset, no input current, infinite gain, and the like are desirable characteristics. However, a little practical experience quickly demonstrates that infinite bandwidth is not an unmixed blessing". We have some customers that use our micropower op-amps *because* they are slow, giving reduced high frequency noise and good tolerance to stray capacitance. But the whole electronic world seems to be moving toward higher speed, and op-amps are definitely no exception.

Obviously, the basic concept for obtaining high bandwidth in an op-amp is to minimise the phase delay through the amplifier. Increasing the stage currents is a fairly straightforward method of reducing the phase delay, but unfortunately amplifiers such as the LM101 which use lateral PNP transistors for level shifting cannot be made faster than about 2 MHz, because of the large transit time of these devices. Two techniques are currently used to circumvent this problem; one is feedforward compensation, the other is to include a faster PNP transistor.

Feedforward compensation involves bypassing the lateral PNP transistor at high frequencies by means of a capacitor. Fig.4 shows the most used technique for achieving this, as found on products such as the LM118, OP-27 and LT1028. C_1 is the dominant pole compensation capacitor, while C_2 provides a bypass function around the entire second stage. C_3 serves to reduce "peak-back" of the output stage at high frequencies.

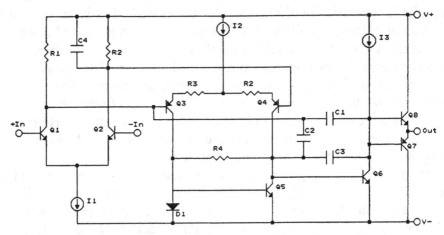

Figure 4: Feedforward compensation.

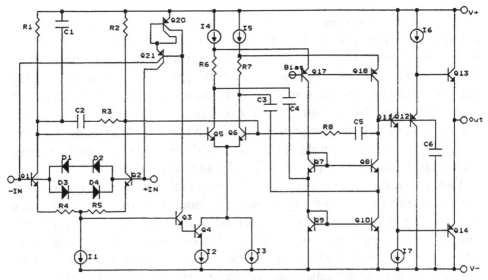

Figure 5: Simplified OP-470 schematic.

Unfortunately, this type of compensation produces a pole-zero pair in the open-loop response. This leads to excessive settling times, and some ingenious solutions to this problem have been proposed [6]. However, this pole-zero doublet can be avoided altogether by using charge conservation techniques.

Fig.5 shows the feedforward technique used on the OP-470 series of amplifiers. Following the input stage, Q_5 and Q_6 comprise the second stage input pair driving a folded cascode pair, Q_{17} and Q_{18}. Two capacitors, C_3 and C_4 provide a bypass function, but they feed into low impedance points created by the load transistors Q_8 and Q_9. At low frequencies, most of the signal currents in the input pair pass through Q_{17} and Q_{18}, but at high frequencies the signal path is through C_3 and C_4. The point is that at all times the difference in collector currents of Q_5 and Q_6 somehow gets to the collector of Q_8, even if the internal nodes have not settled. Using this scheme, settling tails at the 12-bit level are not apparent.

Several manufacturers have decided to explore the alternative route of including a high-speed PNP transistor on their process. Radiation Incorporated, now part of Harris Semiconductor, were the pioneers in this respect with their dielectric isolation process. Amplifiers from Harris generally used a single gain stage with a foldback cascode load, and this has set a general trend for amplifiers where highest possible bandwidth is needed. Today, many complementary bipolar processes are in use, falling basically into two categories. Those intended for 36 V operation have F_t's of up to 600 MHz, with lower voltage processes often having F_t's of several GHz. Using such processes, amplifiers with several hundred megahertz of bandwidth are becoming common.

4. Large Signal Performance

Even when a high small signal bandwidth is achieved, a further speed limitation is the slew rate of the amplifier. This can cause drastic limitations on the large signal bandwidth and settling time of an op-amp. For an amplifier with a simple bipolar input stage and dominant pole compensation, it is easy to show that the slew rate is only around 1 V/µs for each 3 MHz of small signal bandwidth. This would mean, for example that a 100 MHz amplifier driving 5 V r.m.s. would have a full-power bandwidth of less than 1 MHz. Traditional methods of improving the slew rate have concentrated on reducing the input stage transconductance so that smaller compensation capacitors can be used [7]. This unfortunately reduces DC accuracy and increases noise. Historically there have been some new architectures devised to improve this situation, one notable example being the class AB input stage of W. Hearn [8], which eventually turned into the Signetics NE531. More recently, some other interesting techniques have been devised.

Fig.6 shows an ingenious example using complementary technology [9].

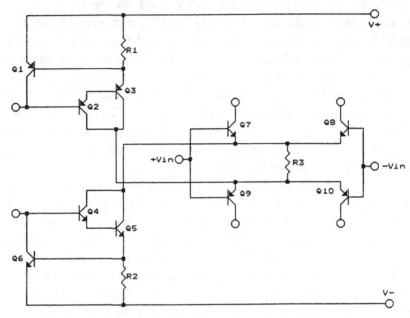

Figure 6: Slew-enhanced complementary input stage.

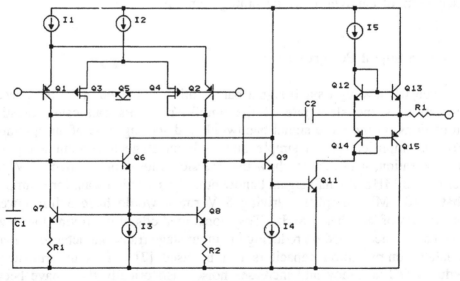

Figure 7: Simplified OP-275 schematic.

Q_7 -Q_{10} are used as a double input stage (actually only one needs to be used to convey the signal, the output from the other being discarded) biased to a V_{be} proportional current by the remaining transistors. R_3 is scaled to rob some of the biasing current, and partly controls the small signal bandwidth. With large differential inputs, R_3 increases the total currents in both pairs, greatly improving the slew rate. Another surprisingly simple technique is reflected in the simplified schematic of the new OP-275, designed by Jim Butler at PMI (Fig.7). The input stage consists of both a JFET and a bipolar transistor pair in parallel. The tail currents are ratioed 8:1, resulting in both pairs having roughly equal transconductance.

The overall transconductance versus differential input voltage is shown in Fig.8. It can be seen that for small inputs the transconductance is high, but it is still finite with large differentials. This enables the amplifier to have the high slew rate associated with a FET input amplifier, combined with noise and precision approaching that of a purely bipolar design. If ultra high slew rate is needed, it is difficult to beat the class of amplifiers known as *current feedback*.

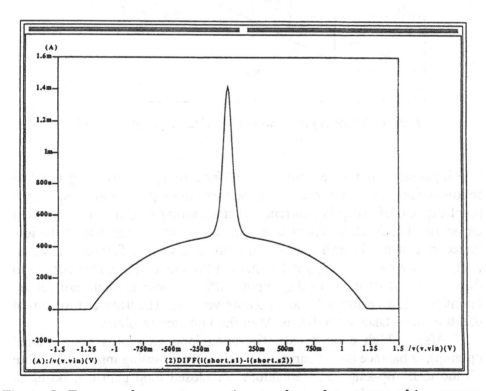

Figure 8: Transconductance versus input voltage for compound input stage.

5. Current-Feedback Amplifiers

Current feedback op-amps differ from voltage feedback types in that the current supplied to the compensation capacitor is equal to the current flowing in the inverting input. Fig.9 shows the simplified current feedback op-amp model.

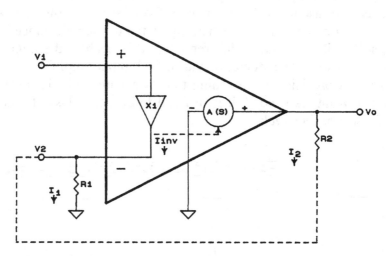

Figure 9: Conceptual current feedback op-amp model.

The input stage is now a unity-gain buffer forcing the inverting input to follow the non-inverting one. Thus, the inverting input is a low impedance point capable of accepting current and transferring it to the compensation capacitor. Feedback is always treated as a current and because of the low impedance at the inverting terminal, there is always a feedback resistor, even at unity gain. Voltage imbalances at the inputs thus cause current to flow in or out of the inverting input buffer. These currents are sensed internally and transformed into an output voltage. The transfer function of this transimpedance amplifier is A(s); the units are in ohms.

If A(s) is high enough (like the open loop gain of a conventional op-amp), at balance no net current flows into the inverting input. Therefore all the usual op-amp static equations for closed loop gain can be used. However, the dominant pole now has its time constant set by the product of the *feedback resistor* and an internal compensation capacitor value. If a fixed feedback resistor is used, the closed loop bandwidth is therefore theoretically independent of gain. In practice finite output impedance in the buffer reduces the bandwidth at very high gains, but it is still a considerable improvement over the conventional op-amp.

The other big advantage of the current feedback architecture lies in its large signal characteristics. No matter how large the input step becomes, the buffer theoretically follows it. This forces the difference between the input and output voltages directly across the feedback resistor. Thus the current available for slewing is proportional to the error voltage, which is exactly the same as saying that the large-signal and small-signal bandwidths are identical. In practice, a variety of second order effects does give a finite slew rate limit, but it is often thousands of Volts per microsecond.

Fig.10 shows a method for implementing such an amplifier, and in fact the majority of commercially available current feedback op-amps are variations on this topology.

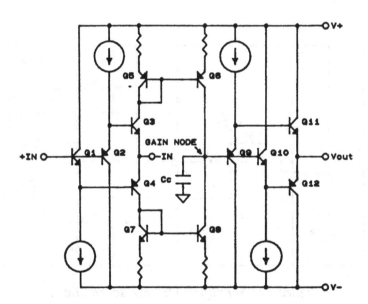

Figure 10: Basic implementation of the current feedback op-amp.

Q_1-Q_4 form the input buffer which forces the inverting terminal to the potential of the non-inverting one. Any imbalances in the collector currents of Q_3 and Q_4 are summed at the gain node, via current mirrors Q_5-Q_8, causing the output buffer, Q_9-Q_{12} to move. Negative feedback via the feedback resistor corrects the imbalance thus forcing a constant error current (ideally zero) into the inverting input. One problem with this implementation is that Early effects in Q_1-Q_4 cause poor common mode rejection, as low as 50 dB on some available devices. This can be corrected by extensive cascoding but this tends to reduce the AC performance, particularly in the non-inverting mode. Another problem is that good NPN to PNP V_{be} matching is required to obtain low offset voltage. This can be alleviated by several design tricks, one being to add diodes of the opposite polarity in series with the emitters of each of the input transistors.

6. Voltage Feedback Revisited

If a second input buffer is added to a current feedback op-amp a voltage feedback amplifier results with similar slewing characteristics to a current feedback type. Fig.11 illustrates the concept, used on the PMI OP-467.

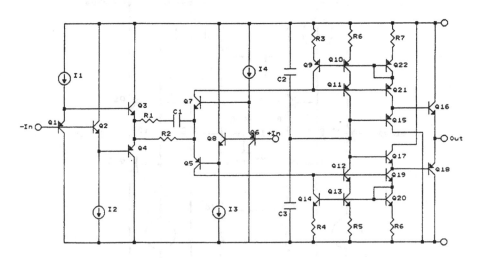

Figure 11: Simplified OP-467 schematic.

Q_5-Q_8 form the equivalent of the current feedback input buffer, while R_2 is the equivalent of the feedback resistor. This resistor is preceded by a second buffer consisting of Q_1-Q_4, identical except for the fact that its output current is discarded. As previously, the full error voltage can be impressed across R_2 and all of this is available to charge the compensation capacitors, C_2 and C_3 (C_1 and R_1 compensate for the inductive behaviour of the input buffers). A side benefit of this arrangement is that the errors of the buffers tend to cancel, resulting in good DC performance. This topology has been used to create an op-amp capable of operating on 36 V supplies with 35 MHz of bandwidth, 200 V/µS of slew rate and 2 mA of supply current. Noise is 6 nV/√Hz, and offset is below 500 µV.

7. Single-Supply Amplifiers

As more systems are being powered from single supplies, the need for op-amps with wide voltage range capability becomes critical. To meet this need, a number of op-amps have appeared with signal handling capability including the negative supply rail, and occasionally both supply rails.

Early single-supply amplifiers used Darlington PNP input stages, as there is ample room to fit a conventional active load beneath the collectors. More recent amplifiers tend to use a single PNP input pair with resistive loads. Care must be taken to keep the drop across these resistors to much less than a normal diode-drop in order to avoid forward biasing the collector-base junctions when the inputs are at ground.

The question of the output stage is less straightforward. The conventional emitter-follower output stage cannot swing much better than about one Volt to the supply rails, so alternatives have to be found. Some ingenious common-collector output stages have been designed [10],[11] which can usually swing to better than 100 mV, but even this is not close enough for many applications. The only really suitable device for improving this is the MOSFET, and for this reason we have turned to our 36 V CBCMOS process to produce a new generation of single-supply amplifiers. This process features vertical PNP and NPN transistors as well as CMOS devices, and opens up a cornucopia of design possibilities.

Fig.12 shows an output stage with fairly high speed properties which can operate single or dual supply. When the output is not near the negative rail, D_1 and D_2 bias Q_4 and Q_5 into the conventional class AB mode. As the negative supply is approached, M_1 turns Q_4 completely off and M_3 can pull the output completely to the rail. One drawback of this circuit is that the gain is not very high, and this problem has been alleviated with a new bootstrapped input stage (Fig.13).

Q_1 and Q_2 form the input pair and are loaded by R_1, R_2 and a foldback cascode consisting of Q_3 and Q_4. This in turn is loaded by a Wilson current source consisting of D_1, Q_6, Q_7 and Q_8. The output impedance of the current source is high, but not high enough to provide precision performance into a MOSFET type output stage.

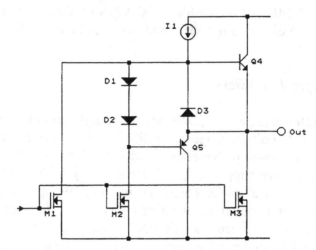

Figure 12: Single-supply BiCMOS output stage.

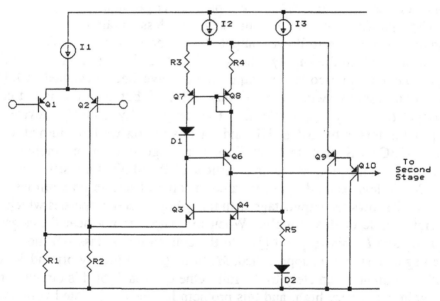

Figure 13: Bootstrapped input stage.

To raise the output impedance, bootstrap transistors Q_9 and Q_{10} are added. If I_2 is chosen such that the current in Q_9 is equal to the total mirror current, then the stage theoretically has no β-induced errors, because the base current of Q_{10} cancels the second-order terms in the Wilson current mirror. A voltage gain in excess of 10^5 has been achieved with this input stage.

Some applications require voltage swing to both rails and the output stage of Fig.14 has been developed for this application. The function of M_2 as the pulldown device is fairly obvious but the upper signal path merits some explanation. First note that M_5 is always on and can be initially ignored. This means that the gate drive to M_3 is controlled by M_5 via D_2 and Q_3. This loop stabilises when the current in M_1 is equal to I_2, so the quiescent current in M_2 and M_3 is equal to I_2 multiplied by the ratio of M_2 to M_1. Capacitor C_1 improves stability when driving capacitive loads.

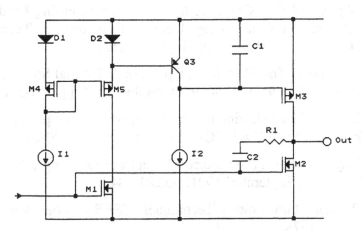

Figure 14: Rail-to rail output stage.

Driving loads in the positive direction, M_1 and M_2 turn off and I_2 applies full gate drive to M_3. A subtle problem exists when driving in the negative direction, however. In this case, the preceding stage applies full gate drive to M_2, which enters the linear region. M_1 is still in the saturated region though, and potentially draws a large current under these conditions. M_5 prevents this by limiting the current in M_1 to about five times I_2, and this in turn means that the output can be driven hard into either rail without incurring excessive supply current.

References

[1] J.R. Ragazzini, R.H. Randall, and F.A. Russell, "Analysis of Problems in Dynamics by Electronic Circuits", *Proc. IRE*, May 1947.

[2] R.J. Widlar, "I.C. op amp with improved input-current characteristics", *EEE*, December 1968, pp.38-41.

[3] G. Erdi, "A Low Drift, Low Noise Monolithic Operational Amplifier", Fairchild Semiconductor Application Brief #136, July 1969.

[4] D. Soderquist and G. Erdi, "The OP-07 Ultra-Low Offset Voltage Op Amp" PMI Application note #13.

[5] R.J. Widlar, "Design Techniques for Monolithic Operational Amplifiers", *IEEE J. Solid-State Circuits*, vol. SC-4, August 1969, pp.184-191.

[6] R.J. Apfel and P.R. Gray, "A Monolithic Fast-Settling Feedforward Operational Amplifier using Doublet Compression Techniques", *ISSCC Dig. Tech. Papers*, 1974, pp.134-155.

[7] J.E. Solomon, "The Monolithic Op Amp: A Tutorial Study", *IEEE J. Solid-State Circuits*, vol. SC-9, December 1974, pp.314-332.

[8] W.E. Hearn, "Fast slewing monolithic operational amplifier", *IEEE J. Solid-State Circuits*, vol. SC-6, February 1971, pp.20-24.

[9] D. Bray, "A High Performance Micropower Op-Amp", *Proceedings of the BCTM*, September 1991, pp.281-284.

[10] R.J. Widlar, "Low Voltage Techniques", *IEEE J. Solid-State Circuits*, December 1978.

[11] J.H. Huijsing and D. Linebarger, "A low-voltage operational amplifier with rail-to-rail input and output ranges", *IEEE J. Solid-State Circuits*, December 1985.

Biography

Derek F. Bowers is staff vice-president of design at Precision Monolithics Inc., where he has worked for more than 10 years. He has been closely involved in the development of a wide variety of analog devices including op-amps, reference sources, and D/A and A/D converters. Derek Bowers holds a B.Sc. in physics and mathematics from the University of Sheffield, England. In his spare time he enjoys music and brewing.

Design of
Low-voltage Bipolar Opamps

Jeroen Fonderie, J.H. Huijsing

Signetics Company, Sunnyvale, USA
Delft University of Technology, The Netherlands

Abstract

This paper describes design techniques for low-voltage bipolar Operational Amplifiers (OpAmps). The common-mode (CM) input voltage of the OpAmp must be able to have any value that fits within the supply voltage range. An input stage capable of realizing this is discussed. The OpAmp should also have a rail-to-rail output voltage range. The multi-path-driven output stage, which is discussed here has a high bandwidth and a high gain. Miller compensation is used to split the pole frequencies of the OpAmp. Finally, two realizations are discussed and a Figure of merit is defined to enable a comparison of the OpAmps' performances.

1. Introduction

The trend towards very low supply voltages forces new demands on the design of accurate analog building blocks. An important building block is the OpAmp. OpAmps that run on a supply voltage of 1 V have already been in existence for over a decade [1], but until now, the properties of these OpAmps have not allowed accurate signal processing. In this paper techniques to design high-quality low-voltage bipolar OpAmps are highlighted. A more comprehensive discussion can be found in [2].

Why use a low supply voltage? First of all, there is a tendency towards chip components of smaller dimensions. A component with these smaller dimensions is subject to breakdown at lower voltages. Secondly, battery-powered portable equipment should be able to function at low supply voltages, using just one single battery cell up to the end of its life span. Finally, if low supply voltages are allowed, applying the OpAmp in system design is less critical; the circuit can, for instance, be made to operate from residual voltages within the system.

The question might arise as to whether the OpAmp could preferably be composed of MOS transistors. After all, MOS transistors can potentially be made with a very low threshold voltage. A disadvantage of MOS transistors is, however, that the transconductance of the devices is much smaller than that of bipolar transistors. Furthermore, the minimum drain-source voltage of MOS transistors is usually much larger than the corresponding collector-emitter voltage of a bipolar transistor. This paper is therefore restricted to the use of bipolar transistors.

The design of low-voltage OpAmps is dealt with in the following Sections of the paper. The design of input stages is discussed in Section 2, Section 3 deals with output stages and Section 4 briefly describes the design of intermediate stages. In Section 5 aspects of the high-frequency design of the OpAmp are illustrated and finally, in Section 6, two implementations are presented and a Figure of merit is defined to compare the performance of low-voltage OpAmps.

2. Input stages

The differential input stage that is shown in Fig.1 has a rail-to-rail CM-input voltage range. In this way a reasonable ratio between the signal and additive interferences, such as noise, is obtained. Further, a rail-to-rail CM-input voltage range also enhances the general-purpose nature of the OpAmp.

If the CM-input voltage is near the negative supply rail, current source I_{B1} activates the p-n-p pair Q_3, Q_4, provided that the supply voltage is high enough. This p-n-p pair is now able to handle the input signal. If the CM-input voltage is now raised above the reference voltage V_{R1}, transistor Q_5 takes away the current from current source I_{B1}, and through Q_6 and Q_7, supplies the n-p-n pair Q_1, Q_2. The p-n-p pair Q_3, Q_4 is now switched off and the signal operation is performed by the n-p-n pair Q_1, Q_2. Even in the turnover range of Q_5, the sum of the tail currents of the input pairs, and therefore also the total transconductance of the input stage, is kept constant. The value of the reference voltage source V_{R1} is 0.8 V, high enough to

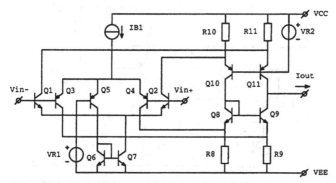

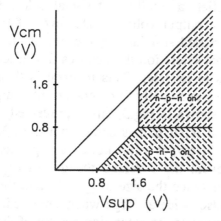

Figure 1: Input stage with rail-to-rail common-mode input-voltage range and a constant transconductance over the full common-mode range.

Figure 2: The operational regions of the input stage shown in Fig. 1. Only one of the input pairs is active.

enable the n-p-n pair to function properly when the CM-input voltage exceeds this value. The transistors Q_8-Q_{11} and resistors R_8-R_{11} sum the collector currents of the input pairs.

Fig.2 shows the CM range of the input stage as function of the supply voltage, the so-called operational regions. It can be seen that either the p-n-p pair or the n-p-n pair is active, but not both. For supply voltages below 1.6 V (the sum of V_{R1}, the base-emitter voltage of Q_5 and the saturation voltage of I_{B1}) the n-p-n pair is completely shut off and only the p-n-p pair is operative.

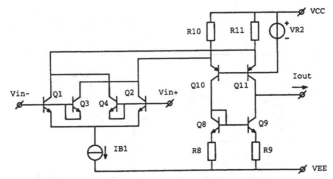

Figure 3: Input stage with diodes to prevent reversal of the output signal.

The CM-input voltage range extends beyond the supply rails for approximately 400 mV. This feature is very useful when the OpAmp is used as an inverting amplifier; a small input signal below the negative supply still yields a positive output voltage. However, if the CM-input voltage is increased or decreased still further, the transistors of the input pairs become saturated. This saturation then causes the collector currents of the input-pair transistors to reverse. This reversal is then transferred through the amplifier, leading eventually to reversal of the output signal of the OpAmp. In the circuit shown in Fig.3, this is prevented. For reasons of simplicity, only the n-p-n input pair is shown but the addition to the circuit, which is described below, is valid both for n-p-n as well as for p-n-p input pairs. To prevent the reversal, the diodes Q_3 and Q_4 are inserted. These diodes have twice the size of Q_1, Q_2, and their collector-base current are therefore also twice as large when saturated. This larger current is used to over-compensate the saturation current of the input-stage transistor by electrically cross-coupling the sensing diodes Q_3, Q_4 with the input-pair transistors Q_1, Q_2. Thus the signal polarity remains correct. This results in the ability to overdrive the circuit input to a CM voltage of at least one diode voltage beyond the supply rails.

Fig.4 shows the region of normal operation, the region that extends 400 mV above the supply rail and the region where the reversal diodes are active. If a p-n-p input pair was added to the circuit shown in Fig.3, the area indicated by "I_{B1} saturates" would also be included in the operational regions.

Although not examined here, an input stage with a fully rail-to-rail CM-input range that operates at a minimum supply voltage of 1 V can also be used [3].

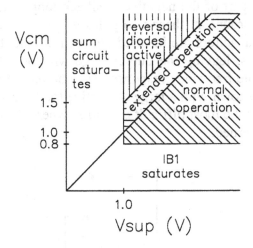

Figure 4: Operational regions of the input stage depicted in Fig. 3.

3. Output stages

The output stage of a low-voltage OpAmp should be able to deliver an output voltage signal that is as large as possible, preferably from one supply-voltage rail to the other. After all, the supply-voltage range is by definition not very great, so we want to be able to make full use of it. This rules out the use of emitter followers as output transistors, because such a use would result in the loss of one diode voltage. The actual output transistors of the output stages discussed in this Section therefore have a common-emitter configuration.

The maximum bandwidth of the OpAmp is determined by the current flowing through the output transistors, which determines their transconductance, and by the capacitive load it has to drive. Any transistors preceding the actual CE output transistors more or less reduce the obtainable bandwidth. This reduction should be kept as small as possible.

The output stage should be able to push and pull output currents in the order of 10 mA, which means that its current gain should be in the order of 10^3. One output transistor in CE configuration does not meet this condition and therefore emitter followers are normally placed in front of it. This contradicts, however, with the previous demand.

Initially, current-driven output stages with collector outputs have two dominant pole frequencies, one at the input and one at the output. Because the output stage is part of the negative feedback loops within the OpAmp and probably also around it, these two poles will give rise to oscillations. Therefore Miller capacitors are inserted between the input and the output. They split the poles, leaving the output stage with one dominating pole, which yields a stable frequency response [4]. The output stages discussed in this Section are shown with a large capacitive load and no resistive load, because the latter is the most difficult condition where the compensation of the circuits is concerned.

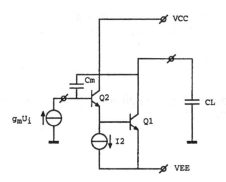

Figure 5: Darlington output stage (n-p-n side).

To illustrate the consequences of the demands stated above, two examples of low-voltage output stages are examined. To increase the current gain of one single CE stage, it can be preceded by an emitter follower. The Darlington output stage that then emerges is shown in Fig.5. Transistors Q_1 and Q_2 are the CE stage and emitter follower, respectively, and the pole splitting is realized with C_m. Current source I_2 ensures a minimum current through Q_2 at low output currents. The current source $g_m U_i$ represents the output of the preceding intermediate stage. The minimum supply voltage for this output stage is about 1.7 V.

Fig.6 shows the frequency response of the Darlington output stage. Because ω_0 always has to be chosen smaller than ω_2 in order to keep a positive phase margin, ω_2 ultimately sets the bandwidth of the output stage and, consequently, of the complete OpAmp. The position of ω_2 can be calculated as:

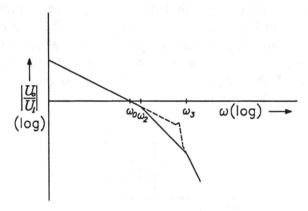

Figure 6: The magnitude of the small-signal voltage gain versus the frequency of the Darlington output stage. The dashed line shows the "output bump" when the output current increases and the poles become complex.

$$\omega_2 = \frac{g_o}{C_L \left(1 + \dfrac{C_{be2}}{C_m}\right)} \tag{1}$$

where g_o is the transconductance of the output stage, C_L is the load capacitor, C_m is the Miller capacitor and C_{be2} is the base-emitter capacitance of transistor Q_2. This leads to the important conclusion that the bandwidth of the OpAmp with a current-driven output stage is determined by the quiescent current through that output stage and by the maximum load capacitor that has to be driven.

If the output current of the output stage increases, ω_2 and the high-frequency pole ω_3, which is situated at the base of Q_1, move towards each other and finally collide. The complex poles that then arise cause the famous "output bump" [4], as is shown by the dashed line in Fig.6. These complex poles may have a negative influence on the overall performance of the OpAmp. The presence of complex poles causes damped oscillations in the transient response.

The emitter follower that is placed in front of the CE output transistor thus introduces an extra pole in the frequency response and this deteriorates the high-frequency behavior of the Darlington output stage. However, the demands on the current gain prohibits the use of just one simple CE stage. The circuit shown in Fig.7, the multi-path-driven (MPD) output stage [5], [6], combines the advantages of the simple CE stage with a higher current gain, and therefore bypasses the trade-off between bandwidth and gain. Furthermore, the MPD output stage is also able to operate at a supply voltage as low as 1 V.

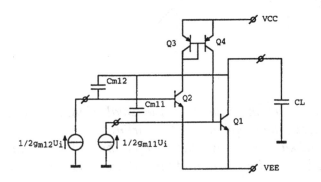

Figure 7: The n-p-n multi-path-driven output stage.

The transistors Q_2, Q_3 and Q_4 drive the CE output transistor Q_1 and parallel to this path there is a feedforward path directly from the intermediate stage to transistor Q_1. Basically, the path through Q_2-Q_4 supplies the necessary current gain, while the direct path to Q_1 guarantees good high-frequency behavior. The intermediate stage has to supply the output stage with two identical, but uncoupled, input signals. This can for instance be realized with two differential stages in parallel, both driven by the same input voltage [7]. The two outputs of the intermediate stage are symbolized in Fig.7 by the two input current sources $\frac{1}{2}g_{m11}U_i$ and $\frac{1}{2}g_{m12}U_i$. The poles at the output and at both inputs are split with Miller capacitors C_{m11} and C_{m12}. The current gain from the bases of Q_1 and Q_2 to the output is $\frac{1}{2}\beta_1(\beta_2+1) \approx \frac{1}{2}\beta_1\beta_2$, half the value of the Darlington output stage and sufficient to drive large output currents.

The frequency response of the MPD output stage is shown in Fig.8. The dashed and dotted lines represent the response of the gain and feedforward paths, respectively; the gain path has a large gain, but a smaller bandwidth, and the feedforward path has a small gain, but larger bandwidth. The slopes of the two responses coincide if the Miller capacitors C_{m11} and C_{m12} as well as the transconductances of the intermediate transistors and therefore the current through these transistors are equal:

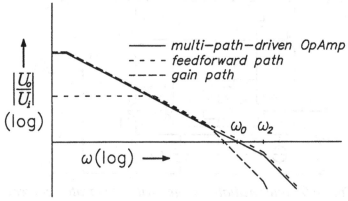

Figure 8: The principle of the multi-path-driven strategy. The overall transfer has the gain of the gain path, and the bandwidth of the feedforward path.

$$\frac{g_{ml1}}{C_{ml1}} = \frac{g_{ml2}}{C_{ml2}} \tag{2}$$

A mismatch of the Miller capacitors and of the intermediate stage transistors introduces a pole-zero doublet. Fortunately, the matching can be done very accurately, because a pole-zero doublet gives a slow settling time of the transient response [8]. Fig.8 shows the frequency response of the intermediate and output stage. As can be seen from Fig.8, no pole-zero doublet occurs and complex poles also do not occur. The bandwidth is, thus, only limited by the current flowing through Q_1 and by the load capacitor C_L.

4. Intermediate stages

A simple OpAmp can consist of an input stage, directly coupled to an output stage. Although the simplicity of this topology is attractive, a major disadvantage is that the DC open-loop gain is relatively low, in the order of 80 dB. If a larger gain is desired, an additional differential amplifier stage should be inserted between the input and output stage. The design of this intermediate stage should be done carefully, however, because, when the supply voltage of the OpAmp is much higher than the minimum value, the voltages on the matching parts of the intermediate stage diverge.

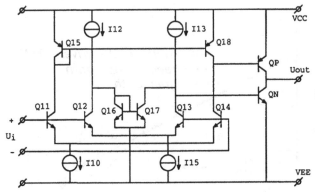

Figure 9: Intermediate stage with separate current mirrors and tail-current sources, to compensate the Early effect on the intermediate-stage transistors.

Because of the Early effect, this voltage difference causes a disturbance of the biasing of the transistors of the OpAmp. To compensate the Early effect, the intermediate stage can be extended by the addition of folded cascodes, much in the same way as this is done in the input stage. The p-n-p cascodes do, however, deteriorate the frequency performance of the intermediate stage. A more simple solution to the problem is shown in Fig.9. The intermediate stage shown in Fig.9 consists of two parts. Transistors Q_{11}, Q_{14} supply the p-n-p output stage, and have current mirrors Q_{15}, Q_{18} that are connected to the positive supply rail, just as the p-n-p output transistor is. The other differential stage, Q_{12}, Q_{13} is connected to the n-p-n output stage and has n-p-n current mirrors Q_{16}, Q_{17}, that are connected to the negative supply rail, just as the n-p-n output transistor is. In order to be able to fully drive the differential stage Q_{12}, Q_{13}, the current sources I_{12} and I_{13} should be equal to the tail-current sources I_{10} and I_{15}. Although this may seem an unnecessary dissipation of supply power, the consumption is less than when the intermediate stage is equipped with folded cascodes.

5. Overall frequency compensation

The input, output and intermediate stages of low-voltage OpAmps, as described in the previous Sections, all have collector outputs and hence contribute dominating poles to the transfer from input to output. If there are two or more dominating poles, the phase margin of the frequency response is negative, and overall feedback of the OpAmp is not possible. To change this, frequency compensation techniques must be applied that give the OpAmp one dominating pole frequency with a straight 6-dB/octave frequency roll-off. The most effective way to compensate the OpAmp is by applying the technique of pole splitting [4].

First, simple pole spitting, using one Miller capacitor is discussed and then nested Miller compensation, for OpAmps consisting of three stages, is examined.

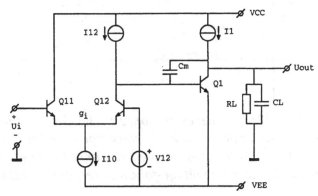

Figure 10: Two-stage Opamp with two poles, which are being split by one Miller capacitor.

The most simple configuration of a low-voltage OpAmp is a differential input stage followed by an output transistor, as shown in Fig.10. This OpAmp has initially two dominant pole frequencies, one at the output of the output stage, and one at the output of the input stage. To give the OpAmp a stable frequency response, these poles can be split with Miller capacitor C_m. Calculations show that the bandwidth of the Miller-compensated OpAmp $\omega_0^{(m)}$ should be chosen:

$$\omega_0^{(m)} = \frac{g_i}{C_m} = \frac{1}{2} \frac{g_o}{C_L\left(1 + \dfrac{C_{bel}}{C_m}\right)} \tag{3}$$

or in words: the ratio of the transconductance of the input stage g_i and the Miller capacitor C_m should be chosen such that the open-loop bandwidth is half the value of the second pole in the open-loop transfer. This second pole is determined by the output stage transconductance g_o and the load capacitor C_L, as is discussed in Section 3. If these conditions are met, the voltage-follower function, i.e. the OpAmp with total external feedback applied to it, has the maximum bandwidth without peaking and its poles are then in Butterworth positions.

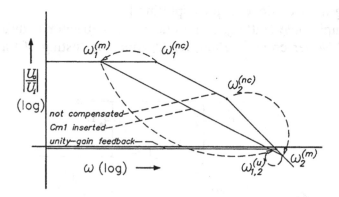

Figure 11: Frequency response of the OpAmp shown in Fig.10. Indicated are: the uncompensated response (top), the Miller-compensated response (middle), and the unity-gain or voltage-follower response (bottom).

Fig.11 shows the effect of the pole movements on the frequency response of the OpAmp. First the poles are split from the non-compensated *(nc)* positions to the Miller *(m)* positions and if then unity-gain feedback is applied to the OpAmp, the poles again move towards each other and coincide at the frequency $\omega_0^{(u)}$.

	non-compensated $\omega^{(nc)}$	Miller compensated $\omega^{(m)}$	unity-gain feedback $\omega^{(u)}$
ω_0	23 MHz	16 MHz	23 MHz
ω_1	160 kHz	16 kHz	23 MHz, +45°
ω_2	3.2 MHz	32 MHz	23 MHz, -45°
A_0	60 dB	60 dB	0 dB

Table 1: Calculated pole positions for a two-stage OpAmp with an n-p-n output stage.

To illustrate the theory we will apply it to an OpAmp with an output stage consisting of 5-GHz n-p-n transistors with a current gain of 100. The quiescent current through the output transistor is 500 μA, the load consists of a 100 pF capacitor in parallel with a 10 kΩ resistor, and the Miller capacitor is chosen 10 pF. Table 1 can now be filled out using (3) and the Equations given in [2]. From (3), it is also found that the tail current I_{10} in Fig.6 must be 100 μA, giving a transconductance g_i of 1 mmho. It is interesting to see from Fig.7 and Table 1 that if unity-gain feedback is applied to the OpAmp, all the 60 dB gain is "transformed" into the maximal obtainable bandwidth, i.e. 23 MHz, the same as the bandwidth of the noncompensated OpAmp.

If the output stage introduces complex poles in the frequency response, the stability of the OpAmp can still be guaranteed, but the obtainable bandwidth should be greatly reduced when compared to the optimal value given in (3). The use of output stages without these complex poles is therefore emphasized.

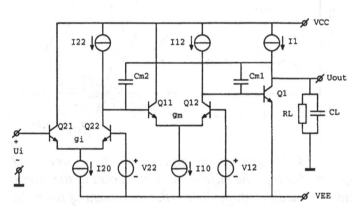

Figure 12: Three-stage OpAmp with three poles, which are being split by two nested Miller capacitors.

Fig.12 shows a three-stage OpAmp, comprising an input stage, an intermediate stage and an n-p-n output transistor. The intermediate stage and the output transistor are compensated with Miller capacitor C_{m1}, and the combination of these two, together with the input stage are in turn compensated with C_{m2}. The open-loop bandwidth of the OpAmp $\omega_0^{(m2)}$ can again be calculated:

54

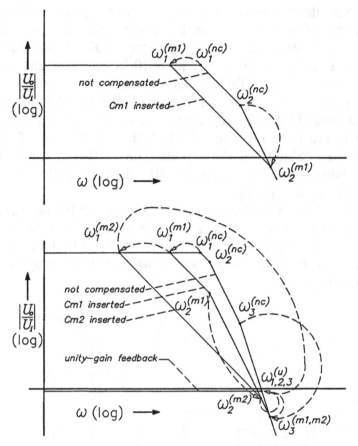

Figure 13: Frequency response of the three-stage OpAmp. On top, the movement of the poles of the output and intermediate stage, below the movement of the poles of the complete OpAmp.

$$\omega_0^{(m2)} = \frac{g_i}{C_{m2}} = \frac{1}{4} \frac{g_o}{C_L \left(1 + \frac{C_{bel}}{C_{ml}} \right)} \tag{4}$$

The ratio of the transconductance of the input stage g_i and the Miller capacitor C_{m2} should now be chosen at one quarter of the value of the output pole, which is determined by the output transconductance g_o and the load capacitor C_L. If unity-gain feedback is applied to the OpAmp, this results in a flat frequency response, and the poles then move into the third-order Butterworth positions.

Fig.13 shows the movement of the poles as a result of the insertion of the Miller capacitors C_{m1} and C_{m2}. The upper half of Fig.13, showing the movement of the poles of the intermediate and output stage, is identical to Fig.11. The lower half of Fig.13 shows the poles of the complete OpAmp, without compensation, after C_{m1} is inserted, after C_{m2} is inserted and when unity-gain feedback is externally applied to the OpAmp.

	non-compensated $\omega^{(nc)}$	Miller compensated $\omega^{(m1)}$	twice Miller compensated $\omega^{(m2)}$	unity-gain feedback $\omega^{(u)}$
ω_0	16 MHz	11 MHz	8.0 MHz	16 MHz
ω_1	160 kHz	16 kHz	320 Hz	16 MHz, +60°
ω_2	320 kHz	320 kHz	23 MHz, +45°	16 MHz, 0°
ω_3	3.2 MHz	32 MHz	23 MHz, -45°	16 MHz, -60°
A_0	88 dB	88 dB	88 dB	0 dB

Table 2: Calculated pole positions for a three-stage OpAmp with an n-p-n output stage, biased at 500 μA and loaded with 100 pF in parallel with 10 kΩ.

In accordance with the preceding, Table 2 illustrates the theory pertaining to the poles of the three-stage OpAmp. The values that are used can be calculated from (3), (4) and the Equations found in [2]. The transistors in this example are again assumed to be the same as above: the load is 100 pF in parallel with 10 kΩ, the transconductances of the input and intermediate stages are 0.25 mmho and 0.1 mmho, respectively, and C_{m1} and C_{m2} are 10 pF and 5 pF, respectively. The voltage follower "transforms" the 88 dB gain of the OpAmp into a -3-dB bandwidth of 16 MHz, equal to the bandwidth of the non-compensated OpAmp.

If the bandwidth of this three-stage OpAmp is compared with that of the two-stage OpAmp discussed in the previous Section, we see that the increase of the DC gain has to be paid for with a reduction of the bandwidth by a factor of two.

56

6. Realizations

The stages described in the previous Sections can be combined in one circuit which is discussed here as an example of a low-voltage OpAmp. The circuit has been made in a 12-V BiCMOS process of Philips Nijmegen, The Netherlands, with 3-GHz n-p-n transistors and 1-GHz vertical p-n-p's.

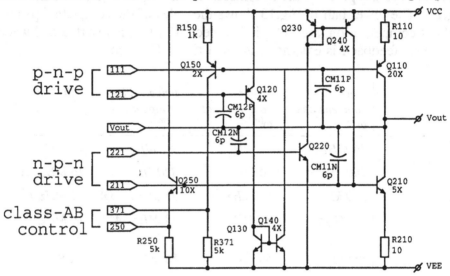

Figure 14: Circuit implementation of the complementary MPD output stage.

The MPD output stage shown in Fig.14 embodies two complementary parts, Q_{110}-Q_{140} and Q_{210}-Q_{240}. Both parts are identical to the output stage shown in Fig.7, except that the current mirrors Q_{130}-Q_{140} and Q_{230}-Q_{240} are scaled 1:4 to increase the gain. The small emitter resistors R_{110} and R_{210} prevent breakthrough of the output transistors at high current and voltage levels, but they do not limit the output-voltage range in normal operation; when driving light loads, the output voltage is able to reach within 100 mV of the supply rails. The output transistors Q_{110} and Q_{210} are able to sink or source a maximum output current of 15 mA. The quiescent current through the output transistors is kept at 320 µA, and the minimum current through either one of these transistors is 160 µA.

Fig.15 shows the input stage, the intermediate stage and the class-AB current control. The intermediate stage now has four outputs to drive the four inputs of the output stage. The feedback class-AB current control embodies transistors Q_{150} and Q_{250}, shown in Fig.14, and Q_{310} through

Q_{380}, shown in Fig.15 [9], [10]. The differential amplifier Q_{310}-Q_{360} consists of two parts, each connected to the matching part of the intermediate stage, again to compensate for the Early effect. The input stage Q_{510}-Q_{540} has a transconductance with a constant value of 0.25 mmho over the full CM-input range. The circuit Q_{580}-Q_{595} adds the currents of the differential input stages. The 0.25 mmho input-stage transconductance, and the capacitors C_{m2} and C_{p2} of 6 pF each, gives the OpAmp with MPD output stage a unity-gain bandwidth of 3.4 MHz. The specifications of this OpAmp, referenced as OpAmp 1, are listed in Table 3.

To show that the concept shown in Figs.14 and 15 also gives a good performance at higher frequencies, a second OpAmp with different component values has been designed, but now in the QUBiC process of the Signetics Company, Sunnyvale, CA. The QUBiC process is a 7-V BiCMOS process, with oxide-isolated transistors. The n-p-n transistors have a transit frequency of 13 GHz and the F_T of the lateral p-n-p's is 200 MHz. This OpAmp design is intended as a system-library building block. Because of this, the maximum output current of the OpAmp could

symbol	OpAmp 1	OpAmp 2	unit
T_{OP}	-55/125	-55/125	°C
V_{CC}	1-10	1-7.5	V
I_{CC}	700	390	µA
V_{OS}	0.6	0.5	mV
I_B	140	45-120	nA
V_{CM}	V_{EE}-0.25/ V_{CC}+0.25	V_{EE}-0.25/ V_{CC}+0.25	V
$CMRR$	100	100	dB
I_L	±15	±2	mA
A_{VOL}	117	90	dB
V_{OUT}	V_{EE}+0.1/ V_{CC}-0.1	V_{EE}+0.1/ V_{CC}-0.05	V
S_R	1.1	3.0	V/µs
BW	3.4	10	MHz
θ_M	61	64	°
$NOISE$	23	15	nV/√Hz

Table 3: Specifications of the OpAmps. V_{sup}= 1 V, T= 27 °C, R_L= 10 kΩ, C_L= 100 pF for OpAmp 1, 10 pF for OpAmp 2.

be chosen smaller than in the design presented above, and the maximum load capacitor is also smaller. However, the bandwidth of the circuit should be as high as 10 MHz, and the total current consumption should be lower. The specifications of this OpAmp, listed as OpAmp 2, can also be found in Table 3.

To be able to compare the performance of low-voltage OpAmps, simply measuring their bandwidth is not enough; the dissipated current in the output stage should also be taken into consideration. Since we strive for the lowest supply voltage, this too is of importance. A Figure of merit is thus defined as the ratio of the bandwidth and the power the OpAmp consumes, viz., the product of the minimum supply voltage and the supply

OpAmp	BPR (MHz/mW)	B_W (MHz)	V_{CC} (V)	I_{CC} (μA)
OpAmp 2	26	10	1	390
OpAmp 1	4.9	3.4	1	700
NE5234	1.8	2.5	2	690
LM10	0.27	0.08	1.1	270
LM741	0.15	1	4	1700

Table 4: Bandwidth-to-Power ratio of the two discussed OpAmps, compared to the BPR of existing (low-voltage) OpAmps.

current. In the formula the bandwidth-to-power ratio (*BPR*) of the OpAmp is:

$$F_M = \frac{B_W}{V_{CC,\text{min}} \cdot I_{CC}} \qquad (5)$$

This Figure of merit is the inverse of the well-known power-delay product in digital circuits. Table 4 gives the *BPR* of the two OpAmps that have been discussed above and compares them with those of low-voltage OpAmps currently on the market. The *BPR* of the 10-MHz OpAmp, OpAmp 2, is by far the highest, however, this OpAmp can only be loaded with a capacitor of 10 pF. OpAmp 1 and the NE5234, which is also discussed in [2], have a *BPR* that is considerably higher than that of the LM10 [1], the previous generation low-voltage OpAmp. For comparison, the LM741 is also listed in Table 4, although this OpAmp is not a low-voltage OpAmp.

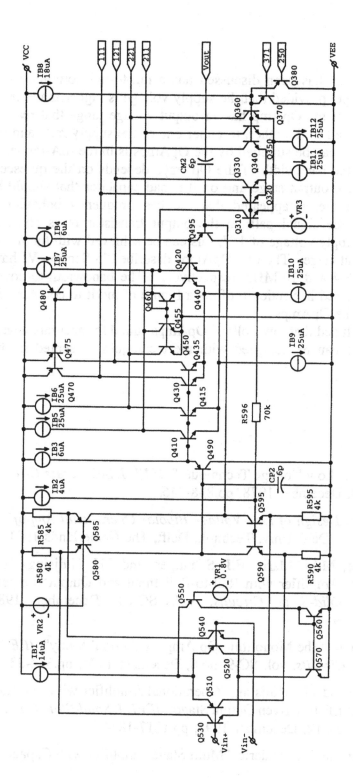

Figure 15: Circuit implementation of the input stage, intermediate stage and class-AB control.

7. Conclusions

The OpAmp that has been discussed has a rail-to-rail common-mode input-voltage range, provided that the supply voltage is higher than 1.8 V. The output stage of the OpAmp has an output-voltage range that reaches from one supply rail to the other, to either one of the supply rails and the output current that can be supplied by the OpAmps is in the mA-range. It was shown that the bandwidth of the OpAmps depends on the quiescent current through the output stage and on the load capacitor that should be driven. Furthermore, it appeared that the best frequency behavior is accomplished if the signal path to the output transistor is as short as possible. At a supply voltage of 1 V, this can be achieved with the multi-path-driven output stage. The two OpAmps described in Section VI have a bandwidth of 3.4 and 10 MHz, respectively. The bandwidth-to-power ratio that was defined in Section 6 offers a Figure of merit to compare the performance of the OpAmps.

It can be concluded that low-voltage OpAmps suited for accurate analog signal processing can be designed with the techniques presented in this paper.

References

[1] R.J. Widlar, "Low Voltage Techniques", *IEEE J. Solid-State Circuits*, vol. SC-13, December 1978, pp.838-846.

[2] J. Fonderie, *Design of Low-Voltage Bipolar Operational Amplifiers*. Ph.D. Thesis, Delft Univ. Technol., Delft, The Netherlands 1991.

[3] J. Fonderie, M.M. Maris, E.J. Schnitger, and J.H. Huijsing, "1-V Operational Amplifier with Rail-to-Rail Input and Output Ranges", *IEEE J. Solid-State Circuits*, vol. SC-24, December 1989, pp.1551-1559.

[4] J.E. Solomon, "The Monolithic Op Amp: A Tutorial Study", *IEEE J. Solid-State Circuits*, vol. SC-9, no.6, December 1974, pp.314-332.

[5] J. Fonderie and J.H. Huijsing, "Operational Amplifier with 1-V Rail-to-Rail Multi-Path-Driven Output Stage", *IEEE J. Solid-State Circuits*, vol. SC-26, no.12, December 1991, pp.1817-1824.

[6] J.H. Huijsing and J. Fonderie, "Multi Stage Amplifier with Capacitive

Nesting and Multi-Path-Driven Forward Feeding for Frequency Compensation", U.S. Pat. Appl. Ser. No. 654.855, filed February 11, 1991.

[7]   J.H. Huijsing, J. Fonderie, "Current Compensation Circuit", Eur. Pat. Appl. Ser. No. 9000351, filed February 14, 1990.

[8]   B.Y. Kamath, R.G. Meyer, and P.R. Gray, "Relationship Between Frequency Response and settling time of Operational Amplifiers", *IEEE J. Solid-State Circuits*, vol. SC-9, December 1974, pp.347-352.

[9]   E. Seevinck, W. de Jager, and P. Buitendijk, "A Low-Distortion Output Stage with Improved Stability for Monolithic Power Amplifiers", *IEEE J. Solid-State Circuits*, vol. SC-23, June 1988, pp.794-801.

[10] J.H. Huijsing and D. Linebarger, "Low-Voltage Operational Amplifier with Rail-to-Rail Input and Output Ranges", *IEEE J. Solid-State Circuits*, vol. SC-20, December 1985, pp.1144-1150.

Biographies

Jeroen Fonderie was born in Amsterdam, the Netherlands, on July 27, 1960. He received a M.Sc. in Electrical Engineering from the Delft University of Technology, Delft, the Netherlands in 1987 and a Ph.D. from the same university in 1991. The title of his Ph.D thesis is: "Design of Low-Voltage Bipolar Operational Amplifiers". In 1992, Jeroen Fonderie joined the Signetics Company, Sunnyvale, Ca. where he is involved in automotive circuit design.

Johan H. Huijsing was born in Bandung, Indonesia, on May 21, 1938. He received the ingenieurs (M.Sc.) degree in Electrical Engineering from Delft University of Technology, Delft, The Netherlands, in 1969, and the Ph.D. degree from the same university in 1981 for work on operational amplifiers. Since 1969 he has been a member of the research and teaching staff of the Electronic Instrumentation Laboratory, where he is now a full professor of electronic instrumentation. His field of research is: operational amplifiers, analog to digital converters, signal conditioning circuits and integrated smart sensors. He is author and co-author of more than fifty scientific papers and holds twelve patents.

Opamp Design towards Maximum Gain-Bandwidth

M. Steyaert, W. Sansen

K.U. Leuven, ESAT-MICAS
Heverlee
Belgium

Abstract

One of the major parameters of an opamp is its gain-bandwidth (GBW). Because nowadays the analog systems are pushed to higher frequencies, the design of opamps with a high GBW becomes very important. An overview of design techniques to achieve the maximum GBW in a CMOS technology is presented. An analysis of different feedforward techniques and their effect on the settling time is studied.

1. Introduction

In the design of analog circuits the operational amplifier (opamp) is an indispensable building block. Opamps allow the user to realize with high accuracy special functions such as inverting amplifiers, summing amplifiers, integrators and buffers. All this can be achieved with an opamp and only a few extra integrated passive circuit elements. The combination of those functions can result in very complex circuits such as higher order switched capacitor filters, telecommunication circuits and very sensitive amplifiers for medical purposes. However the first order calculations of those building blocks are based on feedback structures using an ideal opamp an ideal voltage dependent voltage source with an infinite gain factor. Practical opamps can reach only high gain factors at low frequencies. Usually the gain decreases for frequencies above a few Hertz. The product of the -3 dB point (the open loop bandwidth) and the low frequency open loop gain

(A_o) is called the gain bandwidth (GBW). Because the accuracy of a feedback network (settling error, distortion, power supply rejection ratio, gain accuracy, bandwidth, ...) depends on the loop gain of the opamp, and because the loop gain at a given frequency is a function of the GBW, the higher the GBW can be designed, the better system performances can be obtained. Therefore, the design strategy to achieve the maximum gain bandwidth for a given CMOS technology is analysed. Several structures, including feedforward techniques are analysed.

2. MOS Transistor in Strong Inversion

In high frequency circuit designs the parasitic poles/zeros are generated by the finite f_t of the transistors. In order to achieve high frequency performances, let us first study the f_t of MOS transistors and how to design transistors with a high f_t. This means that we will look into a more complex model than the well-known square-law relationship of MOS transistors. This is necessary because for high frequency design, short channels are very often used in combination with large gate-source voltages. Hence the degradation effects of high electrical fields in the MOS transistor have to be included as well. The I_{DS} versus V_{GS} relationship, including these effects, is given by [1]:

$$I_{DS} = \frac{\mu \cdot C_{ox}}{2\left(1 + \theta \cdot (V_{GS} - V_T) + \xi \dfrac{V_{DSat}}{L}\right)} \cdot \frac{W}{L} \cdot (V_{GS} - V_T)^2 \cdot (1 + \lambda V_{DS}) \tag{1}$$

with W and L the effective transistor width and length. The effect of the gate electrical field is calculated by the θ term, while the source-drain electrical field is modelled by the ξ term ($\theta = \mu/v_{max}$). Those terms are very important for high frequency designs, because, as will be shown later on, large $(V_{GS} - V_T)$ and V_{DSat} values are very common. In practice V_{DSat} can be approximated by $V_{DSat} = (V_{GS} - V_T)$, which will be further assumed in this text. In analog circuit designs, the small signal parameters are more important. From relationship (1) the transconductance (g_m) can be calculated to be:

$$\frac{d\,I_{DS}}{d\,(V_{GS}-V_T)} =$$

$$g_m = \mu \cdot C_{ox} \cdot \frac{W}{L} \cdot (V_{GS}-V_T) \frac{\left(1+\frac{1}{2}(\theta + \frac{\mu}{v_{max}L})\cdot(V_{GS}-V_T)\right)}{\left(1+(\theta+\frac{\mu}{v_{max}L})\cdot(V_{GS}-V_T)\right)^2} \qquad (2)$$

This equation can be very well approximated by (max. error < 6%)

$$g_m = \mu C_{ox} \cdot \frac{W}{L} \cdot \frac{V_{GS}-V_T}{\left(1+2(\theta+\frac{\mu}{v_{max}L})\cdot(V_{GS}-V_T)\right)} \qquad (3)$$

It is important to note that for small $(V_{GS}-V_T)$ values and long transistor lengths the well-known g_m-relationship is found. For very large $(V_{GS}-V_T)$ values the transconductance "saturates" to a maximum value given by

$$g_m = \frac{\mu \cdot C_{ox} \cdot \frac{W}{L}}{2\left(\theta+\frac{\mu}{v_{max}L}\right)} \qquad (4)$$

However, in modern processes the gate electrical field effect is less important compared to the source-drain electrical field effect. For example θ is typical 0.1 1/V while $\xi = \mu/v_{max} \approx 0.3$ μm/V (see also table 1). This means that for large $(V_{GS}-V_T)$ values the relationship can be simplified into

$$g_m = \frac{C_{ox}Wv_{max}}{2} \qquad (5)$$

If now in the transistor only the gate-source capacitance $(C_{gs} = \frac{2}{3}\,C_{ox}WL)$ is included, the f_{to} of the transistor can be defined as

$$f_{to} = \frac{g_m}{2\pi\, C_{GS}} = \frac{\mu}{2\pi\frac{2}{3}L^2} \cdot \frac{V_{GS} - V_T}{\left(1 + 2\left[\theta + \frac{\mu}{v_{max}L}\right]\cdot(V_{GS} - V_T)\right)} \tag{6}$$

It is this relationship which is so important for high frequency designs: the higher $(V_{GS} - V_T)$ and the shorter L are, the higher f_{to} parameter can be achieved. This is well-known, but due to the electrical fields, the increase of f_{to} becomes less than is usually expected.

Parameter	NMOS	PMOS	Units
V_T	0.7	0.8	V
KP	82	27	$\mu A/V$
γ	0.6	0.54	$\sqrt{V}$
C_{ox}	1.6	1.6	$fF/\mu m^2$
C_{go}	0.34	0.34	$fF/\mu m$
C_j	0.32	0.54	$fF/\mu m^2$
C_{jsw}	0.37	0.57	$fF/\mu m$
DW	0.	0.	μm
$LD-DL$	0.2	0.2	μm
θ	0.06	0.11	$1/V$
ξ	0.3	0.14	$\mu m/V$
f_{to} ($V_{GS} - V_T = 1V/V_{DS} = 2.5V$)	6.5	2.5	GHZ
L_{edge}	3.6		μm
$C_{poly\text{-}sub.}$	0.02		$fF/\mu m^2$
α_{np}	2.4		-

Table 1: Typical process parameters of an 1.2µm CMOS process.

The transconductance relationship (2) can be rewritten in the form:

$$g_m = \frac{2I_{DS}}{V_{GS}-V_T} \cdot \frac{1+\frac{1}{2}\left(\theta + \frac{\mu}{v_{max}L}\right)\cdot(V_{GS}-V_T)}{1+\left(\theta + \frac{\mu}{v_{max}L}\right)\cdot(V_{GS}-V_T)} \approx \frac{2I_{DS}}{V_{GS}-V_T} \tag{7}$$

From this relationship it becomes clear that high frequency CMOS designs $(V_{GS}-V_T > 0.1$ V$)$ are very power hungry: for example, if $(V_{GS}-V_T) = 1$ V, g_m becomes equal to $g_m \approx 2I_{DS}$. This is in great contrast with 'normal' CMOS design where usually the transistors are designed at the boundary of strong inversion $(V_{GS}-V_T \approx 0.2$ V$)$ or $g_m \approx 10I_{DS}$. If it is compared with a bipolar design, the contrast is even higher: $g_{m,bipolar} \approx 40I_C$! Or with other words, for the same transconductance, the bipolar transistor consumes approximately twenty times less current (or power) than a CMOS transistor.

Until now we have only considered the gate-source capacitance. In practical transistor designs extra capacitors exist, such as drain-bulk, source-bulk and overlap capacitors. They all decrease the ideal f_{to}. To calculate these effects in a circuit design, the ratio of those capacitors to the gate-source capacitance is calculated. Let us first analyse the overlap capacitances. They are the capacitances between the gate and the source or drain. Hence they can very well be modelled as

$$C_{gd,ovl} = C_{gs,ovl} = C_{go}\cdot W \tag{8}$$

with C_{go} a technology dependent parameter. Hence the ratio becomes

$$\alpha_{gdo} = \alpha_{gso} = \frac{C_{go}\cdot W}{\frac{2}{3}C_{ox}\cdot W\cdot L} = \frac{C_{go}}{\frac{2}{3}C_{ox}\cdot L} \tag{9}$$

This parameter is a strong a function of the length. Because in high frequency designs, minimum lengths are usually taken, this parameter can be considered as a constant. For the technology used as example (see also table 2) this parameter becomes about $\alpha_{gso} = \alpha_{gdo} = \alpha_{go} \approx 0.3$.

Secondly, let us study the drain- (source-) bulk capacitance. To do so, it has to be remarked that those capacitances are lay-out dependent. In Fig.1, three commonly used lay-out structures are shown. For each of those structures the following relationship can be found for the drain-bulk

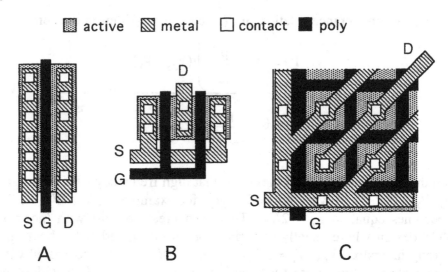

▨ active   ▨ metal   □ contact   ■ poly

Figure 1: Different lay-out style of MOS transistors A: straight forward; B: inter-digit; C: waffle.

capacitances :

$$C_{db,a} \approx W \cdot L_{edge} \cdot C_{dbj} + (W + 2L_{edge}) C_{dbjsw}$$

$$C_{db,b} \approx \frac{1}{2} W \cdot L_{edge} \cdot C_{dbj} + 2L_{edge} \cdot C_{dbjsw} \tag{10}$$

$$C_{db,c} \approx \frac{1}{4} W \cdot L_{edge} \cdot C_{dbj}$$

with L_{edge} the drain (source) diffusion area length, C_{dbj} the junction capacitance and C_{dbjsw} the side-wall junction capacitance. These parameters are all technology dependent, and as a result the designer has only impact on the transistor width. In first order they can be simplified into:

$$C_{db,a} \approx 2C_{db,b} \approx 4C_{db,c} \tag{11}$$

or with a waffle structure approximately four times higher performance can be achieved. However this structure is not so good for noise performances due to the large bulk resistance [2]. Furthermore, to achieve low junction capacitances with this structure, only discrete aspect ratio's can be realized. Therefore it is advised to use the inter-digit structure. If again the ratio factors are calculated, the following relationship can be found :

$$\alpha_{db,a} \approx 2\alpha_{db,b} \approx 4\alpha_{db,c} \approx \frac{L_{edge}}{\frac{2}{3}C_{ox} \cdot L} \cdot C_{dbj} \qquad (12)$$

In fact those parameters are slightly a function of the transistor width (W). In Fig.2 these parameters are plotted as a function of the transistor width.

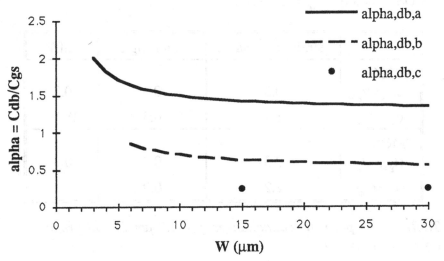

Figure 2: α_{db} of an NMOS transistor as function of lay-out style and transistor width (W): $\alpha_{db,a}$ = straight forward; $\alpha_{db,b}$ = inter-digit; $\alpha_{db,c}$ = waffle.

As can be seen, for wide transistors (and that is usually the case in high frequency designs) they can be assumed to be constant. In table 2 the different parameters can be found for a transistor of $W = 20$ μm and $L = 1$ μm. The reason to introduce those parameters is to make it easy to analyse real circuits. For example, if the -3 dB point of a single transistor has to be calculated, the following relationship can easily be found :

$$f_{-3dB} = \frac{g_m}{2\pi \left(C_{gs} + C_{gso} + C_{db}\right)} = \frac{f_{to}}{1 + \alpha_{go} + \alpha_{db}} \qquad (13)$$

Or with other words, for an NMOS transistor in a straight forward lay-out structure the $f_{-3\ dB}$ drops from the ideal f_{to} = 6.5 GHz down to $f_{-3\ dB} \approx 2.4$ GHz. This is even worse for a PMOS transistor: $f_{-3\ dB}$ drops from f_{to} = 2.5 GHz down to 0.7 GHz. If for example a two-transistor current mirror is studied, the pole associated with this structure becomes:

$$f_{-3dB} = \frac{g_m}{2\pi\,(2C_{gs} + 2C_{gso} + C_{db} + C_{gdo})} = \frac{f_{to}}{2 + 3\alpha_{go} + \alpha_{db}} \tag{14}$$

which gives approximately for an NMOS structure $f_{-3\ dB}$ = 1.5 GHz and for an PMOS structure $f_{-3\ dB}$ = 0.5 GHz.

	straight forward	inter-digit	waffle
NMOS $\alpha_{gdo}=\alpha_{gso}$	0.3	0.3	0.3
$\alpha_{db}=\alpha_{sb}$	1.4	0.5	0.25
PMOS $\alpha_{gdo}=\alpha_{gso}$	0.3	0.3	0.3
$\alpha_{db}=\alpha_{sb}$	2.2	0.9	0.4

Table 2: The parasitic capacitances of a 1.2 μm CMOS process (W=20 μm;L=1 μm).

In order to be totally correct, the interconnecting capacitances have to be included as well. The α-factor can easily be calculated as

$$\alpha_{inter} = \frac{C_{poly-sub} \cdot L_{inter} \cdot W_{inter}}{2\pi\,\frac{2}{3}C_{ox}WL} \tag{15}$$

with L_{inter} and W_{inter} the dimensions of an interconnecting wire. However the effect of this parasitic capacitance can drastically be reduced by designing wide transistors. As a matter of fact, for transistors with $W = 20$ μm, α_{inter} becomes smaller than 0.05 for an interconnecting of $L_{inter} = 2$ μm and $W_{inter} = 150$ μm. Or with other words, this effect can be neglected compared to the other parasitics.

3. Differential pair with active load

A very simple and well-known amplifier structure is a differential pair with an active load (OTA), as presented in Fig.3. The gain-bandwidth (GBW) can easily be calculated as

$$GBW = \frac{g_m}{2\pi\, C_l} \tag{16}$$

with C_l the total capacitance at the output node. Because opamps are used in feedback networks, the stability of the system must be guaranteed. This

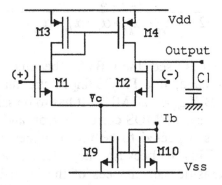

Figure 3: Circuit diagram of the simple amplifier.

is usually described by the parameter phase margin. In order to be able to calculate the phase margin, the position of the second pole should be known. The second pole in this system is a result of the active load or the current mirror. This pole is given by

$$f_2 = \frac{f_{to}}{2 + 3\alpha_{go} + \alpha_{gd} + \alpha_{np}(\alpha_{db} + \alpha_{go})} \tag{17}$$

with α_{np} the ratio between the effective mobility of an NMOS and an PMOS transistor (see table 1). This ratio is also the ratio between the W/L of the NMOS and PMOS devices. This constrain is necessary to ensure that they have the same transconductance. If this is not taken in account, the noise performance of the amplifier can drastically decrease [4].

In order to reach a phase margin of 65° the second pole should be located at a frequency twice the GBW, or

$$f_2 = 2 \cdot GBW = \frac{2g_m}{2\pi\, C_l} = \frac{2f_{to}}{2 + 3\alpha_{go} + \alpha_{gd} + \alpha_{np}(\alpha_{db} + \alpha_{go})} \tag{18}$$

Increasing V_{GS}-V_T increases f_{to} and so also the GBW. However increasing V_{GS}-V_T decreases the output voltage swing and also the input common mode range. For practical use this voltage is limited to approximately V_{GS}-$V_T = 1$V. So, if NMOS input devices are used, the maximum GBW (GBW$_{max}$) is as a result given by

$$GBW_{max} \approx \frac{f_{to,PMOS}}{2(2 + 3\alpha_{go} + \alpha_{gd} + \alpha_{np}(\alpha_{db} + \alpha_{go}))} \approx 230\ MHZ \tag{19}$$

If an inter-digit lay-out is used, the GBW$_{max}$ becomes approximately 330 MHz. On the other hand, if a PMOS input structure is employed, the parameter becomes GBW$_{max} \approx 320$ MHz. It has to be said that although the PMOS input structure has a NMOS current mirror, and as a result the pole at the internal node is expected to be much higher, the GBW$_{max}$ is not significant higher. This is because the internal node capacitance is dominated by the parasitic capacitances of the input PMOS devices. If inter-digit structures are used, the value becomes GBW$_{max} \approx 520$ MHz !

In Fig. 4 the maximum GBW is presented as function of V_{GS}-V_T for inter-digit lay-out structures. As can be seen, by increasing V_{GS}-V_T the GBW$_{max}$ increases too. However the drawback of employing large V_{GS}-V_T values is the reduced output voltage swing and the reduction of the maximum DC gain per transistor. The DC gain per transistor is the ratio of the transconductance (g_m) and the drain-source conductance (g_{ds}). g_{ds} is approximately given by $I_{DS}/V_E \cdot L$ with V_E the Early voltage process parameter. Hence the g_m/g_{ds} becomes

$$\frac{g_m}{g_{ds}} = \frac{2I_{DS} \cdot V_E \cdot L}{(V_{GS} - V_T) \cdot I_{DS}} = \frac{2V_E \cdot L}{V_{GS} - V_T} \tag{20}$$

It can be concluded that the smaller L and the higher V_{GS}-V_T are, the smaller the DC gain becomes. This is just the opposite of which is required to achieve high f_t.

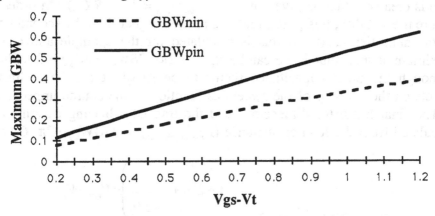

Figure 4: The maximum GBW of the simple amplifier as function of V_{GS}-V_T.

This DC gain problem can be reduced by using cascode transistors or the super gain stage [10,11]. Hence this problem will not be further discussed in this contribution.

Another important conclusion is that higher frequency performances can be obtained with an PMOS input structure. However it has to be remarked that using PMOS input devices can result in very important input capacitances. For example in this case where V_{GS} -V_T is chosen to be 1 V, the capacitance between the inverting and the non-inverting input is approximately 50 fF. It seems to be small, but this parasitic capacitance can drastically decrease the opamp performance. For instance, this is the case if the amplifier is used as unity-gain buffer as is presented in Fig.5.

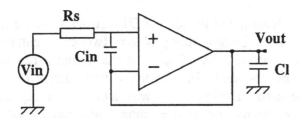

Figure 5: The effect of the input capacitance and the source impedance on the stability.

This is a typical configuration for those applications where the source impedance has to be buffered. Due to the finite input capacitance an extra pole is created which is given by $(C_i > C_{in})$: $f_{in} = 1/(2\pi R_s C_{in})$. In order to ensure the stability, this pole should be at least higher than the second pole of the amplifier itself. If this is calculated for the example above, the maximum source impedance can be $R_s < 1/(2\pi \cdot GBW_{max} \cdot 2C_{in}) \approx 3$ kOhm! Although the input capacitance seems to be small, it can have drastic effects on the stability of the system as function of the source impedance.

The final transistor dimensions and the transistor biasing currents are calculated from the load capacitance ($g_{m,input\ transistors} = GBW_{max} \cdot 2\pi \cdot C_l$) or:

$$W = GBW_{max} \cdot 2\pi \cdot C_l \cdot L \cdot \frac{\left(1 + 2\left(\theta + \dfrac{\mu}{v_{max}L}\right) \cdot (V_{GS} - V_T)\right)}{\mu C_{ox}(V_{GS} - V_T)} \qquad (21)$$

For the example of an inter-digit NMOS input structure, and a total input capacitance of 1 pF the transistor width becomes W $\approx$ 43 μm. The transistor biasing current is given by relationship (7) ($2I_{DS} \approx g_m \cdot (V_{GS} - V_T)$) or

$$I_{DS} = GBW_{max} \cdot 2\pi \cdot C_l \cdot \frac{V_{GS} - V_T}{2} \qquad (22)$$

which is approximately 1 mA for the example above. It is important to note that both W/L (and so also the chip area) and the biasing current (and so also the power drain) are a linear function of the load capacitance.

4. The load compensated OTA (LC-OTA)

The main disadvantage of the simple OTA structure is the limited output voltage swing. To overcome this problem, two extra current mirrors, as presented in Fig.6 are usually added. At the output a rail-to-rail voltage swing can now be obtained. For the upper two current mirrors a current ratio (B) can be used. However this will decrease the higher order poles of the OTA (larger transistor areas $\Rightarrow$ more capacitance) and will increase the total equivalent input voltage noise. Therefore a ratio $B=1$ is better taken as a good compromise and will be further assumed in this text.

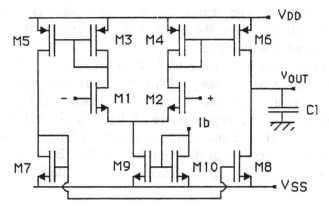

Figure 6: The load-compensated OTA.

Concerning the gain-bandwidth specifications, this structure has the same relationships as the simple OTA:

$$GBW = \frac{g_m}{2\pi C_l} \qquad (23)$$

Due to those extra current mirrors, this structure has mainly two high frequency poles: one as a result of the upper two current mirrors and an extra one due to the bottom current mirror. The first pole (f_a) is the same as the one of the differential pair with active load:

$$f_a = \frac{f_{to,PMOS}}{2 + 3\alpha_{go} + \alpha_{gd} + \alpha_{np}(\alpha_{db} + \alpha_{go})} \qquad (24)$$

The other pole (f_b) is given by

$$f_b = \frac{f_{to,NMOS}}{2 + 3\alpha_{go} + \alpha_{gd} + \alpha_{pn}(\alpha_{db} + \alpha_{go})} \qquad (25)$$

As can be seen from relationship (24) and (25), the main difference, for the same transistor lengths and ($V_{GS} - V_T$) values, is the f_{to}. Hence f_b is typically a factor $f_{to,NMOS}/f_{to,PMOS} \approx 2.4$ higher than f_a. This results in an extra phase shift of about 10°-15°. Hence this amplifier has a slightly smaller GBW (approximately 20-30%) for the same phase margin compared to the differential pair with active load.

A PMOS input structure will result in a higher GBW. However in this case f_b will become smaller than f_a. This will result in a pole-zero pair in the transfer function. The drawback is that a slow settling component exists that deteriorates the step response of the amplifier [5]. Therefore a PMOS input is not advised and will not be studied further on in this text.

To overcome the problem of the reduction in the GBW due to the extra current mirrors, a folded cascode structure, as presented in Fig.7, can be used.

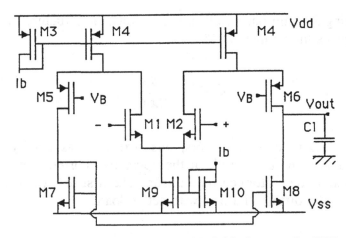

Figure 7: Circuit diagram of a folded cascode OTA.

The pole associated with the upper folded current mirrors is shifted to higher frequencies (less capacitance at the source of the cascode transistors) and given by

$$f_a = \frac{f_{to,PMOS}}{1 + 2\alpha_{go} + 2\alpha_{gd} + \alpha_{np}(\alpha_{db} + \alpha_{go})} \tag{26}$$

which is roughly a factor two higher than the usual current mirror structure. As a result a gain in phase shift of approximately 10°-15° is obtained. So the phase loss due to the bottom current mirror is compensated by the phase gain in the folded cascode structure. Hence, a folded cascode transconductance amplifier has approximately the same GBW (for a given phase margin) as the differential pair with active load. The conclusion is that the maximum GBW that can be obtained with a folded cascode OTA in a given CMOS technology is given by relationship (19). In the technology used as example, this means a $GBW_{max} \approx 330$ MHz.

In Fig.8 the transfer function of a simulated folded cascode amplifier is shown. It can be seen that the achieved GBW is indeed about 330 MHz with a phase margin of 70°.

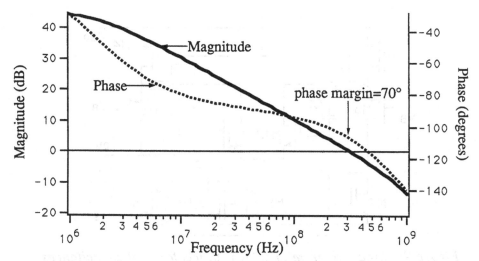

Figure 8: SPICE level 3 simulation of folded cascode OTA (W = 20 μm).

5. Conventional Feedforward Techniques

From the discussion of the folded cascode OTA, it is clear that the maximum GBW is related to the slowest device, in this case the PMOS cascode transistor (see relationship (24) and (26)). Therefore several attempts to overcome this problem can be found in the literature [6,7,8,9]. A simple technique to overcome the problem is using a feedforward capacitance (C_f) as is shown in Fig.9. For high frequencies the AC current of the input devices is flowing through the capacitors C_f and passing-by the PMOS devices. Hence for high frequencies, the circuit acts as an all NMOS design. The second pole is now generated by the bottom NMOS current mirror. Again, if all transistors are designed with maximum $V_{GS}-V_T$ and equal transconductances, the maximum GBW becomes:

$$GBW_{max} = \frac{f_a}{2} \approx \frac{f_{to,NMOS}}{2\left(2 + 2\alpha_{go} + 4\alpha_{gd} + \alpha_{np}\left(\alpha_{db} + \alpha_{go}\right)\right)} \qquad (27)$$

As a result very high gain bandwidths can be achieved. For example in the process used as example, GBW_{max} becomes with an inter-digit lay-out style approximately 450 MHz. However it has to be remarked that in those

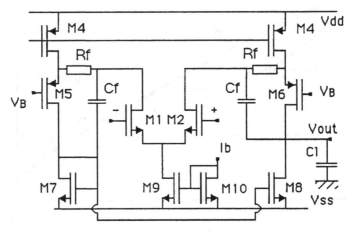

Figure 9: Circuit diagram of OTA with feedforward capacitances.

calculations the effect of any stray capacitance of the feedforward capacitance has been neglected. In practice this feedforward capacitance can become large (1-10 pF) which will indeed result in important stray capacitances, and as a result, it will reduce the maximum achievable GBW.

The limitation in this feedforward structure is the bandwidth limitation of the current mirror. A way to avoid this is using fully differential structures with feedforward capacitances. An example of a total circuit with feedforward capacitances is presented in Fig.10. As can be noted extra transistors are now required to control the common mode output voltage. With this structure a GBW of 850 MHz in a 1.2 μm CMOS process has been reported [8]. It has to be remarked that the feedforward capacitances are connected at the source of the cascode transistors. This will result in a better pole-zero matching as will be shown in the next paragraph.

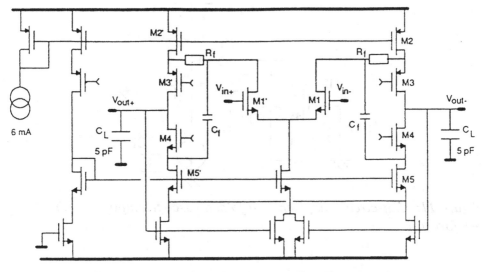

Figure 10: Fully differential wideband amplifier.

6. Advanced Feedforward Techniques

It has already been mentioned that the presence of a pole-zero pair before the GBW of the amplifier can result in a slow settling time on a step-input response. Mathematically it can be shown [5] that the step-input response of an amplifier in unity-gain follower mode is given by

$$V_{OUT}(t) = V_{STEP}\left(1 - \frac{A}{1+A}\exp(-2\pi\,GBW\cdot t) + \frac{f_z - f_p}{GBW}\exp(-2\pi f_z t)\right) \quad (28)$$

with A the low-frequency open loop gain, f_z and f_p the zero and the pole, and V_{STEP} the applied step input voltage. The second term is the well-known settling component. However, due to the pole-zero pair, an extra settling component exists. Unfortunately this is a component with a large time constant ($f_z < GBW$). If the pole-zero pair can very well be matched ($f_z = f_p$) the effect vanishes.

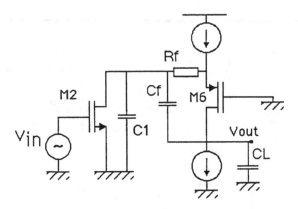

Figure 11: Half-circuit diagram of a folded cascode amplifier with feedforward.

In order to study this effect for the classical feedforward technique, let us analyse the half circuit diagram, as is presented in Fig.11, of the folded cascode amplifier of Fig.9. The pole-zero pair is given by [7]

$$f_p \approx \frac{1}{2\pi R_f \left(C_f + C_1 + \frac{C_f C_1}{C_L} \right)} = \frac{1}{2\pi R_f\, C_f \left(1 + \frac{C_1}{C_f} + \frac{C_1}{C_L} \right)} \qquad (29)$$

and

$$f_z = \frac{1}{2\pi R_f\, C_f} \qquad (30)$$

with R_f and C_f the feedforward resistor and capacitance, C_L the total load capacitance and C_1 the total capacitance to the ground at the internal node. Hence by increasing C_f and C_L, the matching between the pole and the zero becomes better. If the feedforward capacitance is connected to the source of a cascode transistor, as it is the case in Fig.10, the effect of C_L on the pole-zeros can be neglected. This is mathematically equivalent to take $C_L = \infty$. To see the effect on the settling accuracy the ratio of f_z-f_p to the GBW is calculated:

$$\frac{f_z - f_p}{GBW} \approx \frac{1}{2\pi R_f\, C_f\, GBW} \cdot \left(\frac{C_1}{C_f} + \frac{C_1}{C_L} \right) \qquad (31)$$

In order to meet the settling requirement, the relationship should at least satisfy the required accuracy. In Fig.12 the relationship is plotted as function of C_f for the case that $C_1 = 0.2$ pF, $R_f = 5$ kOhm and for the case that $C_L = 1$ pF and $C_L = \infty$.

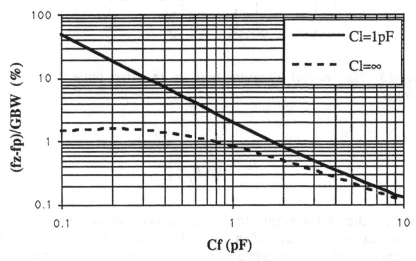

Figure 12: Doublet over GBW ratio: GBW = 450 MHz; $R_f = 5$ kOhm.

As can be noticed, by increasing C_f the settling accuracy decreases too. However, even if C_L is infinite, still large feedforward capacitances are required to achieve high settling accuracies (e.g. $0.1\% \Rightarrow C_f \approx 10$ pF). Even more, for large C_f values the effect of the load capacitance becomes less important. This means that connecting C_f to the sources of the cascode transistors gives no real benefit in the compensation of the pole-zero pair in order to achieve high settling accuracies.

A second compensation technique, based on a resistive biased feedforward technique [7], is presented in Fig.13. Instead of inserting the resistance between the source and the feedforward capacitance, a resistance is placed in series with the gate of the transistor. Analysing the pole-zero pair of the structure, they become

$$f_p = \frac{C_L + C_f}{2\pi R_f C_f (C_L + C_1)} \tag{32}$$

and

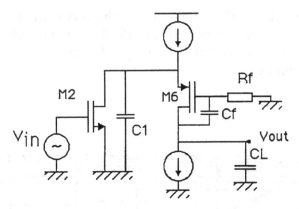

Figure 13: Half-circuit diagram with a resistive biased feedforward technique.

$$f_z = \frac{1}{2\pi R_f C_f} \tag{33}$$

If now C_f is chosen to be equal to $C_f = C_1$, an exact cancellation of the pole and zero occurs. Hence the effect of the slow settling component vanishes. The advantage of this technique is that the required capacitance is very small (in this case only 200 fF). The drawback is that in practice the matching of C_f, which is a double poly capacitance, with C_1, which is a junction capacitance, is not so easy to realize. In Fig.14 the transfer function of an OTA with resistive biased feedforward technique is shown. Due to the perfect cancellation of the pole and zero, no phase dip can be seen in the plot of the phase ($f_z \approx f_p \approx 150$ MHz). However, the best way to see the effect is analysing the step-input response, which is shown in Fig.15. The settling time of the opamp (0.1%) is approximately equal to the one of a first order 450 MHz bandwidth system. Almost no effect of the slow settling component due to the doublet can be noticed. It is important to note that the overshoot of the opamp response is due to the high frequency signal feedthrough (zero in the right-hand s-plane) through the gate-drain capacitances of the input transistors.

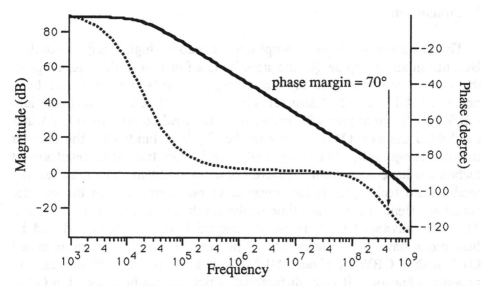

Figure 14: Simulated frequency response of the cascode OTA with enhanced feedforward technique (GBW ≈ 450 MHz).

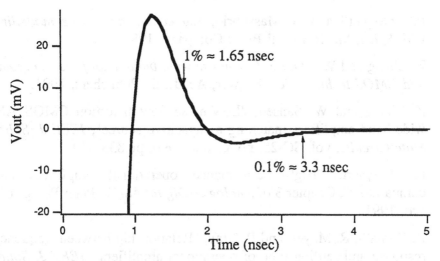

Figure 15: Simulated step-input response of the cascode OTA with enhanced feedforward technique.

7. Conclusion

The design of operational amplifiers towards a high GBW is studied. Because in an opamp design the stability is a function of the internal pole, the design and transistor lay-out strategy towards a high f_t has been presented. Different OTA structures have been analysed towards maximum GBW a differential pair with active load, the load compensated OTA and the folded cascode OTA. Because the GBW is limited by the slowest internal component, which is usually the PMOS transistor, feedforward techniques have been studied to overcome the problem. The drawback of feedforward techniques is the presence of pole-zero pairs in the transfer function. Hence the settling-time of the amplifier can drastically increase. Therefore advanced feedforward techniques have been analysed and its limitations have been presented. With these techniques a single-ended OTA with a GBW of about 450 MHz in a 1.2 μm CMOS process are nowadays feasible. If fully differential structures can be allowed, a GBW of about 850 MHz can be achieved.

References

[1]   P. Antognetti and G. Massobrio, *Semiconductor Device modelling with Spice*, McGraw-Hill Book Company, 1987.

[2]   Z. Chang and W. Sansen, *Low-noise wide-band amplifiers in bipolar and CMOS technologies*, Kluwer, Academic Publishers, 1991.

[3]   Z. Chang and W. Sansen, "Low-noise, low-distortion CMOS AM wide-band amplifiers matching a capacitive source", *IEEE J. Solid-State Circuits*, vol. SC-25, no.3, June 1990, pp.833-840.

[4]   M. Steyaert, "High performance operational amplifiers and comparators", Chapter 3 of *Analogue-Digital ASICs*, Peter Peregrinus Ltd, 1991.

[5]   B. Kamath, R. Meyer, and P. Gray, "Relationship between frequency response and settling time of operational amplifiers", *IEEE J. Solid-State Circuits*, vol. SC-9, no.6, December 1974, pp.347-352.

[6]   R. Apfel and P. Gray, "A fast-settling monolithic operational amplifier using doublet compression techniques", *IEEE J. Solid-State Circuits*, vol. SC-9, no.6, December 1974, pp.332-340.

[7]   Z. Chang and W. Sansen, "Feedforward compensation techniques for

high frequency CMOS amplifiers", *IEEE J. Solid-State Circuits*, vol. SC-25, no.6, December 1990, pp.1590-1595.

[8] F. Op't Eynde and W. Sansen, "A CMOS wideband amplifier with 800 MHz gain-bandwidth", *IEEE Proc. CICC '91*, pp.9.1.1-9.1.4.

[9] P. Gray and R. Meyer, "Recent advances in monolithic operational amplifier design" *IEEE Trans. Circuits Systems*, vol. CAS-21, no.3, May 1974, pp.317-327.

[10] K. Bult and G. Geelen, "The CMOS Gain-Boosting technique", *Analog integrated circuits and signal processing 1*, 1991, pp.119-135.

[11] E. Sackinger and W. Guggenbuhl, "A high swing, high-impedance MOS cascode circuit", *IEEE J. Solid-State Circuits*, vol. SC-25, no.1, February 1990, pp.289-298.

Biography

Michel S.J. Steyaert was born in Aalst, Belgium, in 1959. He received his Ph.D. degree in electrical and mechanical engineering from the Katholieke Universiteit Leuven, Heverlee, Belgium in 1987.

From 1983 to 1986 he obtained an IWNOL fellowship which allowed him to work as a research assistant at the Laboratory ESAT - K.U. Leuven. In 1987 he was responsible for several industrial projects in the field of analog micro-power circuits at the Laboratory ESAT - K.U. Leuven as an IWONL project-researcher. In 1988, he was a visiting assistant professor at the University of California Los Angeles. Since 1989 he has been appointed as a NFWO research associate at the Laboratory ESAT - K.U. Leuven, where he has been an associated professor since 1990. His current research interests are in high frequency analog integrated circuits for telecommunications and integrated circuits for biomedical purposes.

Prof. Steyaert received the 1990 European Solid-State Circuits Conference Best Paper Award, and the 1991 NFWO Alcatel-Bell-Telephone award for innovated work in integrated circuits for telecommunications.

For a biography of *W. Sansen* please refer to page 445.

The CMOS Gain-Boosting Technique

Klaas Bult, Govert J.G.M. Geelen

Philips Research Laboratories
Eindhoven
The Netherlands

This paper was published before in
Analog Integrated Circuits and Signal Processing 1

Abstract

The Gain-Boosting Technique improves accuracy of cascoded
CMOS circuits without any speed penalty. This is achieved
by increasing the effect of the cascode transistor by means of
an additional gain-stage, thus increasing the output
impedance of the sub-circuit. Used in opamp design, this
technique allows the combination of the high-frequency
behaviour of a single-stage opamp with the high DC-gain of
a multi-stage design. Bode-plot measurements show a
DC-gain of 90 dB and a unity-gain frequency of 116 MHz
(16 pF load). Settling measurements with a feedback factor
of ⅓ show a fast single-pole settling behaviour corresponding
with a closed loop bandwidth of 18 MHz (35 pF load) and a
settling accuracy better than 0.03 percent. A more general
use of this technique is presented in the form of a
transistor-like building block: the Super-MOST. This
compound circuit behaves as a normal MOS-transistor but
has an intrinsic gain $g_m \cdot r_o$ of more than 90 dB. The
building block is self biasing and therefore very easy to
design with. An opamp consisting of only 8 Super-MOSTs
and 4 normal MOSTs has been measured showing results
equivalent to the design mentioned above.

1. Introduction

Speed and accuracy are two of the most important properties of analog circuits; optimizing circuits for both aspects leads to contradictory demands. In a wide variety of CMOS analog circuits such as switched-capacitor filters [1] - [3], algorithmic A/D convertors [4], sigma-delta convertors [5], sample-and-hold amplifiers and pipeline A/D convertors [6], speed and accuracy are determined by the settling behaviour of operational amplifiers. Fast-settling requires a high unity-gain frequency and a single-pole settling behaviour of the opamp, whereas accurate settling requires a high DC-gain.

The realization of a CMOS operational amplifier that combines high DC-gain with high unity-gain frequency has been a difficult problem. The high-gain requirement leads to multi-stage designs with long-channel devices biased at low current levels, whereas the high unity-gain frequency requirement asks for a single-stage design with short-channel devices biased at high bias current levels.

Future processes with submicron channel-length will enable us to realize higher unity-gain frequencies. However, the intrinsic MOS transistor gain $g_m \cdot r_o$ will then be lower [7], and the problem of achieving sufficient DC-gain becomes even more severe.

There have been several circuit approaches to circumvent this problem. Cascoding is a well known means to enhance the DC-gain of an amplifier without degrading the high-frequency performance. The result is a DC-gain which is proportional to the square of the intrinsic MOS transistor gain $g_m \cdot r_o$. In modern processes with short-channel devices and an effective gate-driving voltage of several hundreds of milliVolts the intrinsic MOS transistor gain $g_m \cdot r_o$ is about 20 - 25 dB, resulting in a DC-gain of the cascoded version of about 40 - 50 dB. This is however in many cases not sufficient [1], [8], [9].

Dynamic biasing of transconductance amplifiers [10] - [12], was one of the first approaches reported to combine high DC-gain with high settling speed. In this approach the bias current is decreased, either as a function of time during one clock period [10], [11] or as a function of the amplitude of the input signal [12], resulting in a higher DC-gain at the end of the settling period. This decreases the unity-gain frequency which makes the last part of the settling very slow.

In [13] a triple-cascode amplifier has been implemented where the gain is proportional to $(g_m \cdot r_o)^3$. This approach has two significant disadvantages. First, every transistor added in the signal path introduces an extra pole in the transfer-function. In order to obtain enough phase-margin, the minimum load capacitance has to be increased resulting in a lower unity-gain frequency. Secondly, each transistor reduces the output-swing by at least the effective gate-driving voltage.

In [8] positive feedback is used to enhance the gain of an amplifier. This approach however is limited by matching. A gain enhancement of about 16 dB is reported. Starting from an original gain of 50 dB, one could obtain 70 dB DC-gain. For high-Q, high-frequency Switched Capacitor filters, this is not sufficient. Even a moderate Q of 25 and a maximum deviation of 1-percent requires a minimum opamp gain of 74 dB [1], [8]. Therefore, we aim at a DC-gain of at least 80 dB combined with a unity-gain frequency of 100 MHz.

In [14], a regulated-cascode stage was reported which increases the DC-gain of a normal cascode stage. In [15] the performance of this circuit with respect to output-swing and output-impedance was analyzed. The extension of this circuit into a general gain-boosting technique was presented in [16,17], showing a complete opamp design with a measured DC-gain of 90 dB and a unity-gain frequency of 116 MHz. It was shown that this technique enhances the DC-gain of a cascoded amplifier several orders of magnitude *without* any penalty in speed or output-swing.

This paper gives a complete overview of the gain-boosting technique and presents a general use of this technique in the form of a transistor-like building block: the Super-MOST.

In section 2 the principle of the gain-boosting technique is explained. Section 3 deals with the high-frequency behaviour. Section 4 discusses the optimization towards fast settling behaviour. In section 5, some remarks on output swing are made. In section 6 the circuit implementation of an opamp is presented, and in section 7 the measurement results of this opamp are shown. Then, in section 8, the Super-MOST is presented, along with measurement results of the device on its own as well as of an opamp built with this building block.

2. Gain-Boosting Principle

We start this section with the functioning of the simple cascode stage shown in Fig.1. We show that the DC-gain can be expressed as the product of an *effective transconductance* $g_{m,eff}$ and the output-impedance. This explanation leads to the gain-boosting principle. The repetitive application of this principle leads to a decoupling of the DC-gain and the unity-gain frequency of an opamp.

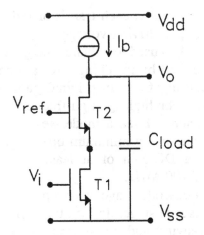

Figure 1: Cascoded gain stage.

2.1. Cascode stage

The transfer function of the cascode stage can be found as follows. Suppose an ideal voltage source is connected to the output of the stage. If a voltage ΔV_i is applied to the input, a current ΔI_o will flow into the voltage source. For low frequencies:

$$\frac{\Delta I_o}{\Delta V_i} = g_{m1} \frac{(g_{m2}r_{o1} + r_{o1}/r_{o2})}{(g_{m2}r_{o1} + r_{o1}/r_{o2} + 1)} = g_{m,eff} \tag{1}$$

This *effective transconductance* is almost equal to g_{m1}, the reduction is caused by the feedback via the drain of the input transistor. Removing the voltage source causes a change $\Delta V_o = \Delta I_o R_{out}$, where ΔI_o is calculated using (1) and R_{out} is the output resistance of the total circuit. The voltage-gain A_o is:

$$A_o = g_{m,eff} R_{out} \tag{2}$$

The output impedance of the total circuit can easily be calculated to be:

$$R_{out} = (g_{m2}r_{o2} + 1)r_{o1} + r_{o2} \tag{3}$$

which leads to the following expression for the DC-gain:

$$A_o = g_{m1} r_{o1} (g_{m2} r_{o2} + 1) \qquad (4)$$

The behaviour of the circuit, in the vicinity of the unity-gain frequency, is similar with the voltage-source connected to the output. Now the load-capacitor C_{load} forms the short-circuit to ground. At the unity-gain frequency we use the effective transconductance as given by (1), and this results in the following expression for the Gain-Bandwidth product (GBW):

$$GBW = \frac{g_{m,eff}}{C_{load}} \qquad (5)$$

From equations (2) and (5) we may conclude that the only way to improve A_o without reducing the GBW is to *increase the output impedance*. Note that this is the effect of the cascode transistor itself with respect to the non-cascoded situation: the cascode transistor shields the drain of the input transistor from the effect of the signal swing at the output.

As we can see from (2), the DC-gain is proportional to the output impedance of the circuit. This implies that we may consider the gain of such a stage as the output-impedance normalized on $1/g_{m,eff}$. We will use this approach in section 4.

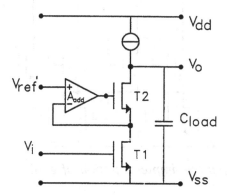

Figure 2: Cascoded gain-stage with gain enhancement.

2.2. Gain-boosting principle

The technique presented here is based on increasing the cascoding effect of T2 by adding an additional gain-stage as shown in Fig.2. This stage further reduces the feedback from the output to the drain of the input transistor. Thus, the output impedance of the circuit is further increased by the gain of the additional gain-stage A_{add} :

$$R_{out} = \left(g_{m2}r_{o2}(A_{add}+1)+1\right)r_{o1}+r_{o2} \tag{6}$$

Moreover, the effective transconductance is slightly increased:

$$\frac{\Delta I_o}{\Delta V_i} = g_{m1}\frac{g_{m2}r_{o1}(A_{add}+1)+r_{o1}/r_{o2}}{g_{m2}r_{o1}(A_{add}+1)+r_{o1}/r_{o2}+1} = g_{m,eff} \tag{7}$$

Hence, the total DC-gain now becomes:

$$A_{o,tot} = g_{m1}r_{o1}\left(g_{m2}r_{o2}(A_{add}+1)+1\right) \tag{8}$$

Section 3 deals with the high-frequency behaviour of this circuit and discusses what happens if the gain of the additional stage decreases as a function of frequency.

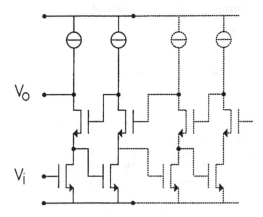

Figure 3: Repetitive implementation of gain enhancement.

2.3. Repetitive implementation of gain-boosting

If the additional stage is implemented as a cascode stage, the gain enhancement technique as described above can also be applied to this additional stage. In this way, a *repetitive* implementation of the gain enhancement technique can be obtained as shown in Fig.3. The limitation on the maximum voltage gain is then set by factors such as leakage currents, weak avalanche and thermal feedback. The implementation of the additional stage is discussed in section 6.

From the above, we may conclude that the repetitive usage of the gain-enhancement technique yields a *decoupling* of the opamp DC-gain and unity-gain frequency.

3. High-Frequency behaviour

In this section, we discuss the high-frequency behaviour of the gain-enhanced cascode stage of Fig.2. It is shown that for a first-order roll-off, the additional stage need not be fast with respect to the unity-gain frequency of the overall design.

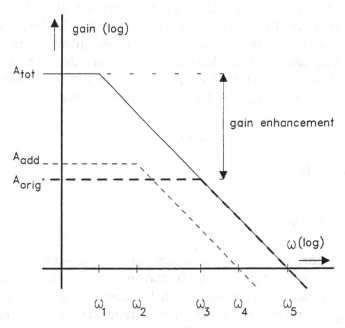

Figure 4: Gain Bode-plots of the original cascoded gain-stage (A_{orig}), the additional gain-stage (A_{add}), and the improved cascoded gain-stage (A_{tot}).

In Fig.4, a gain Bode-plot is shown for the original cascoded gain-stage of Fig.1 (A_{orig}), the additional gain-stage (A_{add}) and the improved cascoded gain-stage of Fig.2 (A_{tot}). At DC, the gain enhancement A_{tot} / A_{orig} equals approximately $[1+A_{add}(0)]$, according to (4) and (8). For $\omega > \omega_1$, the output impedance is mainly determined by C_{load}. In fact, we have to substitute $(R_{out} /\!/ C_{load})$ in equation (2) for R_{out}. This results in a first order roll-off of $A_{tot}(\omega)$. Moreover, this implies that $A_{add}(\omega)$ may have a first order roll-off for $\omega > \omega_2$ as long as $\omega_2 > \omega_1$. This is equivalent to the condition that the unity-gain frequency (ω_4) of the additional gain-stage has to be larger than the 3-dB bandwidth (ω_3) of the original stage, but it can be much lower than the unity-gain frequency (ω_5) of the original stage. The unity-gain frequencies of the improved gain-stage and the original gain-stage are the same.

From the above, to obtain a first order roll-off of the total transfer-function, the additional gain-stage does not have to be a fast stage. In fact, this stage can be a cascoded gain-stage as shown in Fig.1, with smaller width and non-minimal length transistors biased at low current levels. Moreover, as the additional stage forms a closed loop with T2, stability problems may occur if this stage is too fast. There are two important poles in this loop. One is the dominant pole of the additional stage and the other is the pole at the source of T2. The latter is equal to the second pole, ω_6, of the main amplifier. For stability reasons, we set the unity-gain frequency of the additional stage lower than the second pole frequency of the main amplifier. A safe range for the location of the unity-gain frequency ω_4 of the additional stage is given by:

$$\omega_3 < \omega_4 < \omega_6 . \tag{9}$$

This can easily be implemented.

4. Settling behaviour

In this section, the settling behaviour of the gain-enhanced cascoded amplifier-stage is discussed. It is shown that a single-pole settling behaviour demands a higher unity-gain frequency of the additional stage than a simple first-order roll-off in the frequency domain requires as discussed in the previous section. The reason for this is the presence of a closely spaced pole and zero (doublet). In section 4.1 the occurrence of a doublet in the gain-enhanced cascode amplifier-stage is discussed. Section 4.2 presents a graphical representation of the settling behaviour of an opamp, and finally in section 4.3, the condition for one-pole settling is presented.

4.1. Doublet

From (6), the gain-enhancement technique increases the output impedance, Z_{out}, by a factor approximately equal to $(A_{add} + 1)$. The gain of the additional stage, A_{add}, decreases for frequencies above ω_2 (Fig.4) with a slope of -20 dB/decade. For frequencies above ω_4, A_{add} is less than one, and the normal output impedance Z_{orig} of a cascode stage without gain-enhancement remains. This is shown in Fig.5. Also shown is the impedance of the load capacitor Z_{load} and the parallel circuit which forms the total impedance, Z_{tot}, at the output node. A closer look at this plot reveals that a doublet is present in the plot of the total output impedance near ω_4.

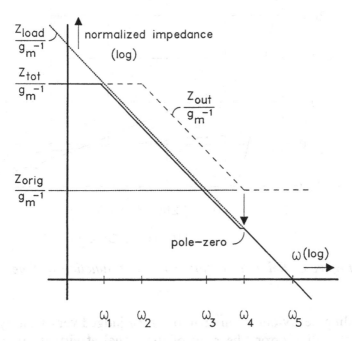

Figure 5: The Normalized output impedance as a function of frequency.

Non-complete doublet cancellation can seriously degrade the settling behaviour of an opamp [18]. If a doublet is present in an opamp open loop transfer-function at ω_{pz} with a spacing of $\Delta\omega_{pz}$, and the opamp is used in a feedback situation with feedback factor β, a slow settling component is present with time-constant $1/\omega_{pz}$. The relative magnitude (with respect to the total output signal) of the slow-settling component is given by [18]:

$$\frac{\Delta V_{out,slow}}{\Delta V_{out,total}} = \frac{\Delta\omega_{pz}}{\omega_{pz}} \cdot \frac{\omega_{pz}}{\beta\omega_{unity}} = \frac{\Delta\omega_{pz}}{\beta_{unity}} \tag{10}$$

For a single-pole settling behaviour, the relative magnitude of the slow settling component, given in (10), has to be smaller than the ultimate settling accuracy $1/\beta A_{tot}(0)$. Note that for opamps with a very high DC-gain this requirement can be very difficult to realize.

4.2. Graphical representation of settling behaviour

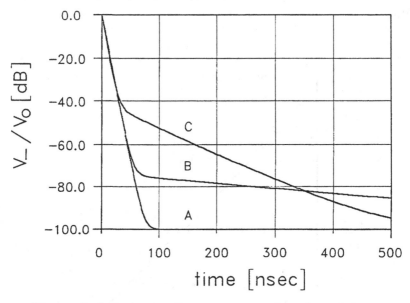

Figure 6: Relative settling error as a function of time.

The settling behaviour of an opamp can be judged very well by plotting the relative settling error (the ratio of the signal at virtual ground of the opamp and the output-step) versus time, as shown in Fig.6. Here an ideal single-pole settling behaviour is shown as a straight line. In Fig.6, the simulated result is shown of an opamp in unity-feedback with a unity-gain frequency of 25 Mhz and a DC-gain of 100 dB. Curve A shows the result without the presence of a doublet: a straight line down to -100 dB with a steep slope corresponding to one small time constant. Curve B shows the result of an opamp with the same DC-gain and unity-gain frequency but now with a doublet at 500 KHz and 1 percent spacing. The slow settling component causes a severe deviation from the straight line. According to (10), the slow-settling component has a relative magnitude of -74 dB in this situation, which is in close agreement with the result shown. The slope of this line is 50 times smaller than the slope of the fast settling component. Curve C shows the result of a simulation with an opamp with again the same DC-gain and unity-gain frequency but now with a doublet at 2.5 MHz and a relative spacing of 10 percent. As seen from this figure, slow-settling components can easily be detected in this way. In section 7 where we show the measurement results, we use this representation to judge the settling behaviour of our design.

4.3. Optimization of settling behaviour

To determine the spacing in the doublet in the transfer-function of the gain-enhanced cascode stage, consider again the impedance-plot of Fig.5. The total impedance at the output of the amplifier is the parallel connection of the load capacitance and the output impedance of the circuit. At ω_2, the output impedance of the circuit begins to decrease as a function of frequency due to the roll-off of the additional stage. This can be modelled as a small capacitor in parallel with the output resistor. The ratio between this small capacitor and the load capacitor is ω_1/ω_2 (=ω_3/ω_4), as can be seen in Fig.5. This small capacitor is simply parallel connected to the (large) output capacitance and gives a small shift of the total impedance at the output. At ω_4 however, the effect of this small capacitor disappears due to the 1 in the (A_{add} + 1) term. At this frequency, a small shift in the total output impedance occurs, back to the original line (in Fig.5) determined by the load-capacitor only. It is also at this frequency where the doublet is located in the total transfer-function. From the above we can conclude that the relative spacing of the doublet is approximately ω_3/ω_4. This results in a relative magnitude of the slow-settling component according to (10) of:

$$\frac{\Delta V_{out,slow}}{\Delta V_{out,total}} = \frac{\omega_3}{\beta \omega_5} \tag{11}$$

which is equal to the inverse of the feedback factor β multiplied by the DC-gain of the original cascode stage without gain-enhancement! Thus we may conclude that the pole-zero cancellation is not accurate enough. Our approach here is to make this "slow"-settling component fast enough. If the time-constant of the doublet $1/\omega_{pz}$ is smaller than the main time-constant $1/\beta \omega_{unity}$, the settling-time will not be increased by the doublet. This situation is achieved when the unity-gain frequency of the additional stage is higher than the -3 dB bandwidth of the closed-loop circuit. On the other hand, for reasons concerning stability, the unity-gain frequency must be lower than the second-pole frequency of the main amplifier as indicated by (9). This results in the "safe" area for the unity-gain frequency of the additional stage

$$\beta\omega_5 < \omega_4 < \omega_6 \tag{12}$$

as shown in Fig.7. Note that this safe area is smaller than given by (9). A satisfactory implementation however is still *no problem*, even if $\beta=1$, because the load capacitor of the additional stage which determines ω_4, is much smaller than the load capacitor of the opamp, which determines ω_5.

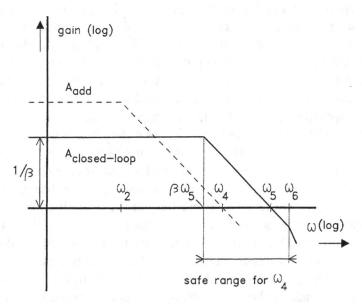

Figure 7: The "safe" range for the unity-gain frequency of the additional stage.

5. Output swing

The output swing is limited by the requirement that all transistors have to remain in the saturation region, otherwise a severe decrease in gain resulting in large distortion will occur. The edge of saturation of an MOS-transistor occurs when the gate-drain voltage equals the threshold-voltage at the drain [19]:

$$V_{D,\min} = \frac{V_G - V_{to}}{1 + \alpha} \qquad (13)$$

which is the effective gate-driving voltage divided by $(1+\alpha)$ representing the body effect, where α is determined by processing [19]. Note that a large body-effect is advantageous here. In this design, an effective gate-driving voltage of 250 mV was chosen. With $\alpha=0.3$ this results in a minimum drain-source voltage of 190 mV. Applying this result to the amplifier stage in Fig.2 leads to a minimum output voltage of 380 mV and to a large signal output swing of V_{dd} - 2 * 380 mV. With a 5.0 V supply, this results in a maximum output swing of 4.2 V. Note that a different criterion is used for the maximum swing at the output than in [15]. To be able to obtain this large output swing, the additional stage has to have an input common-mode range close to the supply voltage V_{ss}.

6. Opamp Circuit Implementation

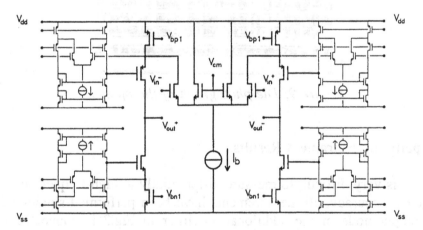

Figure 8: Complete circuit diagram of the opamp.

In this section, the implementations of the main opamp and the additional gain-stages are discussed. The main stage is a folded-cascode amplifier [2]. The simplest implementation of the additional stage is one MOS-transistor [14, 15, 20]. We have chosen a cascode version because of its high gain and the possibility of repetitive usage of the gain enhancement technique as discussed in section 2. The input stage design of the additional amplifier is determined by the common-mode range requirement which is close to the V_{ss} as discussed before. As a consequence, a folded-cascode structure with PMOS input transistors is chosen for the additional amplifier in Fig.2. To realize a very high output impedance, the current source in Fig.2 is also realized as a cascoded structure with an additional gain stage. A fully differential version (Fig.8) has been integrated in a 1.6 μm CMOS process. The two input transistors connected to V_{cm} have been added to control the common-mode bias voltage at the output. Using this scheme, the circuit can also be used as two single-ended opamps. The die photograph of Fig.9 clearly shows that the additional stages are much smaller in chip area than the main opamp.

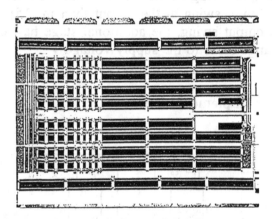

Figure 9: Die photograph of the opamp.

7. Opamp Measurement Results

As it is very difficult to measure differentially at high-frequencies with sufficient accuracy, all measurements have been performed single-ended. The output node of the additional amplifier in Fig.2 is connected to a bonding pad in order to be able to switch off the gain enhancement technique. Results of gain measurements both with and without gain enhancement are shown in Fig.10. A DC-gain enhancement of 45 dB was measured without affecting the gain or phase for higher frequencies, resulting in a total DC-gain of 90 dB combined with a unity-gain frequency of 116 MHz. This shows a good agreement with Fig.4. Note that when measured differentially, both the DC-gain and the unity-gain frequency are expected to be twice as high. The settling behaviour is measured according to Fig.11 by applying a step, ΔV_i, at the input. The resistors are needed for DC-biasing of the opamp and have no influence on the settling behaviour.

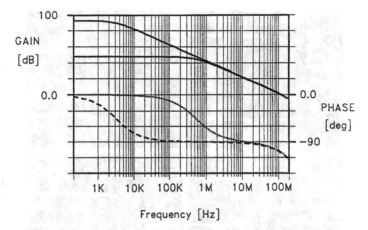

Figure 10: Results of gain and phase measurements both with and without gain enhancement.

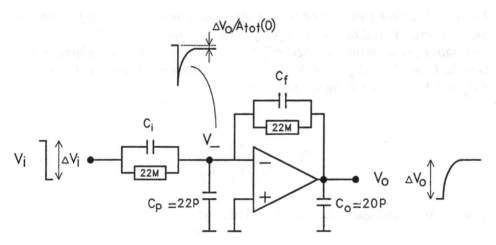

Figure 11: Scheme for measuring settling behaviour.

The error signal V_- at the opamp input, and the output signal V_o are shown in Fig.12. In Figs.12a and 12b, $\Delta V_o = 1$ V which is small enough to avoid slewing. With the gain enhancement switched off, an error signal of 4.75 mV is measured after settling (Fig.12a). This corresponds to the measured DC-gain of 46 dB. Switching on the gain enhancement reduces the error signal to a value smaller than 0.1 mV (Fig.12b), which corresponds to a DC-gain higher than 80 dB. Settling speed can be calculated as follows. In Fig.11 the feedback factor β is given by:

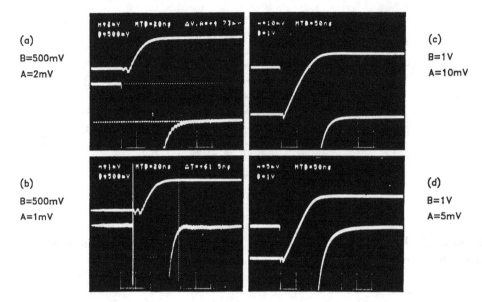

(a)

B=500mV

A=2mV

(c)

B=1V

A=10mV

(b)

B=500mV

A=1mV

(d)

B=1V

A=5mV

Figure 12: Settling-measurement results. The output signal (upper trace) and the error signal at the opamp input (lower trace) with: (a) $\Delta V_o = 1$ V and gain enhancement switched off, (b) $\Delta V_o = 1$ V and gain enhancement switched on, (c) $\Delta V_o = 4$ V and gain enhancement switched off, (d) $\Delta V_o = 4$ V and gain enhancement switched on.

$$\beta = \frac{C_f}{C_i + C_p + C_f} \tag{14}$$

whereas the unity-gain frequency is given by:

$$\omega_{unity} = g_m \frac{C_i + C_p + C_f}{C_o(C_i + C_p + C_f) + (C_i + C_p)C_f} \tag{15}$$

The theoretical settling time-constant, τ, can now be calculated:

$$\tau = \frac{C_p + C_i + C_o + (C_i + C_p) \cdot C_o / C_f}{g_m} \tag{16}$$

In Fig.11, $C_o = 20$ pF and $C_p = C_i = C_f = 22$ pF, which is relatively large due to probing. With $g_m = 0.012$ A/V, we find $\omega_{unity} = 54$ MHz, $\beta = \frac{1}{3}$ and $\tau = 8.8$ ns. Settling to 0.1 percent takes $7\tau = 62$ ns and corresponds to a 1 mV error at the output. With a feedback factor $\beta = \frac{1}{3}$, this corresponds to an error signal of $V_- = 0.33$ mV. From Fig.12b the measured settling time for 0.1 percent accuracy is 61.5 ns which is in agreement with the theory.

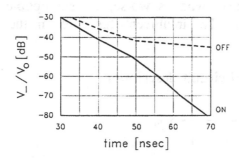

Figure 13: Measured relative-settling error as a fuction of time for both with and without gain enhancement.

Gain enh.	on	off
DC gain	90 dB	46 dB
Unity-gain freq.	116 MHz	120 MHz
Load cap.	16 pF	16 pF
Phase margin	64 deg.	63 deg.
Power cons.	52 mW	45 mW
Output swing	4.2 V	4.2 V
Supply voltage	5.0 V	5.0 V

Tabel 1: Main characteristics of the opamp.

In Fig.13, the measured settling behaviour is shown as a function of time as discussed earlier. During the entire settling process, each 10 dB increase in settling accuracy takes approximately 8 ns. This clearly shows that there are no slow settling components. In Figs.12c and 12d, $\Delta V_o = 4$ V, $C_i = 33$ pF and $C_f = 15$ pF, showing a normal slewing behaviour and a large output swing. The main measured characteristics of the opamp are summarized in table 1.

8. The Super-MOST

A disadvantage of the above shown implementation of the opamp is the complexity of the design and layout. A disadvantage of cascoded amplifiers in general is the number of required biasing voltages resulting in long wires across the chip. These long wires consume a considerable amount of space and, what is worse, are susceptible to cross-talk and therefore instability. To circumvent this problem, the Super-MOST was developed.

8.1. Basic idea and circuit description

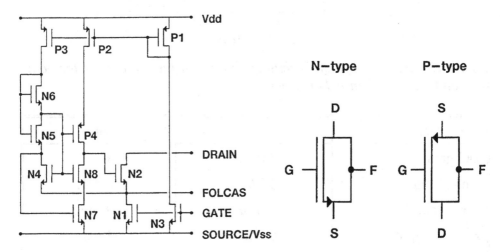

Figure: 14   a) N-type Super-MOST.
b) Super-MOST symbols.

The Super-MOST is a compound circuit which behaves like a cascoded MOS-transistor and has, like a normal MOST, a source, a gate and a drain terminal. The Super-MOST however has an extremely high output impedance due to implementation of the gain-boosting technique. Moreover, it does not require any biasing voltage or current other than one single power supply. The circuit of an N-type Super MOST is shown in Fig.14a. The circuit consists of three parts:

1. Transistors N1 and N2 are the main transistor and its cascode transistor and form the core of the circuit. Their size determines the current-voltage relations and the high frequency behaviour of the "device".

2. Transistors N7, N8, P2 and P4 form the additional stage for the gain-boosting effect.

3. Transistors N3, P1, P3, N4, N5 and N6 are for biasing purposes and ensure that N2 is always biased in such a way that N1 is just 50 - 100 mV above the edge of saturation, independent of the applied gate-voltage. This ensures a low saturation voltage of the Super-MOST.

The size of the transistors of the additional stage and of the biasing branch is as small as a few percent of the main transistors.

Fig.14b shows the symbol for the Super-MOST, S is the source terminal, G is the gate terminal and D is the drain terminal. Furthermore an extra low-impedance current-input terminal F is available which can be used for folded-cascode structures.

8.2. Measured "device" characteristics

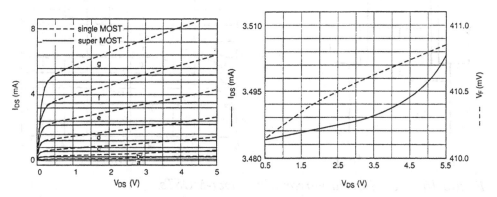

Figure 15:   *a) Measured I_{ds} - V_{ds} characteristics of single N-type MOST and Super-MOST for V_{gs} ranging from 0.7 V (a) to 1.0 V (g). b) Enlargement of curve f of figure 15a.*

Fig.15a shows the measured characteristics of a single N-type MOST and a Super MOST for V_{gs} ranging from 0.7 V (a) to 1.0 V (g). The single MOST shows an early voltage of approximately 5 V whereas the Super MOST has an Early voltage which is several orders of magnitude higher. Note that the "device" is already saturated at a voltage only slightly above the saturation voltage of one single MOST, indicating a large possible output swing. In Fig.15b an enlargement of Fig.15a is shown for V_{gs} =

0.95 V. From the I_{DS} curve an early voltage of 1750 Volts can be calculated which is an increase of 350 times compared to a single MOST.

8.3 Fundamental limitation of the gain-boosting technique

This measurement (Fig.15b) also reveals the upper limit for the gain-boosting technique. The voltage V_F at the F-terminal changes 0.8 mV as the drain voltage varies 5 V. This would imply an increase in early voltage and output impedance of 6250 times. The lower measured early voltage is due to a weak avalanche current which flows directly from drain to substrate and is therefore not influenced by the gain boost technique. In fact this weak avalanche current imposes an upper limit to the output impedance achievable with the gain boosting technique.

8.4 Applications of the Super-MOST

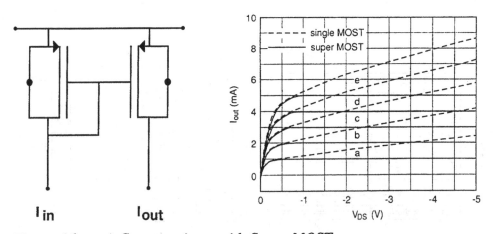

Figure 16:   *a) Current mirror with Super-MOSTs.*
         b) Measured output current of a P-type current mirror with
         I_{in} ranging from 1.0 mA (a) to 5.0 mA (e).

The Super MOST can be used as a normal MOS-transistor. As an example a current mirror has been realized as shown in Fig.16a. Fig.16b shows the measured results for a P-type current mirror at several input currents and output voltages. The mirror accuracy is independent of output voltage and is only determined by matching.

The F-terminal of the Super MOST provides a very low ohmic current input which can be used for folded-cascode structures [2]. In Fig.17

measurement results of a P-type Super MOST show an input impedance lower than 1 ohm for low frequencies.

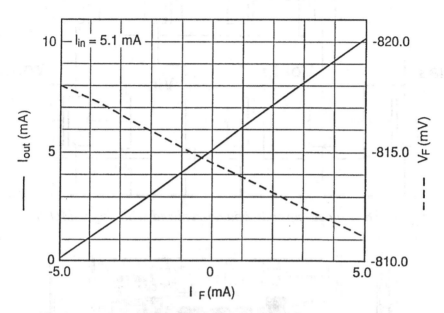

Figure 17: Measured output current I_{out} and terminal voltage V_f as a function of I_f for a P-type Super-MOST.

As an example Fig.18 shows a straight-forward circuit topology of a folded-cascode Opamp but now realised with Super-MOSTs. Note that the tail current of the differential input-pair also consists of a Super-MOST. In this way the common-mode rejection ratio CMRR is increased several orders of magnitude for low frequencies. For the input-pair normal transistors have been used. Measured results of this straight-forward design are equivalent to the results shown in table 1.

Fig.19 shows the die photograph of the chip, containing 1 Opamp, 1 N-type Super-MOST and 2 P-type Super-MOSTs. As can be seen, a Super-MOST consumes only 20 percent more chip area compared to a normal cascoded transistor. With a more careful lay-out this can even be further reduced. Because there is no need for extra biasing, the total opamp chip area is more than 25 percent smaller as compared to the design of Fig.9.

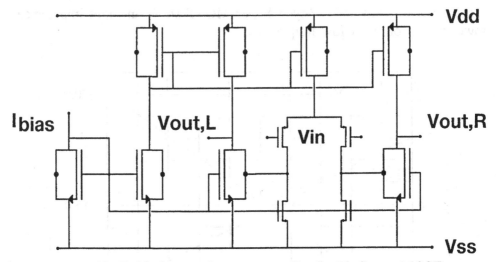

Figure 18: Folded-cascode opamp realized with Super-MOSTs.

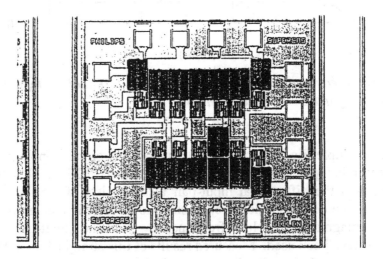

Figure 19: Die photograph of Super-MOST opamp.

8.5. Advantages of the Super-MOST

There are several advantages in using Super-MOSTs as a building block in circuit design :

1. The design of high-quality Opamps, OTAs, current sources, etc., can be split into two parts. First the design of the Super-MOSTs, in which the

knowledge of proper biasing, gain-boosting, optimal settling and stability is used. Much attention can be paid to an optimal lay-out also, as this sub-circuit is going to be used in many designs at several points in the comprising circuit. Secondly, the design of the comprising circuit, which now becomes rather straight-forward. In this higher level design no knowledge of the gain-boosting principle or of optimal biasing of a cascode transistor is required. This eases the design of such circuits and shortens the design time considerably.

2. The design of Super-MOSTs is very suitable for parameterized automatic generation.

3. As each Super-MOST is self-biasing, no long biasing wires are required on chip, leading to a design which is much less susceptible to cross-talk and therefore also for instability problems.

4. As wiring usually consumes a relatively large part of the chip area, the approach presented here consumes considerably less chip area.

Comparison of the die-photographs of figs. 9 and 19 clearly shows the difference in design strategy with and without Super-MOSTs.

9. Conclusions

A technique is presented which decouples the DC-gain and unity-gain frequency of an opamp. A very high DC-gain can be achieved in combination with any unity-gain frequency achievable by a (folded-) cascode design. With this technique, an opamp is realized in a standard 1.6 μm CMOS process which has a DC-gain of 90 dB together with a unity-gain frequency of 116 MHz. The opamp shows one-pole roll-off and a single-pole settling behaviour. This technique does not cause any loss in output voltage swing. At a supply voltage of 5.0 V an output swing of about 4.2 V is achieved without loss in DC-gain.

The advantages above are achieved with only an increase in chip area of 30 percent and an increase in power consumption of 15 percent.

The Super-MOST presented here eases the design of high-gain amplifiers considerably. As the building block is completely self-biasing, there are no long wires in the design, reducing cross-talk and instability, and chip area.

References

[1] K. Martin and A.S. Sedra, "Effects of the opamp finite gain and bandwidth on the performance of switched-capacitor filters," *IEEE Trans. Circuits Syst.*, vol. CAS-28, August 1981, pp.822-829.

[2] T.C. Choi et al., "High-frequency CMOS Switched-Capacitor Filters for Communications Application" *IEEE J. Solid-State Circuits*, vol. SC-18, no.6, December 1983, pp.652-664.

[3] T.C. Choi and R.W. Brodersen, "Considerations for High-Frequency Switched-Capacitor Ladder Filters" *IEEE Trans. Circuits Syst.*, vol. CAS-27, no.6, June 1980, pp.545-552.

[4] P.W. Li et al., "A Ratio-Independent Algorithmic Analog-to-Digital Conversion Technique", *IEEE J. Solid-State Circuits*, vol. SC-19, no.6, December 1984, pp.828-836.

[5] P. Naus et al., "A CMOS Stereo 16-bit D/A Convertor for Digital Audio", *IEEE, J. Solid-State Circuits*, vol. SC-22, no.3, June 1987, pp.390-395.

[6] S.H. Lewis and P.R. Gray, "A Pipelined 5-Msample/s 9-bit Analog-to-Digital Convertor", *IEEE J. Solid-State Circuits*, vol. SC-22, no.6, December 1987, pp.954-961.

[7] S. Wong and C.A.T. Salama, "Impact of scaling on MOS analog performance", *IEEE J. Solid-State Circuits*, vol. SC-18, no.1, February 1983, pp.106-114.

[8] C.A. Laber and P.R. Gray, "A Positive-Feedback Transconductance Amplifier with Applications to High-Frequency, High-Q CMOS Switched-Capacitor Filters", *IEEE J. Solid-State Circuits*, vol. 23, no.6, December 1988, pp.1370-1378.

[9] B.-S. Song, "A 10.7-MHz switched-capacitor bandpass filter," *IEEE J. Solid-State Circuits*, vol. SC-24, April 1989, pp.320-324.

[10] M.A. Copeland and J.M. Rabaey, "Dynamic amplifier for MOS technology", *Electron. Lett.*, vol. 15, May 1979, pp.301-302.

[11] B.J. Hosticka, "Dynamic CMOS amplifiers", *IEEE J. Solid-State Circuits*, vol. SC-15, October 1980, pp.887-894.

[12] M.G. Degrauwe, J. Rijmenants, E.A.Vittoz, and H.J.DeMan, "Adaptive biasing CMOS amplifiers", *IEEE J. Solid-State Circuits*,

vol. SC-17, June 1982, pp.522-528.

[13] H. Ohara et al., "A CMOS programmable self-calibrating 13-bit eight-channel data acquisition peripheral", *IEEE J. Solid-State Circuits*, vol. SC-22, December 1987, pp.930-938.

[14] B.J. Hosticka, " Improvement of the gain of MOS amplifiers", *IEEE J. Solid-State Circuits*, vol. SC-14, no.6, December 1979, pp.1111-1114.

[15] E. Sackinger and W. Guggenbuhl, "A High-Swing, High-Impedance MOS Cascode Circuit", *IEEE J. Solid-State Circuits*, vol. SC-25, no.1, February 1990, pp.289-298.

[16] K. Bult and G.J.G.M. Geelen, "A Fast-Settling CMOS Opamp with 90-dB DC gain and 116 MHz Unity-Gain Frequency", *ISSCC Dig. Tech. Papers*, February 1990, pp.108-109.

[17] K. Bult and G.J.G.M. Geelen, "A Fast-Settling CMOS Opamp for SC Circuits with 90-dB DC Gain", *IEEE J. Solid-State Circuits*, vol. SC-25, no.6, December 1990, pp.1379-1384.

[18] B.Y. Kamath, R.G. Meyer, and P.R. Gray, "Relationship between frequency response and settling time of operational amplifiers," *IEEE J. Solid-State Circuits*, vol. SC-9, December 1974, pp.347-352.

[19] H. Wallinga and K. Bult, "Design and Analysis of CMOS Analog Signal Processing Circuits by Means of a Graphical MOST Model", *IEEE J. Solid-State Circuits*, vol. SC-24, no.3, June 1989, pp.672-680.

[20] H.C. Yang and D.J. Allstot, "An Active-Feedback Cascode Current Source", *IEEE Trans. Circuits Syst.*, vol. CAS-37, no.5, May 1990, pp.644-646.

Biographies

Klaas Bult was born in Marienberg, The Netherlands, on June 26, 1959. He received the M.S. degree in electrical engineering from the University of Twente, Enschede, The Netherlands, in 1984, on the subject of a design method for CMOS opamps. In 1988 he received the Ph.D. degree from the same university on the subject of analog CMOS square-law circuits. He is now with Philips Research Laboratories, Eindhoven, The Netherlands. His main interest are in the field of analog CMOS integrated circuits. He is the recipient of the Lewis Winner Award for outstanding conference paper of ISSCC 1990.

Govert J.G.M. Geelen was born in Heythuysen, The Netherlands, on November 20, 1957. He received the M.S. degree in electrical engineering from the University of Eindhoven, Eindhoven, The Netherlands, in 1983. In 1984, after his military service, he joined the Philips Research Laboratories, Eindhoven, where he is currently engaged in the design of analog CMOS integrated circuits. He is co-recipient of the Lewis Winner Award for outstanding conference paper of ISSCC 1990.

CMOS Buffer Amplifiers

Rinaldo Castello

Università di Pavia
Dipartimento di Elettronica
Pavia
Italy

Abstract

This paper gives a tutorial presentation on the design of buffer amplifiers in CMOS technology. These are circuits that must drive a load made up of either a large capacitor or a small resistor or both. The core of the paper deals with special purpose buffer amplifiers intended for a specific application most often within a mixed analog/digital system integrated on a single chip. Several architectures for both input class A/B and output push-pull stages are discussed and compared. Issues like quiescent current control, frequency compensation, open loop and closed loop linearity are analysed in detail with the help of many examples.

1. Introduction

The tremendous increase in chip size and complexity that has occurred over the last decade has made possible to integrate complete subsystems, containing both analog and digital parts, on the same chip. This has resulted in an increased attention of the scientific community toward the subject of analog design using MOS technology.

The workhorse of most of these analog circuits is the operational amplifier. Over the last decade there has been a tremendous amount of development in the field of output buffers which has been extensively reported in the literature [4-9,11-15,20]. Also there has been a shift toward lower supply voltage operation, i.e. typically 5 Volts today and probably less in the future.

All of these developments have produced a wide variety of circuit solutions to the various design problems arising in the context of buffer amplifiers. However, contrary to the case of transconductance amplifiers, for the buffer amplifiers there seem to be no clear design criteria or generally accepted topologies. There is, therefore, the need for a systematic review of the state of the art in this field. This paper attempts to fill this gap focusing on the trade-offs existing in the design of CMOS buffer amplifiers.

The paper is organized as follows: in Section 2 the performance requirements and main design problems for special purpose buffer amplifiers are discussed. In this Section the basic topology of a buffer is also introduced. In Section 3 the requirements for the input stage of a buffer amplifier are discussed. Several input class A/B circuits are described together with criteria to compare them. In Section 4 capacitive and/or resistive drivers are examined. In these sections the various problems associated with push-pull output stages are discussed like quiescent current control, linearity, frequency stability, etc.

2. Buffer amplifier performance requirements

One of the main objectives of a buffer amplifier is to be able to drive off-chip loads. As a consequence, a load consisting of a large capacitance (up to several nanofarads) and/or a small resistance (down to 50 Ω or less) can be expected. The presence of a heavy load at the output (especially a small resistance) requires to use a multistage structure as shown conceptually in Fig.1. The preamplifier stage is typically a transconductance amplifier which provides differential input and a large amount of gain. The output stage is typically a push-pull circuit that provides low output impedance and a large current driving capability.

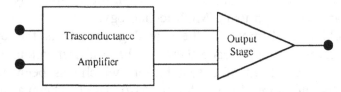

Figure 1: Buffer amplifier block diagram.

As larger and more complex systems are integrated on the same chip, special purpose buffer amplifiers are required for a wide range of diverse applications with a broad set of specifications. Special purpose buffers are often operated under some favourable and generally well defined conditions. In particular the supply voltage is specified within a small range around some nominal value (typically 5 V ± 10%). Almost always the driver is operated in a well defined feedback configuration (possibly programmable). Finally, in some cases, due to the presence of voltage amplification in front of the buffer, noise, offset and overall gain are not particularly critical.

The favourable operating conditions of special purpose buffer amplifiers often have the counterpart of some difficult performance requirements. In particular a very large output swing is often required. The output load is sometimes very hard to drive, due to the presence of either a small resistor or a large capacitor or both. As an example the output driver of a SLIC chip may have several nanofarads of capacitive load. High linearity and a low output resistance over a large frequency range may be required. As an example in an ISDN U interface chip. The driver requires more than 70 dB of linearity up to 40 kHz or more. Another objective often required and difficult to obtain is a very high ratio between the quiescent current level and the peak current to be delivered to the load. Finally, a good PSRR, especially at high frequency, is often required if the buffer is operating at the periphery of an A/D system. A typical example of this situation is found in an ISDN combo chip.

3. Buffer's input stages

A variety of class A/B input stages will be presented and discussed below. Several criteria to evaluate the merits and drawbacks of a given topology will be introduced and applied to the various circuits presented.

The first circuit considered is shown in Fig.2 [1]. The quiescent current level in the input cross-coupled devices M_5 - M_8 is determined by current source devices M_3, M_{12}. Assuming for simplicity $M_1 = M_5 = M_7 = M_{10}$ and $M_2 = M_6 = M_8 = M_{11}$ are identical devices, than the current in M_5 - M_8 in quiescent conditions ($V_{in} = 0$) is equal to the current in M_3 and M_{12}. In these conditions the current in the output branch (M_{13}, M_{14}) is also determined by the current source devices (M_3, M_{12}) and the current ratio of the two current mirrors (M_4, M_{13} and M_9, M_{14}).

For a large differential input signal V_{in}, e.g. positive at In+ and negative at In-, the voltage drop across transistors M_5, M_6 is increased by V_{in} while the drop across M_7, M_8 is decreased by the same amount.

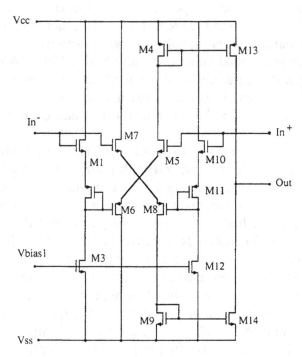

Figure 2: Circuit schematic of input class A/B stage of Ref. [1].

As a consequence the current level through M_{13} is increased by a large amount while the current in M_{14} is reduced close to zero. A current much larger than the quiescent level is therefore available to be delivered to the load.

Some of the specific features of the circuit are listed below. Class A/B operation is achieved without using feedback, so no stability problem arises. The quiescent current level is well defined (typically within ± 20%) and independent from technology parameters and supply variations. This is because the current level is controlled by matching identical (or ratioed) devices. The absolute maximum current available at the output for a given supply voltage is limited by the total voltage drop in the input branch (M_4, M_5, M_6 for a positive input). Current saturation occurs when the current level is such that the voltage drop in this branch is equal to the supply voltage, i.e. $2V_T + 3(V_{gs} - V_T) = V_{supply}$.

The circuit is moderately complicate compared with other class A/B configurations. Furthermore it is particularly suitable for differential applications since it can produce a complementary output by simply adding two additional current mirrors symmetrically with respect to M_4, M_{13} and M_9, M_{14}.

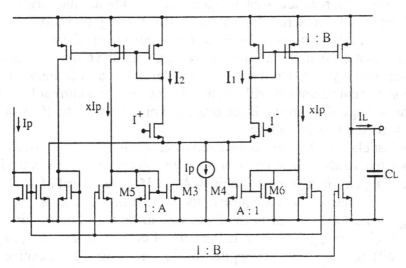

Figure 3: Circuit schematic of input class A/B stage of Ref. [2], [3].

A second class A/B topology is shown in Fig.3 [2,3]. The circuit is such that the current in the tail current source in the input differential pair controls the current level in the entire circuit in quiescent conditions. The maximum sourcing and sinking capability of the circuit can therefore be controlled by modulating the value of this tail current source. In quiescent condition the current through M_3 and M_4 is equal to zero and the differential pair is biased at a current equal to I_p.

As an input signal is applied (suppose positive at In+) I_2 increases. When I_2 becomes larger than I_p a current starts flowing through M_5 and a corresponding current (A times in value) through M_3. This further increases I_2 producing a positive feedback reaction that dramatically increases the total tail current and consequentially the current driving capability of the circuit. This is called adaptive biasing. Because the circuit uses positive feedback to insure stability some conditions must be met [2,3]. First, the current mirror gain X must be greater than 0.5. Furthermore, the product of the two current mirror gains AX must be smaller than 1.

In the circuits of this kind there is a trade-off between the precision in the quiescent current level on the one hand and both the linearity at small signals and the effectiveness of the class A/B operation on the other hand. In fact if the extra current sources (M_3, M_4 in Fig.3) are kept on even in quiescent conditions a large sensitivity of the quiescent current level to mismatches results. On the other hand, the further away from turn-on the extra current sources are kept in quiescent conditions, the larger is the signal amplitude required for the adaptive biasing to become active. Furthermore, large crossover distortion can occur for small amplitude input signals. Notice that for a sufficiently large input voltage the positive feedback will try to make the tail current level to go to infinity. In this situation current saturation will result. This occurs for a current level such that the voltage drop in the input branch, which is equal to $V_T + 3(V_{gs} - V_T)$, equals the supply voltage.

The last class A/B circuit considered is shown in Fig.4 [4]. Also in this case positive feedback is used to increase the tail current source in the input differential pair. In quiescent condition M_{1A}, M_{1B}, M_{2A}, and M_{2B} all carry the same current. As a consequence, the current in M_{3A} and M_{4A} is equal to I_{bias}. The total current in the input pair in this condition is therefore equal to: $(1 + a) I_{bias}$, where a is the gain of current mirror M_{4A}, M_{4B}.

If a differential voltage is applied to the circuit with the positive side connected to In+ the current in M_{2B} increases and that in M_{1B} decreases. This causes a net current to flow out of the input differential stage. This current is forced to flow through M_{3A} and M_{4A}. Current mirror device M_{4B} brings a portion of this signal current back to the input stage creating a positive feedback reaction which increases its biased current level. At the same time current mirror device M_{4C} sends to the output the signal current with a gain generally larger than one. Stability is guarantied if mirror M_{4A}, M_{4B} has a gain smaller than one. Practical considerations, like mismatches, force the mirror current gain to be chosen substantially smaller than one.

The precision of the quiescent current level of the circuit for a given mismatch depends on the value of the above mirror. There is a trade-off between stability and quiescent current control on the one hand and peak current driving capability on the other hand.

The circuit of Fig.4 operates in class A/B only for one input signal polarity (the one assumed before). For an input signal of opposite polarity the auxiliary current source M_{4B} is turned off and only I_{bias} is left to bias the input stage. To achieve class A/B operation for both signal polarities a complementary circuit with the same topology as the one shown in Fig.4 but using p-type input devices must be used in parallel to the one just described. As a consequence a fairly complex topology results. As in the previous examples current saturation occurs when the voltage drop in the input stage — which is equal to $V_T + 3(V_{gs} - V_T)$ — equals the supply voltage.

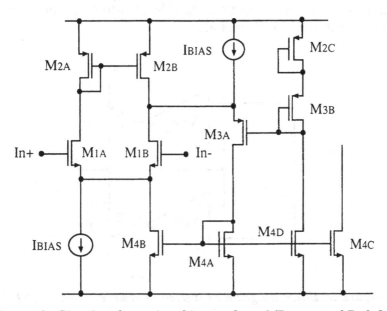

Figure 4: Circuit schematic of input class A/B stage of Ref. [4].

4. Push-pull stages

The key block of a resistor driver is most often the output stage which generally operates in a push-pull mode. The main characteristics of a push-pull output stage are: quiescent current control, frequency behaviour and linearity. The objective of a push-pull stage is to have a large current sourcing and sinking capability with a small quiescent current level. Several different ways to control the quiescent current level of a push-pull stage will be considered first.

The first push-pull topology to be considered is shown in Fig.5 [5]. The two large common source output transistors M_1, M_2 are driven by two amplifiers A_1, A_2. The feedback loop around A_1 (A_2) and M_1 (M_2) ensures low output impedance. This gives a large current drive capability and pushes the output pole to a high frequency even for a large capacitive load.

The two amplifiers A_{V1} and A_{V2} must satisfy three requirements for proper operation of the stage. First they must be broad band to prevent cross-over distortion. This is however difficult to achieve due to stability constraint. Second, they must have a rail-to-rail output swing and a rail-to-rail input common mode range to give good driving in front of the output devices. Finally, they must have a reduced gain (typically less than 10). This is due to the input offset of A_{V1} and A_{V2}. In the case of a large gain this offset can create an unacceptable chip-to-chip variation of the quiescent current level in M_1 and M_2.

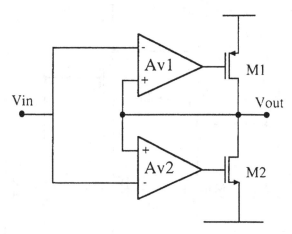

Figure 5: Block diagram of a pseudo source-follower [5].

A modified version of the previous circuit which attempts to address some of its shortcomings is shown in Fig.6 [6]. This circuit is obtained by merging a complementary class A/B source follower with the circuit of Fig.5. The problem of quiescent current control is solved by deliberately introducing an offset voltage in A_1 and A_2 in such a way that M_1 and M_2 are not carrying any current in quiescent conditions. In this situation the output stage current is controlled by M_3 -M_6 so that A_1 and A_2 can have a large gain.

The second function of the complementary source follower M_3 -M_6 is to provide a high frequency feed-forward path from the input to the output of the push-pull stage. This extra path compensates the excess phase shift of the main class B stage due to amplifiers A_1 and A_2. This allows to chose A_1 and A_2 with a large bandwidth while ensuring stability even in the presence of a purely capacitive load. The main disadvantage of this new circuit is its reduced output swing which is limited to a V_{gs} from either supply.

The circuit of Fig.6 can be considered as a special case of a family of push-pull stages. These circuit have the characteristic of being made-up of the parallel combination of a class A/B or even class A circuit and of a class B one [7,8]. The class A/B circuit carries the entire output current in quiescent conditions. A simple class A/B circuit with a moderate current driving capability can be used thus allowing to easily control its quiescent current level. The class B circuit has a very large current driving capability but is kept off in quiescent conditions. When a large current is required by the load the class B circuit takes over the stage operation.

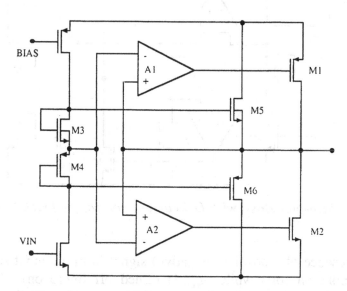

Figure 6: Pseudo source-follower with DC feed-forward path [6].

A key design parameter for this topology is the point where the control of the stage is switched over from the class A/B to the class B circuit and vice versa. Variation on the level of current where this cross-over occurs can result in gross distortion and even instability.

An alternative way to control the quiescent current level in a pseudo source follower like the one of Fig.5 without drastically reducing the gain of the two amplifiers A_1 and A_2 is shown in Fig.7 [9]. In this configuration a DC feedback loop controls the quiescent current in the output branch making it independent from the offset of A_1 and A_2.

The operation of the circuit is as follows: transistor M_9 senses the current through M_6 and feeds it back to the input of amplifier A_2. If, due to an offset in A_1, the current in M_6 is larger (smaller) than the desired value, the feedback action will try to reduce (increase) the current in M_{6A}. In this way the feedback loop tends to keep constant the sum of the current in M_6 and M_{6A}. If no current is required by the load (quiescent condition), the overall amplifier feedback loop will force the current in M_6 and M_{6A} to be the same and equal to the nominal value.

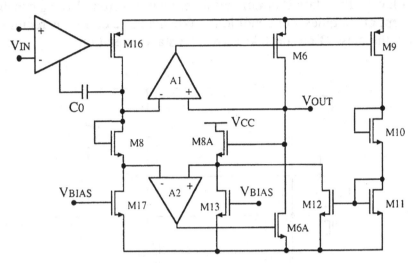

Figure 7: Push-pull stage with DC current control feedback loop [9].

In the presence of a positive (negative) signal from the input stage, M_6 is turned more on (off) while M_{6A} is turned off (more on). The DC feedback tends to further turn on (off) M_6 therefore reinforcing the push-pull action of the stage.

A possible problem with this solution is the undesirable effect of the DC loop on the AC operation of the circuit. In fact an interaction between the DC loop and the overall feedback loop cannot be avoided. This interaction causes cross-over distortion which reduces the maximum achievable linearity.

A straightforward implementation of the topology of Fig.7 is shown in Fig.8 [8,10]. In this solution the input stage directly drives the gate of the n-MOS output device M_2. A scaled version of the current through M_2 is compared with the reference current source I_1 through sensing device M_3 and current mirror M_4, M_5. The output node A of this comparison circuit drives the gate of the large p-MOS output device M_1. In quiescent conditions (no current to the load) the overall feedback loop forces the current through M_1 and M_2 to be the same. When this occurs the voltage at node A is such that M_5 is in saturation. In this situation the current through M_1 is equal to I_1 times n. Where n is the size ratio between M_2 and M_3 (the gain of current mirror M_4, M_5 is assumed to be 1). A precise ratio

n between the current in the output branch and I_1 is therefore maintained independently of process variations.

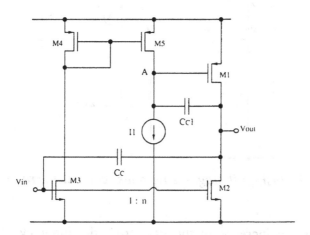

Figure 8: Simple push-pull stage with DC current feedback [8], [10].

Notice that the circuit has one extra gain stage in the p signal path (in front of the p-MOS device) with respect to the n signal path.

One difference between the circuit of Fig.7 and that of Fig.8 is that, while the first one has a DC feedback (unity) around it, the second one uses an AC feedback through capacitor C_C.

A third possible technique to control the quiescent level in a push-pull stage is the use of "quasi current mirrors" in front of the output devices [11,12,13]. This technique is shown conceptually in Fig.9 with reference to the p side of the push-pull circuit. A dual configuration is used for the n side. In quiescent conditions the sum of the currents through M_2 and M_4 is equal to that through M_3 when the voltage at node A is equal to the voltage at node B. In this situation the current through M_1, and therefore the current in the output branch, is given by I_1 times the ratio between the sizes of M_1 and M_5.

The precision of the output current level depends on both the precision of current source I_1 and the precision of the DC current level through M_2 and M_3 in addition to the matching of all the devices in the stage.

As the current through M_2 increases in response to some input signal a corresponding reduction in the current through M_4 has to occur since the current through M_3 is constant. As a result the voltage at node A begins to fall increasing the current through M_1. As the voltage at node A is reduced the impedance level at the same node increases.

124

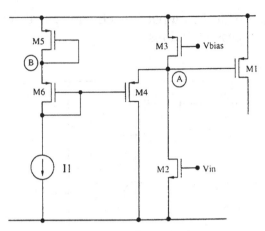

Figure 9: Push-pull stage using quasi current mirror [10-13].

The result is an increase in the gain from the input (V_{in}) to node A. When the current through M_2 is equal to the current through M_3, M_4 shuts off. From this point on the gain from the input to node A becomes equal to that of an actively loaded common source stage. Any current increase beyond this point will result in a very large voltage variation at node A.

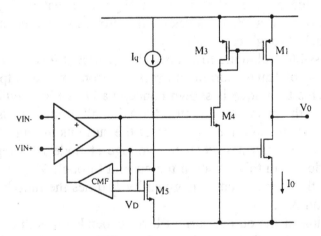

Figure 10: Push-pull using fully-differential driving stage [14].

A last possible way to bias a push-pull stage is shown conceptually in Fig.10 [14,15]. In this circuit the two large output devices M_1, M_2 are preceded by a fully differential preamplifier A_1. The common mode voltage of this preamplifier is controlling the quiescent current level in the output branch. When no current is delivered to the load the differential output signal of A_1 is zero and the current in M_1 and M_2 is the same. This is because M_2 and M_4 are ratioed in the same way as M_1 and M_3.

The desired value of the quiescent output current is controlled by I_q and the area ratios between M_5 and M_2. This is due to the fact that the common mode feedback circuit forces the common mode of the preamplifier to be equal to V_D.

When the circuit is required to source (sink) current to the load the differential output of A_1 becomes different than zero and this has the effect of turning on M_1 (M_2) and off M_2 (M_1). One problem with this circuit is that while sourcing a large current to the load M_2 becomes completely shut-off with its gate very close to ground. This produces cross-over distortion.

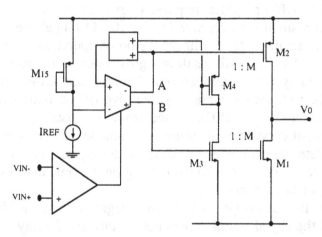

Figure 11: Improved push-pull using fully-differential driving stage [15].

A push-pull circuit based on a similar principle which however does not suffer from this problem is shown in Fig.11 [15]. In this implementation the quiescent output current is controlled by the differential output voltage of the fully differential preamplifier, while the current delivered to the load is controlled by its common mode voltage. This is the opposite of what is done in the previous circuit. The extra feature of this last circuit is that the smallest of the currents through the two large output devices must be equal to the quiescent current in the output branch. This reduces the cross-over effect.

4.1. Sources of distortion in push-pull stages

Closed loop linearity is the combined result of two characteristics of a buffer: its open loop linearity and its gain in the frequency band over which linearity is important. The main source of open loop distortion in a buffer amplifier is the push-pull output stage. The problem of getting large gain over a large band translates into achieving a stable amplifier with a sufficiently large bandwidth. A list of sources of open loop distortion will be given first while the problem of AC stability and compensation in a buffer circuit will be analysed next.

There are two situations that can cause distortion in a push-pull circuit. The first is during high current driving which for a resistive load corresponds to the voltage peaks at the output. The second is at cross-over where the circuit switches from sourcing to sinking and vice versa.

For an MOS buffer amplifier required to give an almost rail to rail swing while driving a small load resistor (in the 100 Ω range) the output MOS devices will generally enter the linear region of operation in the proximity of the signal peaks. This will result in a gross open loop distortion that can only be reduced by the overall feedback. The amount of distortion that this causes can be reduced by increasing the size of the output devices. A practical limit on the size of the device however exists.

Two causes of cross-over distortion are considered. The first one is due to the complete shut-off of some devices during one half of the signal cycle. The second one is due to the turning on of some devices during the wrong phase of the signal.

If a device is shut-off ($V_{gs} = 0$ or even negative for an n-MOS) during one phase of the signal it will experience an unwanted delay in turning on when the output is crossing from one polarity to the other. This results in a cross-over distortion effect as depicted in Fig.12a. In this situation, as the current in transistor M_3 becomes smaller than I, node X will rise very close to V_{DD} shutting off completely M_1. As V_{out} approaches zero Volts from the negative side the current in M_3 increases to turn on M_1. However, before M_1 can turn on node X must fall by one Volt or so. Due to the large parasitic capacitor at node X and the small driving current available, a delay equal to ΔT is necessary before this can occur. The result is a cross-over distortion as shown in Fig.12b.

Another possible mechanism that can create the same effect is capacitive coupling into a very high impedance node as shown in Fig.13a. In this case as M_3 turns off, node X tends to rise till about one threshold below V_{DD} leaving M_1 and M_2 on the edge of conduction. However, as shown in Fig.13b as V_{out} rises from its negative peak towards ground, this voltage variation is coupled into the high impedance node X by capacitor C_C. If C_C is much larger that C_P node X can rise above V_{DD}. This behaviour will cause the same cross-over distortion as in the previous example.

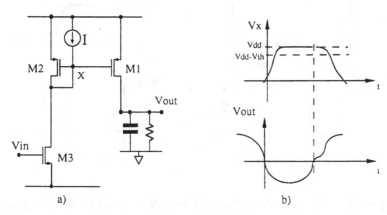

Figure 12: Cross-over distortion due to the circuit structure.

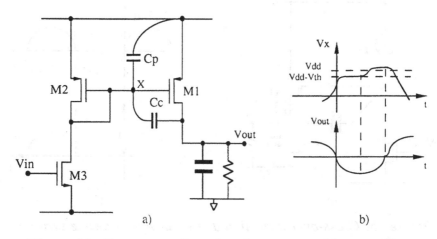

Figure 13: Cross-over distortion due to capacitive coupling.

In order to prevent complete shut-off of some devices, a clamping device can be placed at the critical high impedance node as shown in Fig.14. Device M_C is biased in such a way to be off when M_2 and M_1 are on and therefore does not interfere with the circuit during normal operation. As node X tends to rise however, M_C goes on clamping node X at a voltage just less than a threshold below the positive rail as shown in Fig.14b. This reduces drastically cross-over distortion. Changing the bias voltage at the gate of M_C, results in a trade-off between the effectiveness of M_C as a clamp and its unwanted interference with normal circuit operation.

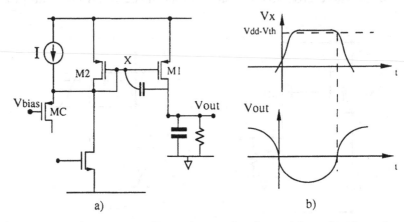

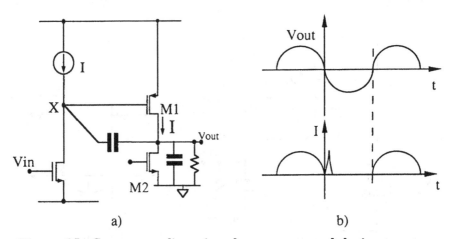

Figure 14: Cross-over distortion reduction with node clamping.

Figure 15: Cross-over distortion due to unwanted device turn on.

Another source of cross-over is due to the unwanted turn on of one of the large output devices during the wrong phase. An example of this effect is shown in Fig.15. In this case as V_{out} goes from zero towards the negative rail transistor M_1 should be off. However the negative going output transient couples into node X through capacitor C_C. If the displacement current through C_C is larger than I, node X begins to fall. This starts to turn on M_1 whose current is subtracted from the load. As a consequence the overall amplifier feedback tends to further turn on M_2 to compensate for the reduced current driving. A current spike between the two supplies through M_1, M_2 therefore results. This has the effect of distorting the output signal and in extreme cases can damage the circuit.

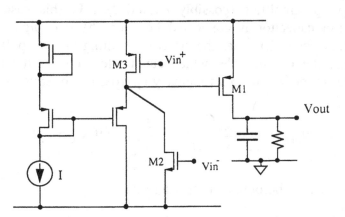

Figure 16: Circuit to prevent unwanted turn-on of the output devices.

A possible way to prevent this effect is to use quasi current mirror structures with a push-pull driver as shown in Fig.16. This solution turns off M_1 with a large current signal provided by M_3. As a consequence the displacement current necessary to create the unwanted turn on of M_1 must be much larger since it must overcome the peak current into M_3 and not its quiescent value.

4.2. Push-pull frequency response

For a given amount of open loop nonlinearities the residual distortion in a closed loop configuration depends on the amount of gain in front of the source of distortion at the frequency of interest. To maximize closed loop linearity at a given frequency amounts to maximizing the unity gain bandwidth of the amplifier. This is generally true although different topologies may have a quite different linearization effect in closed loop for the same unity gain bandwidth as will be discussed later.

Due to the need of driving a resistive load, a buffer almost always uses a multiple stage topology. If a DC feedback is applied around the output stage the problem of stabilizing the buffer is reduced to that of separately stabilizing the output stage and the rest of the amplifier. If an AC feedback is applied around the output stage the compound structure made up of the preamplifier and the output stage both interacting with each other must be stabilized. This is by far a more complicated problem than the previous one and will be addressed next.

The most simple multi-stage topology is shown in Fig.17 where the output stage is a simple common source (pseudo-follower) preceded by a single stage preamplifier (possibly cascoded). In this case a Miller compensation capacitor is connected between the two high impedance nodes in the circuit. This has the effect of splitting the two poles [16,17]. In particular the output pole, which can be low due to the large load capacitance, is moved up to a frequency approximately equal to

$$\omega_{out} = \frac{g_{m2}C_C}{(C_C+C_P)C_L} \approx \frac{g_{m2}}{C_L} \quad (\text{if } C_C \gg C_P) \tag{1}$$

Stability can easily be obtained insuring that

$$\frac{g_{m1}}{C_C} < \frac{1}{2}\frac{g_{m2}}{C_L} \tag{2}$$

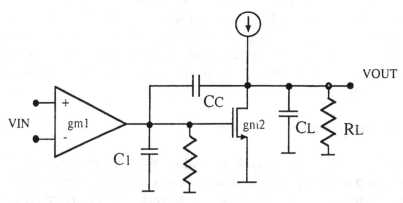

Figure 17: Simplest multi-stage topology with Miller compensation.

If a very large load capacitor is present, the above condition may be difficult to achieve. In this case a modified Miller compensation scheme as shown in Fig.18 can be used [8,18,19]. In the circuit a common gate MOS transistor is placed between one terminal of the Miller capacitor and the high impedance node at the output of the first stage. Furthermore a follower circuit is placed between the first and the second stage of the amplifier.

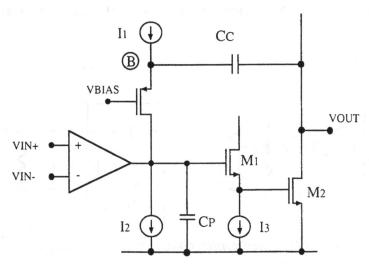

Figure 18: Compensation scheme for large load capacitances [8,18,19].

By returning the Miller compensation to a low impedance point like node B the high frequency gain from the output to node A becomes C_C/C_P as opposed to $C_C/(C_C + C_P)$ in the previous case. Due to the effect of the follower M_1, C_P is small so that a substantial gain (up to 10 or more) from the output to node A can be achieved at high frequency. The output impedance at high frequency is equal to $1/g_{m2}$ divided by the gain from the output to node A. As a consequence the modified circuit can give a much smaller output impedance. This moves the output pole up in frequency to

$$\omega_{out} = \frac{g_{m2}C_C}{C_L C_P} \tag{3}$$

In the case of a very small load resistance the gain of the output stage can become so small that a two stage solution may not give enough DC gain. In this case a three or more stage topology can be used. To stabilize such a topology a nested Miller [19] or even a double-nested Miller compensation must be used. This configuration is shown in Fig.19 for the case of a three stage circuit.

Some conditions that must be verified in order to achieve a stable closed-loop operation for this topology can be identified. First the ratio g_{mi}/C_i for each stage should be at least doubled going from the input to the output stage. This implies that for a given output pole frequency the amplifier unity gain bandwidth is reduced as the number of stages is increased.

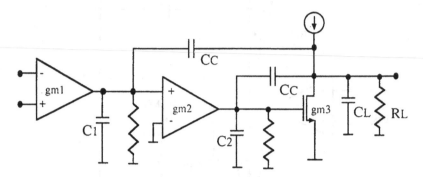

Figure 19: Nested Miller compensation scheme [21].

Second, all stages in front of the output can be similar often being a simple class A circuit. On the other hand, the output stage has typically a different topology (push-pull) and to guarantee stability must have a g_m much larger than that of the previous stages. This condition is however almost automatically verified in a resistive buffer since the output devices must be very large and carry a large quiescent current to give good linearity.

As for the case of a simple Miller topology also for a nested Miller (or double-nested) there exists the possibility of returning the Miller compensation capacitances to a low impedance node. In this case the technique can be used in either one of the two loops or in both.

Increasing the number of stages reduces its unity gain bandwidth. For this reason feedforward is sometimes employed to by-pass some of the additional poles introduced by the extra stages therefore achieving a better frequency response. Both AC and DC feedforward can be used [20, 21, 6].

An example of a multistage configuration with AC feedforward is shown conceptually in Fig.20 [20]. This topology can be seen as the dual of the nested Miller configuration of Fig.19. A nested Miller topology in fact is characterized by 2 nested feedback AC paths from the single output node (the amplifier output) to the inputs of the two previous stages. In the multipath feedforward configuration instead, there are two nested feedforward paths from a single input node (output of the first stage) to the output of the following stages.

A typical problem of a feedforward compensated topology is the creation of pole-zeros doublets in the frequency response. If the doublet is located within the passband it may degrade the step response of the amplifier [22]. The problem becomes more and more severe as both the relative distance between the pole and the zero in the doublet and the settling precision required increase.

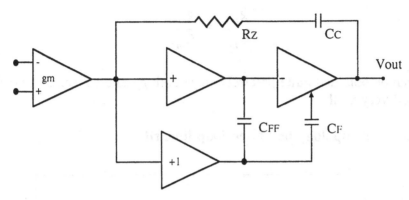

Figure 20: Multi-stage topology with AC feed-forward [20].

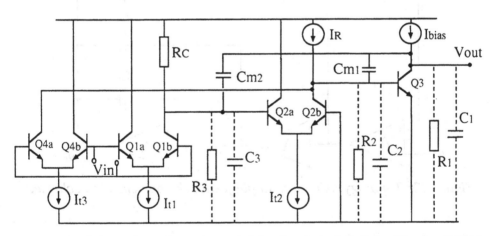

Figure 21: Circuit combining nested feedback and DC feedforward [23].

A topology that combines nested feedback and DC feedforward is shown in Fig.21 [23]. In this circuit an auxiliary differential pair is added in parallel with the main input pair. One of the two outputs of the main differential pair is feeding the second stage of the three stage structure. The corresponding output of the auxiliary pair is instead sent directly to the input of the last stage therefore bypassing the intermediate amplifying stage A_2. Since A_2 is non-inverting the two parallel paths produce in phase signals in front of the output stage.

A nice feature of this circuit is that the pole and the zero (doublet) produced by the feedforward path ideally cancel each other. To ensure this condition the following relationship must hold:

$$\frac{g_{m1}}{C_{m2}} = \frac{g_{m4}}{C_{m1}} \tag{4}$$

This can be obtained quite precisely since both g_m and capacitances can be matched very well.

4.3. Effect of topology on closed-loop linearity

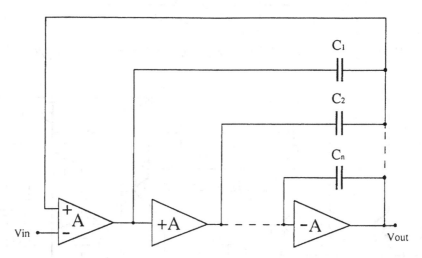

Figure 22: Topology using a multiple feedback compensation scheme.

It has been shown that the closed loop linearity of a buffer stage is depending on the open loop linearity of the circuit and on its gain in the band of interest. In addition to this, the topology of the buffer can have a major impact on the closed-loop linearity. In particular the number of stages of a buffer plays an important role on its closed loop behaviour and this will be discussed next.

Fig.22 shows schematically a general multistage topology using a multiple feedback compensation scheme. For the general structure of Fig.22 open loop distortion is reduced by the product of the gains of all the loops that surround the distortion source (both external and compensation loops) at the frequency of interest. Another way to express the same thing is to say that the open loop distortion is reduced by the total gain that exists between the input and the distortion source at the frequency of interest. This gain for the multiple feedback structure can be obtained opening all the feedback loops. This can be done by cutting the connections between all the compensation capacitors and the output node. The compensation capacitors are then returned to ground to preserve their loading effect on all the nodes.

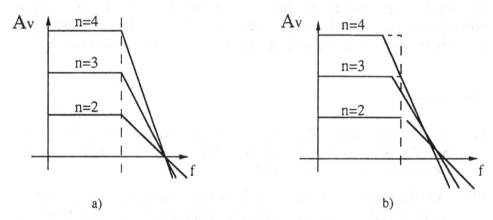

Figure 23: Closed loop linearization factor for different number of stages.

In a buffer amplifier with a small resistor at the output the main distortion sources are all located in the output stage. As a consequence the linearization factor is the product of the gains of all the stages except the output one. This quantity is shown in Fig.23a as a function of frequency for the case of an amplifier with two, three, and four stages. In Fig.23a it has been assumed, for simplicity, that all the stages are identical and that the overall bandwidth of the amplifier can be maintained constant independently of the number of stages. From Fig.23a follows that as the number of stages increases the linearization factor also increases over the entire buffer bandwidth. Furthermore the increase of the linearization factor is the same as the increase in the DC gain up to the frequency of the open loop poles f_x in the figure.

From the above results it may seem that an arbitrarily small closed loop distortion can be obtained by increasing more and more the number of stages. In reality as previously mentioned a larger number of stages necessarily produces a smaller bandwidth especially if a constant power consumption is assumed. As a consequence a more realistic situation is that shown in Fig.23b. For a given frequency band of interest there is therefore an optimum number of stages that gives the largest closed loop linearization factor. Qualitatively speaking the smaller the bandwidth of interest the larger the optimum number of stages that results.

As a final note it should be noticed that although the main open loop distortion sources are in the output stage these are divided by a very large linearization factor when the overall feedback loop is closed. Smaller distortion sources in the previous stages are on the other hand divided by smaller linearization factors. In particular distortion sources associated with the input stage remain unaltered when the loop is closed. As a consequence care must be taken to ensure that no slewing or other input stage distortions are produced if a very linear buffer must be designed.

References

[1] R. Castello and P.R. Gray "A High-Performance Micropower Switched-Capacitor Filter", *IEEE J. Solid-State Circuits*, vol. SC-20, December 1985, pp.1122-1132.

[2] M.G. Degrauwe, J. Rijmenants, E. Vittoz, and H. DeMan, "Adaptive Biasing CMOS Amplifiers", *IEEE J. Solid-State Circuits*, vol. SC-17, June 1982, pp.522-528.

[3] M.G. Degrauwe and W. Sansen, "Novel Adaptive Biasing Amplifier", *Electron. Lett.*, vol.19, no.3, February 1983, pp.92-93.

[4] L.G.A. Callawert and W.M.C. Sansen, "Class AB CMOS Amplifiers with High Efficiency", *IEEE J. Solid-State Circuits*, vol. SC-25, June 1990, pp.684-691.

[5] B.K. Ahuja, P.R. Gray, W.M. Baxter, and G.T. Uehara, "A programmable CMOS dual channel interface processor for telecommunications applications", *IEEE J. Solid-State Circuits*, vol. SC-19, December 1984, pp.892-899.

[6] J.A. Fisher, "A High-Performance CMOS Power Amplifier", *IEEE J. Solid-State Circuits*, vol. SC-20, December 1985, pp.1200-1205.

[7] H. Khorramabady, J. Anidjar, and T.R. Peterson, "A Highly-Efficient CMOS Line Driver with 80 dB Linearity for ISDN U-interface Applications", *ISSCC Dig. Tech. Papers*, February 1992, pp.192-193.

[8] R. Castello, F. Lari, L. Tomasini, and M. Siligoni, "100 V High Performance Amplifiers in BCD Technology for SLIC Applications", European Solid-State Circuits Conference, Grenoble, September 1990, pp.177-180.

[9] K.E. Brehmer and J.B. Wieser, "Large Swing CMOS Power Amplifier" *IEEE J. Solid-State Circuits*, vol. SC-18, December 1983, pp.64-69.

[10] D. Sanderowicz and G. Nicollini, *Private communication.*

[11] S.L. Wong and C.A.T. Salama, "An Efficient CMOS Buffer for Driving Large Capacitive Loads", *IEEE J. Solid-State Circuits*, vol. SC-21, June 1986, pp.464-469.

[12] M.D. Pardoen and M.G.R. Degrauwe, "A Rail-to-Rail Input/Output CMOS Power Amplifier", IEEE Custom Integrated Circuits Conference, 1989, pp.25.5.1-25.5.4.

[13] R. Castello, G. Nicollini, and P. Monguzzi, "A High-Linearity 50-W CMOS Differential Driver Output Stage", *IEEE J. Solid-State Circuits*, vol. SC-24, December 1991, pp.1809-1816.

[14] J.N. Babanezhad, "A Rail-to-Rail CMOS Op Amp", *IEEE J. Solid-State Circuits*, vol. SC-23, December 1988, pp.1414-1417.

[15] F.N.L. Op't Eynde, P.F.M. Ampe, L. Verdeyen, and W.M.C. Sansen, "A CMOS Large-Swing Low-Distortion Three-Stage Class AB Power Amplifier", *IEEE J. Solid-State Circuits*, vol. SC-25, February 1990, pp.265-273.

[16] P.R. Gray and R.G. Meyer, "MOS Operational Amplifier Design - A Tutorial Overview", *IEEE J. Solid-State Circuits*, vol. SC-17, December 1982, pp.969-982.

[17] J.E. Solomon "The monolithic op amp: a tutorial study" *IEEE J. Solid-State Circuits*, vol. SC-9, December 1974, pp.314-332.

[18] R.D. Jolly and R.H. Mc Charles "A Low-Noise Amplifier for Switched-Capacitor Filters", *IEEE J. Solid-State Circuits*, vol. SC-17, no.6, December 1982, pp.1192-1194.

[19] B.K. Ahuja, "An Improved Frequency Compensation Technique for CMOS Operational Amplifiers", *IEEE J. Solid-State Circuits*, vol. SC-18, no.6, December 1983, pp.629-633.

[20] D.M. Monticelli, "A Quad CMOS Single-Supply Op Amp with Rail-to-Rail Output Swing", *IEEE J. Solid-State Circuits*, vol. SC-21, December 1986, pp.1026-1034.

[21] J.H. Huijsing and D. Linebarger, "Low-Voltage Operational Amplifier with Rail-to-Rail Input and Output Ranges", *IEEE J. Solid-State Circuits*, vol. SC-20, December 1985, pp.1144-1150.

[22] B.Y. Kamath, R.G. Meyer, and P.R. Gray, "Relationship between Frequency Response and Settling Time of Operational Amplifier", *IEEE J. Solid-State Circuits*, vol. SC-9, December 1974, pp.347-352.

[23] R.G.H. Eschauzier, L.P.T. Kerklaan, and J.H. Huijsing, "A 100 MHz 100 dB Operational Amplifier with Multipath Nested Miller Compensation Structure", *ISSCC Dig. Tech. Papers*, February 1992, pp.196-197.

Biography

Rinaldo Castello was born in Genova, Italy, in 1953. He received the degree of Ingegnere (summa cum laude) from the University of Genova, Italy, in 1977. In 1979 he began his graduate study at the University of California, Berkeley, where he received the M.S.E.E. degree in 1981 and the Ph.D. degree in 1984. While at Berkeley he was Teaching Assistant and a Research Assistant. Both in 1983 and 1984 he was a Visiting Professor during part of the academic year at the University of Genova. From 1984 to 1986 he was a Visiting Assistant Professor at the University of California, Berkeley. Since 1987 he has been an Associate Professor with the Department of Electrical Engineering at the University of Pavia, Pavia, Italy, and a consultant with ST SGS-Thomson Microelectronics, Milan, Italy. His main interest is in circuit design, particularly in the area of telecommunications and analog/digital interfaces.

Dr. Castello has been a member of the Program Committee of the European Solid-State Circuits Conference (ESSCIRC) since 1987 and was the Program Chairman of ESSCIRC '91. Since 1992 he is a member of the Program Committee of the International Solid-State Circuits Conference (ISSCC). Dr. Castello was Guest Editor for the Special Issue of the Journal of Solid-State Circuits devoted to ESSCIRC '91.

Analog to Digital Conversion

Introduction

Analog-to-digital converters provide the link between the analog world of transducers and the digital world of signal processing, computing and other digital data collection or data processing systems. Numerous types of converters have been designed which use the best technology available at the time a design is made. High performance bipolar and MOS technologies result in high-resolution or high-speed converters which can be applied in digital audio or digital video systems. Furthermore, the advanced high-speed bipolar technologies show an increase in conversion speed into the Giga Hertz range. Applications in these areas are, for example, digital oscilloscopes. In this chapter papers will be presented which address very high conversion speeds and very high resolution implementations using sigma-delta modulation architectures.

In the first paper flash architectures are examined. The drawbacks of the flash architecture, such as high-power consumption, large die size and large input capacitance to mention a few, can be overcome by using a pipeline configuration. However, the pipeline configuration requires an accurate sample-and-hold circuit. The sample-and-hold amplifier on the other hand can be used to overcome dynamic errors of flash type converters. A simple model is given to predict the limits of these types of amplifiers.

In the second paper a folding and interpolation architecture will be described which combines the digital sampling of the full flash architecture with the small component count of the two-step converter configuration. As a result a low-power high performance converter can be designed. A simple model which predicts dynamic errors in flash type converters is presented. Using this model high-speed converters can be optimized for minimum distortion at maximum analog input frequency.

In the third paper an overview of oversampled converter architectures is presented. A comparison is made between the single loop modulator and the cascaded or multistage architecture. In a single loop higher order modulator stability is an important issue. First and second-order coders are inherently stable, while higher order single loop modulators need a careful design or a special configuration to overcome stability problems. A special configuration which overcomes stability problems by controlling the pole locations of the linearized loop is presented.

An alternative approach to higher-order noise shaping functions is to cascade first or second-order sigma-delta modulators. Cascading of first or second-order coders avoids the stability problem while at the same time the overall noise-shaping function is improved. A discussion of future noise-shaping coder designs completes this paper.

In the fourth paper examples of cascaded higher order modulators are given which result in a 14-bit equivalent dynamic range using a low oversampling ratio of 16. Furthermore an operational transconductance amplifier (OTA) using a minimum number of transistors in the signal path is described. Such an amplifier is seen as a crucial part to solve high-speed high-resolution coder implementations.

In the fifth paper an implementation of a multi-bit noise-shaping digital-to-analog converter is shown. Multi-bit noise shaping coders reduce the amount of out of band noise resulting in a much simpler output filter. In the multi-bit D/A converter a dynamic self-calibration technique is used which stores charge on the gate-source capacitance of MOS devices to calibrate the current sources in the converter. Furthermore a sign-magnitude architecture combined with the self-calibration technique is introduced which results in a very high performance low-noise reconstruction of the quantized signals. A dynamic range of 115 *dB* over the audio band with an oversampling ratio of 128 is obtained.

The last paper in this section describes the use of oversampling coders in bandpass analog-to-digital conversion for application in high performance radio systems. The advantage of oversampling coders in bandpass applications is found in the inherently better linearity of the system resulting in a larger spurious free dynamic range. Such a specification is very important in radio applications. Three examples of bandpass implementations are presented. The choice of sampling frequency with respect to incoming signal frequency allows even the use of subsampling in the converter. To increase the dynamic range of the system an interpolative solution is used to obtain a multi-bit D/A configuration.

Rudy J. van der Plassche

High Speed Sample and Hold
and
Analog-to-Digital Converter Circuits

John Corcoran

Hewlett-Packard Laboratories
Palo Alto, California
U.S.A.

Abstract

Architectures for analog-to-digital (A/D) conversion above 100 Msamples-per-second are described. The well known pipeline and flash configurations are examined, including variations which reduce complexity and power consumption. Static and dynamic errors common to flash architecture A/D circuits are enumerated. The advantages of sample and hold (S/H) circuits in reducing dynamic errors in flash converters are described. A simple model for sample and hold circuits is used to predict S/H performance limits and sources of error.

1. Introduction

High speed analog-to-digital converters (ADCs) with sample rates above 100 MHz have numerous potential applications, including those in digitizing oscilloscopes, spectrum and network analyzers, radar systems, medical imagers, and communications equipment. The combination of A/D converters with digital signal processing in these products allows new levels of performance and offers features previously unavailable. To a great extent, the penetration of ADCs into these applications will depend on the power, cost, and performance of the available converters. Typically, high speed A/D converters require several watts of power [1,2,3], and commercially available circuits (at this writing) typically cost many hundreds of US dollars. Furthermore, it is difficult to achieve accuracy equal to the nominal resolution of the converter, at least for converters with six or more bits of resolution. This is due to various error mechanisms in the ADC, some of which are correlated with the high sample rate, but most of which are associated with the high input signal frequencies that these converters typically process.

In this tutorial paper, we will describe the most common architectures used to implement high speed A/D converters, and why high power is typically required. Improvements on these architectures which reduce power and device count will then be examined. Next we will focus on error mechanisms in high speed ADCs, and describe some techniques for reducing their effects. Finally, we will describe the effectiveness of sample and hold (S/H) circuits in reducing dynamic errors in ADCs, and we will propose a simple model for understanding sample and hold performance limits and errors.

2. High Speed A/D Converter Architectures

2.1. Flash Convertors

The most commonly used architecture for high speed ADCs is the parallel or "flash" architecture (Fig.1). In the flash architecture, 2^n clocked comparators are used to simultaneously compare the input with a set of reference voltages generated with a resistor divider (n is the nominal resolution of the ADC). The "thermometer code" (logic ones for comparators with references below the input signal, zeroes above it) is then converted to a 1-of-N code, which is subsequently encoded to n bits (usually binary weighted) to produce the output.

Flash converters have been implemented most commonly in bipolar IC technology, where the excellent V_{BE} matching allows design of comparators accurate to eight bits or better. In MOS technology, calibration cycles are typically required to eliminate comparator offset, which reduces the maximum available clock rate. In this paper, we will focus on bipolar circuit implementations, since (to date at least) they have yielded the highest speed converters.

The principal problem with the flash architecture is the high complexity and power associated with the 2^n clocked comparators. A related problem is the high input capacitance associated with these comparators, which must be driven at high input frequency.

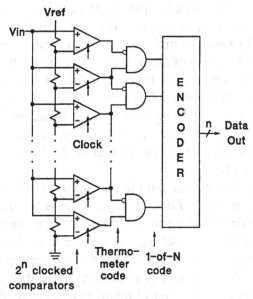

Figure 1: Simplified flash A/D block diagram

2.2. Pipeline Convertors

Both of these problems are alleviated by the use of the pipeline architecture, which converts one or more bits at a time in each of several simultaneously acting pipeline stages. In Fig.2, a one-bit per stage pipeline is illustrated. The input signal is captured by a sample and hold and compared to zero. If the input was above (below) zero, a reference level is subtracted (added) to the input, and the result amplified by two and passed to the next stage. This one-bit pipeline architecture thus requires only n comparators (versus 2^n for the flash.)

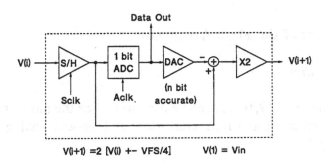

$$V(i+1) = 2 [V(i) +- VFS/4] \qquad V(1) = Vin$$

Figure 2: One-bit per stage pipeline ADC (1 of n required stages shown)

A disadvantage of the one-bit per stage pipeline is that n sample and hold circuits are required, versus one (or none) in the flash architecture. Since a high performance sample and hold typically requires significant power, the power advantage of the pipeline over the flash architecture is not as great as might be expected. Furthermore, the critical path in the pipeline architecture consists of the sum of sample and hold acquisition, comparator decision time, and DAC and amplifier settling time, much longer than the critical path in a flash converter. In fact, the maximum clock rate for the pipeline architecture may be 1/3 or 1/4 that of the flash architecture within a given technology [4].

The high power devoted to sample and hold circuits in the one-bit per stage pipeline can be reduced in two ways. One is to eliminate all the S/H circuits except the first, leading to a "ripple-through" converter [5,6]. This is of course slower than a pipeline, since there is no longer simultaneous processing of multiple input samples, but there is still the advantage of low comparator count. Another way is to digitize more than one bit per stage, and use fewer stages [7,8]. The ultimate extension of this is a two step pipeline, which digitizes $n/2$ bits per stage. Although this architecture requires only two S/H circuits, the comparator count is no longer negligible (e.g., 128 are required for a 12-bit converter).

Clearly, the optimum architectural choice (flash or pipeline, number of bits per stage in the pipeline) will depend on the available IC technology and the sample rate and resolution goals. For the purpose of narrowing the focus of this paper, we will assume that our sample rate objective precludes the use of any variation of the pipeline architecture, and that a flash (or variation of the flash) architecture is required. In the following section we will describe some ways in which the power and area penalties of the flash architecture can be reduced. These techniques will also be of interest for pipeline converters, to the extent that flash converters represent building blocks for pipeline ADCs.

2.3. Variations of the Flash Architecture

Analog Encoding

Analog encoding [9,10,11] can be used to reduce the number of clocked comparators required in a flash converter. A four-way analog encoder is shown in Fig.3.

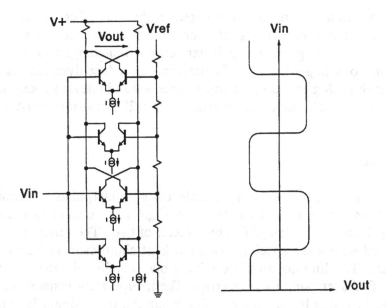

Figure 3: Four-way analog encoder circuit and DC transfer function

This circuit consists of four differential pairs with their collectors alternately cross-connected. The input signal is tied to one side of all differential pairs, with four reference voltages connected to the other side. The output of this circuit makes four transitions across zero when the input voltage covers its full range. If the output of this circuit is digitized by a single clocked comparator, we have essentially duplicated the function of four clocked comparators and the encoding logic required to produce one bit with the pattern vs. input voltage shown in Fig.3.

By logically combining the outputs of multiple four-way analog encoders (each with unique connections to the reference ladder) we can produce bits which vary with the input signal in a binary (or Gray) coded fashion. Roughly speaking, the use of four-way analog encoders reduces the number of clocked comparators required in an ADC by a factor of four, producing a considerable savings in power and device count. Although there are still 2^n differential pairs required in the analog encoders, these can be thought of as the preamp pairs usually used just ahead of the clocked comparators in a traditional flash architecture, and thus don't represent additional circuitry.

Higher levels of analog encoding (8-way or 16-way) offer the promise of even larger reductions in device count. However, several practical limits make this difficult: The finite g_m of bipolar devices means that full scale transitions will not occur at the output of the encoder if the reference voltages within a cell are too close together; the ratio of common mode to differential signal output signal from the encoder gets larger with higher-level encoding, requiring better resistor matching; and the higher capacitance of a larger number of collectors can reduce dynamic accuracy.

Although analog encoding produces a reduction in device count, it does not reduce the ADC input capacitance, since 2^n transistors must still be driven.

Interpolation

Lower input capacitance can be achieved with interpolation techniques. he simple interpolating circuit shown in Fig.4 can be used to reduce he input capacitance of a flash ADC by a factor of two. The reference divider is designed with a tap spacing twice the least significant bit LSB) of the converter. The limiting amplifiers have ow enough gain that the active regions of adjacent amplifiers overlap. The interpolating resistors produce a voltage V_I which has a zero crossing when the input signal is half way between he reference divider taps J and K. Clocked comparators digitize the voltages on the outputs of the amplifiers and interpolating resistors.

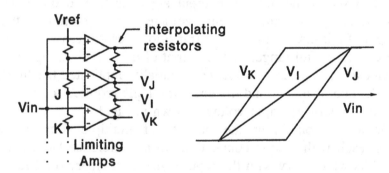

Figure 4: Interpolation circuit and DC transfer function

Although interpolation reduces capacitance on the input line, it does not intrinsically reduce the number of clocked comparators. Folding converters do achieve such a reduction, and have often been used in conjunction with interpolating techniques [12,13,14]. The reader is referred to the article by J. van Valburg, elsewhere in this volume, for a detailed description of folding and interpolating converters.

3. Error Sources in Flash Converters

Most variations of the flash architecture suffer from the same types of errors. These errors can be separated into two categories, static and dynamic, depending on whether they occur with a DC or a rapidly changing input signal. In this section we will describe the most common error sources in flash converters, and describe some ways in which these errors can be reduced.

3.1. Static Errors

Nonlinearity

Most static errors can be described by the integral and differential nonlinearity curves of the converter. Integral linearity errors often occur due to the base current from comparators affecting the tap voltages on the reference divider network. The effect is generally to produce a parabolic bow in the reference divider outputs as shown in Fig.5. This may be fixed in numerous ways, but one of the most effective is to have an additional reference divider, assumed to not be affected by base current errors. The outputs of this divider drive op-amps which correct the reference divider voltages at one or more places along the chain. A single center tap correction (Fig.5) will reduce bow errors by a factor of four.

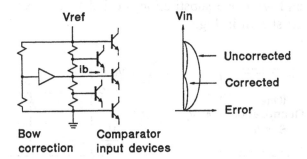

Figure 5: Reference divider bow errors and one scheme for correction

Differential linearity errors arise most commonly from V_{BE} or resistor mismatch. These errors are generally characteristic of the process, but can be reduced by the circuit designer by using sufficiently large geometry devices.

Metastability Errors

Metastability errors are another form of static error. Metastability errors arise when the input signal is very close to a comparator threshold, and the output of the comparator does not regenerate to a full logic level within the allotted time (generally half the clock period of the converter). If this bad logic level drives two or more gates in the encoder network and if one gate interprets this level as a "one" and another as a "zero", large errors can occur in the encoding. These errors can be up to half full scale in a binary converter.

The probability of a metastable error occurring can be derived fairly simply. In Fig.6 we represent one quantization band at the collectors of an ECL clocked comparator as $A*V_q$, where A is the transparent gain of the comparator, and V_q is the input referred quantization voltage (weight of one LSB). At the two quantization band edges we find error bands of width $V_e/2$, which are defined to be the voltage bands within which inadequate regeneration will occur. In this model, the probability of a metastable error can then be written as $V_e/(A*V_q)$.

When the clock occurs, ECL clocked comparators increase their output level in an exponential fashion due to the positive feedback nature of the circuit. This regeneration will be adequate to avoid metastability errors if the output after the available time T exceeds V_a, where V_a is the "ambiguity voltage" of the encoder [2]. (Put another way, a comparator output of V_a or above could not be interpreted differently by two gates in the encoder). We can then say that an input level of $V_e/2$ must regenerate to a voltage V_a after time T, and with one substitution are led to the expression for the probability P, as shown in Fig.6.

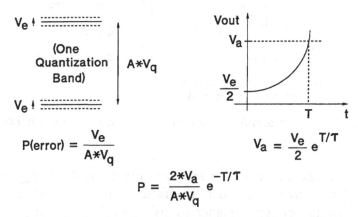

$$P(\text{error}) = \frac{V_e}{A*V_q}$$

$$V_a = \frac{V_e}{2} e^{T/\tau}$$

$$P = \frac{2*V_a}{A*V_q} e^{-T/\tau}$$

Figure 6: Derivation of metastable error probability

Metastable errors occur within $V_e/2$ of the quantization band edge. In the expression for P, the term in front of the exponential has a value typically near one (e.g. , V_a= 25 mV, A= 5, V_q= 10 mV). Thus the probability of error is virtually equal to the value of the exponential term. For a typical 10 GHz f_T process, a well designed comparator might have τ equal to 70 ps. For a 1 GHz clock, the predicted error rate is then about 1 x 10^{-3}, for most applications a totally unacceptable value.

To reduce the metastable error rate substantially, additional latch stages can be employed after the comparators (before the encoder) to increase the time available to regenerate. One additional latch stage essentially squares the probability to 1×10^{-6} in the example above. Additional latches provide further improvements. However, especially in a full flash converter with 2^n comparators, this is an expensive solution.

Another approach is to encode the comparators into Gray code (or a code similar to Gray) which has the property that adjacent codes differ in only one bit position. This property means that each comparator affects only one output bit (in a properly designed encoder), and so if a metastability occurs it only produces a one LSB error. If Gray code is not acceptable for the output code of the converter, additional latches can be added after the Gray encoder to reduce the metastable errors to an acceptable level, then the Gray code can be converted to binary. The advantage of this approach of course is that only n latches are needed per stage of additional pipelining, versus 2^n latches per stage when the pipelining is done before the encoder, as in the example above.

3.2. Dynamic Errors

Dynamic errors are defined here as those arising from a rapidly changing input signal. Classes of dynamic errors include those that cause encoding errors, those that produce distortion, and those that add noise to the output data.

Encoding Errors

Encoding errors often result from what is referred to as a "ragged thermometer code" or a "bubble code". These describe deviations from the normal thermometer code in a flash converter (i.e., ones below a transition point, and zeros above that). Fig.7 depicts the normal thermometer code and the corresponding 1-of-N code, and a case where the thermometer code has a "bubble", a zero surrounded by ones. In this case the 1-of-N code

150

actually has two ones, as would be generated by the AND gates of Fig.1. Depending on the design of the following encoder, this situation can produce very large errors. For example, if the encoder is implemented as a diode ROM, then the output from the encoder may be a bit-by-bit OR of the codes that would have resulted from each of the active input lines. This could produce half full scale errors.

Bubble code errors can be produced by rapidly changing input signals if the circuit layout produces different arrival times of the comparator input signals relative to their clocks. This can cause two comparators to essentially sample a different value of the input signal when the clock occurs. "Seams" in the circuit layout (due to construction of the circuit as a side-by-side assembly of several identical blocks for example) often produce a timing discontinuity which can result in a bubble error.

Normal Thermometer Code	1–of–N Code	Thermometer Code With "Bubble"	1–of–N Code
0		0	
0	0	0	0
0	0	1	1
0	0	0	0
1	1	1	1
1	0	1	0
1	0	1	0

Figure 7: Thermometer codes and corresponding 1-of-N codes

The impact of bubble errors can be reduced by coding techniques. This is not nearly as good a solution as repairing the layout (or other anomalies) which cause the error in the first place, but reducing the magnitude of errors which persist is nevertheless a worthwhile endeavor. Many coding techniques have been published which reduce the effect of bubble errors. An example of such a technique is shown in Fig.8. Here a three input gate is used to produce the 1-of-N code, and it successfully suppresses one of the two "one" outputs in the example of Fig.7. However, this circuit would not eliminate errors from more deeply buried bubbles in the thermometer code (more than one comparator way from the transition point). The reader is referred to the references [2,3,15,16] for other encoding approaches which reduce the impact of bubble errors.

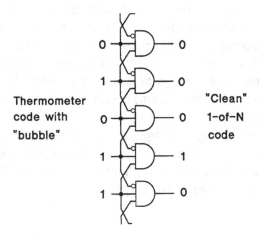

Figure 8: Reducing impact of bubble errors through coding

Distortion

Distortion can occur in several ways when high frequency input signals are applied to a flash converter. One of the most common mechanisms is the distortion produced by nonlinear semiconductor junction capacitances when driven from a finite source impedance. An example would be the input capacitance of the converter, which would typically consist of nonlinear components from the C_{JE} and C_{JC} of the comparator input devices, plus more-or-less linear wiring capacitance. Since this input capacitance can be large, it must often be driven from very low source impedance to avoid distortion at high input frequency. Strategies to minimize this distortion (other than reducing impedance levels which usually costs power) include reducing input signal swing, putting more voltage across base-to-collector junctions, and using architectures which have lower node capacitances.

Another distortion effect occurs due to the fact that ECL comparators have a slew rate dependent delay (in transparent mode) from input to output. This delay is typically less for fast slewing signals. The effect of this slew rate dependent delay is to produce distortion in a sine wave input, since the comparators which digitize the top of the sine wave see low slew rates at their sampling instant, and those that sample near the zero crossing of the sine wave see a high slew rate. The result is to "reshape" the sine wave, and produce distortion. This effect can be reduced somewhat with careful circuit design, but substantial improvements require higher f_T processes.

Comparator aperture effects represent another important distortion source. When comparators are switched into regeneration mode, they do not make this transition instantaneously. In fact, comparators which have only small differential signals stored in their collector capacitances at the clock transition may see the sign of this signal reversed by a high slew rate input. This happens (for one reason) because the current is not immediately removed from the input differential pair at the clock transition. This effect may be viewed as having a comparator whose threshold is a function of signal slew rate and slew direction. The effect once again is to reshape the input sine wave and, as above, this phenomenon is probably best addressed by employing a higher performance process.

Noise

The final dynamic error that we will discuss is noise in excess of the thermal or shot noise that occurs with DC inputs. Time jitter on the input clock can produce excess noise in the converter output data for dynamic inputs. This happens because the clock jitter is directly converted to voltage noise in proportion to the slew rate of the ADC input signal. This problem can become quite severe for very high frequency inputs even with reasonably low jitter clock sources. For example, to keep the error at the zero crossing of a 1 *GHz* sine wave below 1% of full scale requires less than 3 *ps* peak clock jitter. These errors can be minimized in the converter itself by maintaining fast edges in the clock path to minimize the conversion of thermal noise in the gate inputs to time jitter. Furthermore, low phase noise oscillators may be required as clock sources for ADCs with high bandwidth-resolution product.

A general comment that can be made about dynamic errors is that they tend to get worse as the DC resolution of the converter is increased. An obvious example is the distortion due to nonlinear input capacitance. If one more bit of DC resolution is required, the input capacitance typically doubles. This increases the amount of dynamic distortion unless the source impedance can be reduced, which is often not possible. Another example is the errors introduced by propagation delay mismatch between the input and clock lines to the comparators. As the converter resolution is increased, it gets physically bigger, and matching these delays gets more difficult. Thus the converter designer is faced with a difficult tradeoff between DC resolution and AC accuracy. To some extent this tradeoff can be relaxed through the use of a sample and hold circuit, which is the subject of the next section.

4. Sample and Hold Circuits

Sample and hold circuits can be used to reduce most dynamic errors in a flash converter. Encoding errors, dynamic distortion due to nonlinear junction capacitance, distortion due to comparator effects, and errors due to clock and input propagation delay mismatch will all be reduced through the use of a sample and hold (S/H). This is true because in a well designed system the comparators will be clocked at a time when their inputs (the output of the S/H) are essentially unchanging. Operated in this way, only the static errors of the comparators should be important. The dynamic performance of the system then depends entirely on the dynamic performance of the S/H, and its ability to deliver a near zero slew-rate signal to the inputs of the comparators before they are clocked.

There are at least three types of sample and hold circuits suitable for use in a high speed ADC system: *the impulse sampler, the sample and filter* circuit, and the *track and hold*. An impulse sampler captures the input signal by sampling it with a very short sampling pulse. Usually, a pulse of charge is generated with this circuit whose value is proportional to the average of the input signal over the sampling pulse. This charge is then collected on a hold capacitor and converted to a voltage. Because of the averaging of the input signal over the sampling pulse, the bandwidth of such a sampler is inversely proportional to the length of the pulse. For high bandwidth systems, this require a very short pulse, but results in a system whose bandwidth is independent of other component values.

The sample and filter approach [17] is a variation of the impulse sampler. Instead of collecting the charge on a hold capacitor and settling the input line of the converter to the corresponding voltage, this circuit filters and amplifies the impulse of charge into a longer Gaussian-shaped voltage pulse whose height is strictly proportional to the original charge sample. The following ADC is then clocked at the peak of this Gaussian, where the slew rate of the signal is zero. The advantage of this approach is a reduction of the bandwidth of the circuit which drives the ADC input line, since step settling is not required.

The track and hold circuit is the most commonly used sampler. In this circuit, the analog signal is acquired explicitly as a voltage on a hold capacitor via a sampling switch. When the switch is turned off, the input signal present at that time is captured on the hold capacitor and delivered to the digitizer. The advantage of this circuit is that a full scale voltage signal is available for digitization without further amplification.

Although sample and hold circuits typically make significant improvements in converter dynamic performance, there are naturally problems that remain. The requirements for a low jitter clock (for example) are not reduced by use of a S/H. Furthermore, all sample and hold circuits suffer from various linear and nonlinear performance-limiting effects. In the next section we will describe the most important of these phenomena for the commonly used track and hold circuit, and some ways for reducing their effects. For the purposes of the following discussion then, *sample and hold* refers specifically to the *track and hold* configuration.

5. Sample and Hold Error Sources

5.1. Linear Error Mechanisms

Linear performance-limiting mechanisms in sample and hold circuits include bandwidth, pedestal, and feedthrough effects. To define these effects, we will use the simple model of Fig.9. The sample and hold consists of a switch with series resistance *Rs*, capacitance *Cf* across the switch from input to output, and capacitance *Cp* from switch control line to output. The pedestal of the switch is then defined as the change in the voltage on the hold capacitor at the transition from track to hold, due to the capacitance *Cp*. Likewise, the feedthrough is defined as the change in value on the hold capacitor during hold mode, due to the capacitance *Cf*. Both of these parameters are best expressed as a fraction of the full scale signal range *Vfs*.

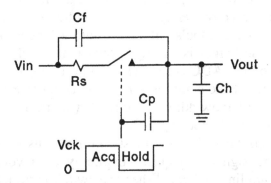

Figure 9: Simple model for linear effects in sample and hold circuits

In Fig.10, the bandwidth, pedestal, and feedthrough are derived from this simple model. The hold capacitor value *Ch* is present in the denominator of all three expressions. Clearly, we can reduce pedestal or feedthrough by using a larger hold capacitor, but this will come at the expense of lower bandwidth. Furthermore, lower bandwidth will mean lower maximum sample rate. This is so because acquisition of a new sample takes place with the same RC time constant that limits the track-mode bandwidth. If acquiring the input signal requires (say) 6 time constants, and if the clock has 50 % duty cycle, it can easily be shown that the maximum sample rate is about one half the track mode bandwidth. In a practical design then, the hold capacitor value would typically be fixed, bandwidth selected by changing the size of the switch devices, and the errors due to pedestal and feedthrough examined.

Bandwidth

$$B = 1 / (2 \pi * Rs\ Ch)$$

Pedestal

$$P = (Vck * Cp) / (Vfs * Ch)$$

$$P \sim Cp / Ch$$

Feedthrough

$$F = Cf / Ch$$

Figure 10: Parameter values derived from sample and hold model

If the feedthrough errors are excessive, they can be reduced with the more complicated series-shunt-series switch configuration shown in Fig.11. The grounded switch shunts the capacitive feedthrough current from the first series device and makes a substantial reduction in feedthrough for all reasonable input frequencies. Unfortunately, since there are now two series switch devices, they must be made twice as large as in the single switch design or the bandwidth objective will no longer be met. The pedestal induced on the hold capacitor by the larger output side device will then be twice as large.

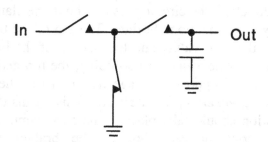

Figure 11: Series-shunt-series sample and hold configuration

To reduce pedestal errors, a complementary switch configuration can be used. Fig.12 illustrates a simple complementary switch using CMOS devices. If the control voltages are equal and opposite, and the N and P devices are substantially identical in other characteristics, some pedestal cancellation will take place. The improvement is limited however by the fact that the *Vgs* of the P and N devices is a function of the input signal. When the gate signals drive the circuit into hold, the remaining voltage transition on the gates *after* the FETs turn off will be different on the p and N devices, creating a residual pedestal. This could be avoided if the "ON" gate control level somehow followed the input signal, and if the gate signals always went through a fixed transition when going into hold.

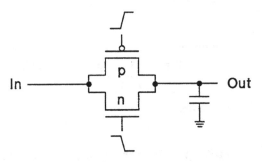

Figure 12: Complementary sample and hold implemented using MOS devices

A desirable circuit is one which combines the series-shunt-series architecture with the signal-tracking complementary switch, and achieves both low feedthrough and pedestal. One such circuit is a diode bridge sampler with feedback clamp [18], as shown in Fig.13. With the currents flowing in the direction shown, the circuit is in track mode; the output voltage is ideally equal to the input. The top of the bridge is at the input plus V_D and the bottom of the bridge is at the input minus V_D. The clamp diodes *D1* and *D2* are reverse biased by one diode drop V_D. When the currents reverse direction, the circuit goes into hold; the clamp diodes carry the currents away from the diode bridge. The top of the bridge goes to a voltage V_D below the held value, and the bottom of the bridge goes to a voltage V_D above the held value. In so doing, the top and bottom of the bridge make equal and opposite voltage transitions at the hold event of magnitude $2V_D$, *independent of input signal*. This means that nearly ideal pedestal cancellation should take place. The two clamp diodes (and the two diodes in series in each leg of the bridge) implement the series-shunt-series switches required for improved feedthrough performance.

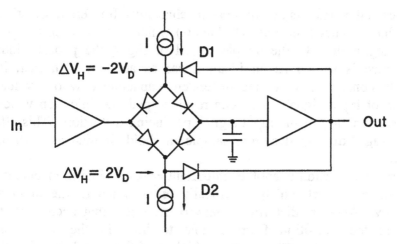

Figure 13: Diode bridge with feedback clamp

Another way to reduce pedestal and feedthrough errors is with an improved switch, i.e. one with either lower Rs, or lower Cp and Cf. These improvements come somewhat naturally with improvements in IC technology. In the case of CMOS switches, the product of Rs and Cp, which we can take to be a figure-of-merit for the switch (smaller is better), is roughly inversely proportional to f_T of the transistor. And f_T of course improves with device scaling. For Schottky diode switches, the product of Rs and Cp can be very low. For GaAs transistors, this figure-of-merit has been reported at below 0.2 *ps* [19], considerably better than for diode-connected transistors in the same technology.

5.2. Nonlinear Error Mechanisms

Nonlinear error mechanisms in sample and hold circuits include *track mode distortion, pedestal distortion,* and *aperture distortion.* Track mode distortion can result from simple DC transfer function curvature, common in the open loop amplifiers which may be necessary in very high bandwidth sample and hold circuits. Track mode distortion can also result from dynamic effects, e.g. when high frequency signals are applied from finite source resistances to nonlinear junction capacitances in the sample and hold circuit. This effect is typically less than in a flash ADC without a sample and hold, since capacitances are typically much higher there, but it can still be important. Track mode distortion represents the lowest distortion level achievable for the overall S/H circuit. Other mechanisms associated with the transition to hold usually dominate the overall performance.

Pedestal distortion is one of those mechanisms. It refers to a static effect (one which occurs even with DC inputs) which occurs when the pedestal is not negligible. In the simple model of Fig.9, the pedestal-inducing capacitance is shown as an ideal, linear capacitance. In fact, it will typically consist of a voltage dependent capacitance whose value is a function of input signal. This can result in pedestal injection which is a nonlinear function of the input signal, producing distortion. This effect is best managed through design of sample and hold circuits with low overall pedestal.

Aperture distortion is another important error-producing mechanism. It is a dynamic effect which is typically proportional to the input signal frequency. Aperture distortion refers to any effect that occurs due to the fact that the transition from sample to hold in the switch is not instantaneous. If the effective turn-off time of the switch is a function of input signal slew rate and/or direction, distortion can occur. It is easy to postulate a slew-direction dependence of turn-off time in an NMOS switch for example, since a falling gate signal will cut off the channel sooner for a rising input signal than a falling one. Aperture distortion can be a very important effect in sample and hold circuits for high frequency inputs. It's magnitude can be reduced by delivering very fast edges to the switch control input of the sample and hold. The ability of the designer to do this largely depends on the speed of the underlying IC technology.

There is limited published data available to verify the assertion that sample and hold circuits do offer an improvement in dynamic performance over flash converters without a sample and hold. Technologies applied have generally been different, and a comparison within a single technology is difficult to find. However there is at least one case where the same A/D was evaluated with [18] and without [9] a sample and hold. Without a sample and hold, 5 effective bit accuracy (in this 6-bit resolution circuit) was achieved with 100 MHz input frequency. With a sample and hold, the same performance was achieved with input signals up to 1 GHz.

Although sample and hold circuits eliminate dynamic errors associated with the nonlinear input capacitance of a flash converter, this capacitance still presents a significant problem. In fact, the main limit to sample rate for a sample and hold driving an ADC may well be the time required to settle the distributed, highly capacitive input line of the digitizer. This problem of course gets worse as DC resolution is increased, and the designer may once again be faced with a difficult choice, this time between DC resolution and sample rate.

6. Summary

The traditional flash A/D converter offers a higher sample rate than other architectures within a given technology, but is complex and requires high power. Furthermore, dynamic accuracy suffers from the large distributed nature of the circuit and its high input capacitance. The pipeline architecture reduces all of these problems, but makes a considerable compromise in sample rate. Variations on the flash architecture such as analog encoding or folding and interpolating can reduce complexity, power, and input capacitance with smaller impacts on maximum sample rate than the pipeline configuration.

Most variations of the flash architecture suffer from similar sources of error. Dynamic errors are particularly difficult to address with circuit design techniques, but can be reduced with a high performance IC process. Increases in static resolution tend to come at the expense of higher dynamic errors.

Sample and hold circuits are particularly effective at reducing dynamic errors in flash converters. However, care must be taken in design of the sample and hold circuit to realize this result. Static errors in sample and hold circuits depend to a significant extent on the on-resistance and off-capacitance of the switching device available within the chosen technology, although circuit design techniques are available to reduce the effects of these parameters. Dynamic performance depends on the quality of the switching device, the speed of the active devices which drive the switch, and the care taken in the design of the overall circuit.

References

[1] Yukio Akazawa, et al., "A 400 MSPS 8b Flash AD Conversion LSI", ISSCC Digest of Technical Papers, 1987, pp.98-99

[2] Bruce Peetz, et al., "An 8-bit 250 Megasample per Second Analog-to-Digital Converter: Operation Without a Sample and Hold", IEEE Journal of Solid-State Circuits, vol. SC-21, no.6, December 1986, pp.997-1002

[3] Yuji Gendai, et al., "An 8b 500MHz ADC", ISSCC Digest of Technical Papers, 1991, pp.172-173

[4] Ken Poulton, Hewlett-Packard Laboratories, private communication

[5] Robert A. Blauschild, "An 8b 50ns Monolithic A/D Converter with Internal S/H", ISSCC Digest of Technical Papers,1983, pp.178-179

[6] Robert Jewett, et al., "A 12b 20MS/s Ripple-through ADC", ISSCC Digest of Technical Papers, 1992, pp.34-35

[7] Stephen H. Lewis, et al., "A Pipelined 5 MHz 9b ADC", ISSCC Digest of Technical Papers, 1987, pp.210-211

[8] David Robertson, et al., "A Wideband 10-bit, 20 MSPS Pipelined ADC using Current-Mode Signals", ISSCC Digest of Technical Papers, 1990, pp.160-161

[9] John J. Corcoran, et al., "A 400MHz 6b ADC", ISSCC Digest of Technical Papers, 1984, pp.294-295

[10] T.W. Henry, et al., "Direct Flash Analog-to-Digital Converter and Method", U.S. Patent 4,386,339

[11] Adrian P. Brokaw, "Parallel Analog-to-Digital Converter", U.S. Patent 4,270,118

[12] Rob E.J. van de Grift, et al., "An 8b 50 MHz Video ADC with Folding and Interpolation Techniques", ISSCC Digest of Technical Papers, 1987, pp.94-95

[13] Rudy van de Plassche, et al., "An 8b 100 MHz Folding ADC", ISSCC Digest of Technical Papers, 1987, pp. 94-95

[14] Johan van Valburg, et al., "An 8b 650 MHz Folding ADC", ISSCC Digest of Technical Papers, 1992, pp.30-31

[15] Yoji Yoshii, et al., "An 8b 350 MHz Flash ADC", ISSCC Digest of Technical Papers, 1987, pp. 96-97

[16] V.E. Garuts, et al., "A Dual 4-bit, 1.5 Gs/s Analog-to-Digital Converter", 1988 Bipolar Circuits and Technology Meeting Proceedings, pp.141-144

[17] Ken Rush, et al., "A 4 GHz 8b Data Acquisition System", ISSCC Digest of Technical Papers, 1991, pp.176-177

[18] Ken Poulton, et al., "A 1-GHz 6-bit ADC System", IEEE Journal of Solid-State Circuits, vol. SC-22, No.6, December 1987

[19] S. Prasad, et al., "An Implant-Free 45 GHz AlGaAs/GaAs HBT IC Technology Incorporating 1.4 THz Schottky Diodes", 1991 Bipolar Circuits and Technology Meeting Proceedings, pp.79-82

Biography

John J. Corcoran received the B.S. degree in electrical engineering from the University of Iowa in 1968, and the M.S. degree in electrical engineering from Stanford University in 1970. Since 1969 he has been with Hewlett-Packard Company, first at Santa Clara Instrument Division, then at Hewlett-Packard Laboratories in Palo Alto, California, where his work has focused on integrated circuit design in numerous technologies including silicon bipolar, BIMOS, NMOS, CCD, GaAs MESFET, and GaAs HBT. He is presently manager of the High Speed Electronics Department of HP Laboratories, where he develops integrated circuits for instrumentation and communication applications.

He has published numerous papers on high speed A/D conversion and A/D testing, and was co-author of the paper "A 1-GHz 6-bit ADC System" which received the Best Paper Award from the Journal of Solid State Circuits for 1987-1988. He received the Best Evening Panel Award from the 1988 International Solid State Circuits Conference for a panel entitled "GaAs vs. Si for High Speed Analog ICs", and was co-recipient of the Editorial Excellence Award from ISSCC in 1992 for the paper entitled "A 12b 20MHz Ripple-through ADC". He has served as Guest Editor for the Journal of Solid State Circuits, and as chairman of the Analog Subcommittee for ISSCC.

High Speed Folding ADC's

Johan van Valburg, Rudy J. van de Plassche

Philips Research Laboratories
Eindhoven
The Netherlands

Abstract

This paper describes the evolution steps used within the Philips company on the development of a high speed analog to digital converter (ADC) with a low power consumption and a small die size. Analog preprocessing of the input signal before conversion to binary information is a key function in all the explained systems.

1. Introduction

The continuously increasing sampling frequencies of analog to digital converters (ADC) and the demand for low power devices ask for ADC structures with a reduced number of power consuming decision stages.

In this paper, several techniques are shown for how to handle this reduction of comparators to come up with state of the art techniques used in the latest ADC designs of Philips.

2. Flash ADC Structure

The full parallel solution in analog to digital conversion, also called flash conversion, is briefly described in the section. To have a good comparison with other structures, all the described techniques will be used to implement an 8 bit ADC.

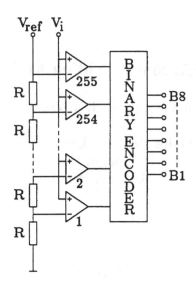

Figure 1: Flash ADC

Flash ADC is a very straightforward technique to convert an analog signal V_i into an 8 bit output code. A resistor ladder is used to generate a set of 255 reference voltages. Those reference voltages are used together with a set of 255 decision stages to compare the input signal V_i with those references. As long as V_i is smaller as the reference voltage, the output of a decision stage will be a logic '0', where a logic '1' occurs at the output if V_i exceeds the V_{ref} of a decision stage.

At the output of the decision stages, a so called linear code, or thermometer code is generated, where the transition from '0'-block to '1'-block in the 255 bit code is related to the value of the input signal V_i.

Very simple techniques are used to derive an 8 bit code out of this thermometer code. In the first place, a differential code is generated using a simple boolean operation on the thermometer code. The differential code has the property to have only one logic '1' which represents the region in the 255 wide decision stage array where the actual transition from '0' outputs to '1' outputs occurs.

The 1 out of 255 code can easily be converted into 8 bit output code using a large but simple analog encoder ROM. This ROM can be simple, using the property of its input code.

The flash structure is a simple technique but uses a lot of chip area to implement the block of decision stages as well as analog encoding ROM. The large chip area can give some problems on a layout like skew in the clock signals, buffering of the sampling clock etc, etc. For higher sampling rates, power consumption can be a practical limit for using this technique in analog to digital conversion. To give an example, SONY presented a paper at the ISSCC 1991, implementing an 8 bit full flash ADC. This 500 *Msample/s* device consumed more then 3 *Watt*, where the chip area of the device was about 21 mm^2.

3. Two Step ADC Structure

A very interesting structure to solve the problem of the large number of comparators is the so called two step structure. In such a structure, the conversion is splitted into two parts, a coarse conversion and a fine conversion.

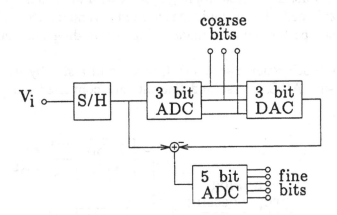

Figure 2: Two Step ADC

During the conversion, the input signal V_i is transformed into a constant DC signal using an sample and hold amplifier (S/H). This provides timing delay variation errors between the coarse and the fine section. In the first place, the analog signal is converted into for instance a 3 bit code using a normal flash ADC. The decision of this coarse ADC is converted back into an analog value and subtracted from the applied original input signal. The residual signal after subtraction is then converted into 32 fine levels using a 5 bit flash ADC.

This solution reduces the number of comparators from 255 to 7+31=38. The disadvantages of the structure for higher sampling frequencies are the sample and hold and the digital to analog converter (DAC). Those blocks are difficult to design for the intended clock frequencies. The SH amplifier should have a very small settling and aperture time and the DAC must have at least an 8 bit linearity for not limiting the performance of the device.

It is clear, that the fine conversion takes place after the coarse conversion is completed. This multistep approach is not limited to two step but the number of steps can be increased to 8 steps, one step for each bit.

4. Folding ADC Structure

The folding structure tries to use the best of the previous explained techniques. The structure does not require a S/H and a DAC but has a comparable reduction rate in the number of comparators.

In this paragraph, the techniques used in the evolution of folding ADC's is described. First the current folding system with the extension to double folding is explained. This technique is used in several devices, but a mayor step to higher performance is made using the folding and interpolation technique.

A general block scheme shows that, in stead of a step by step approach in the two step ADC, coarse and fine bits are converted independent of each other.

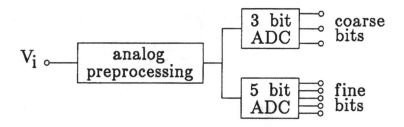

Figure 3: Folding ADC

The comparator reduction can be explained using a simplified model for the DC behaviour of the analog preprocessing. In a flash ADC, each of the 255 comparators is responsible for only one code transition in the input range. A folding ADC has a multiple use of a comparator for instance 8 times which is, as a consequence of that, the reduction rate in the number of comparators.

As visualized in Fig.4 the preprocessing in a folding ADC converts the input signal V_i into a signal with a reduced input range.

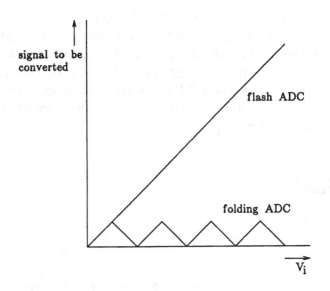

Figure 4: Folding operation

4.1. Current Folding

An implementation of this analog preprocessing is the so called current folder, shown in Fig.5. The input signal V_i is transformed into a current I_{in}, which is continuously compared with the currents I. If the range of I_{in} is defined from 0 to $4I$, the system can be explained as follows.

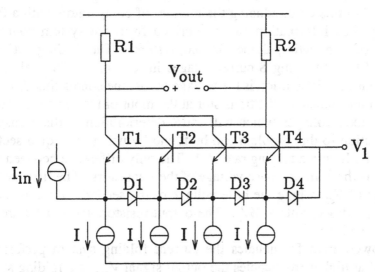

Figure 5: Current folder

Suppose I_{in}=0, the transistors $T1..T4$ will conduct and the current through $R1$ equals the current through $R2$ and the resulting output voltage V_{out} will be 0. If I_{in} increases, the current through $T1$ decreases and as a result of that V_{out} will increase. If I_{in} exceeds the value of I, no current will flow through $T1$ anymore and diode $D1$ will start conducting. All the current flowing through $D1$ will be substracted from the current through $T2$, resulting in a decrease of V_{out}.

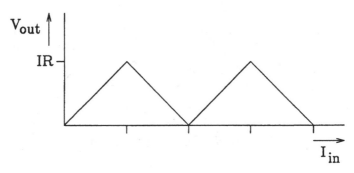

Figure 6: DC transfer of current folder

This process repeats for all the diodes and transistors, resulting in a transfer function as shown in Fig.6. The repetition rate in the output signal is called the folding rate and equals 4 in this system. The signal V_{out} can be converted into the fine bits of the system using a conventional flash ADC and in this case reducing the number of comparators with a factor 4.

The coarse information can be derived from the system counting the number of conducting diodes in the system. The folding rate can be increased to 8 by using 8 current stages instead of 4. The only problem implementing this solution is the voltage across the conducting diodes. The base emitter voltage of the transistor at the input of the system can increase up to 8 diode voltages which will cause deterioration of this transistor.

A solution to this problem can be found by using two equal sections in parallel, both with a folding rate of 4. The only difference between the two sections is the common base voltage of the transistors. The first section has a voltage of V_{cb}, where the base voltage of the second section is increased with half a diode voltage $\frac{1}{2}V_d$. The output resistors $R1$ and $R2$ are shared by both sections.

At lower input frequencies the current folding system performs very well, but at higher frequencies the output signal V_{out} of a folding system is rounded at the top caused by the limited bandwidth of the system.

4.2. Double Current Folding

In a double current folding system, two output voltages are generated using two current folders where a phase shift of 90^0 is obtained between the two output signals of the current folders.

The basic idea for using two current folding systems in parallel is that at the moment the output signal of one of the two systems will come in its nonlinear region, the other output signals will per definition operates in its linear region and vice versa. Coarse bit information is used to determine which of the two systems operates in its linear region.

Another advantage of using parallel current folding system in this double configuration is that the total linear region is increased by a factor two and so the resolution is increased with 1 bit. A total system overview of a double current folding system is shown below.

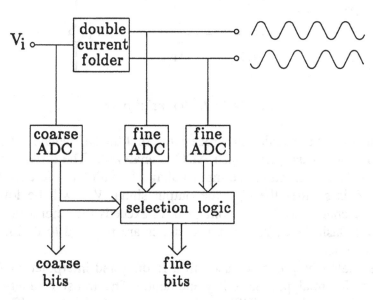

Figure 7: Double current folder

4.3. Folding and Interpolation

The major drawback on the previous explained folding systems is that the absolute value of the output signal is used for the conversion from an analog system to digital information. There will always be an analog input frequency where in case of the double current folding system both current folders will operate in there nonlinear regions which will cause missing codes at the output of the ADC.

The folding and interpolation technique has a complete other approach in designing an analog preprocessing unit. Up to now, this technique of analog preprocessing seems to be the most promising solution for reducing the number of comparators in an ADC.

Therefore, in this section we will go in more detail, compared to the previous system descriptions.

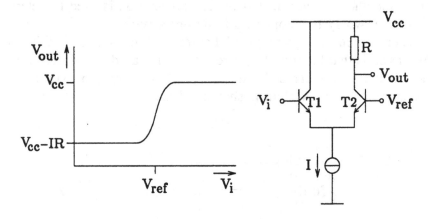

Figure 8: Differential pair

To understand the analog preprocessing, first the DC behaviour of a part of the comparator used in a flash ADC is analyzed. The differential pair, formed by $T1,T2$ compares the input voltage V_i with the reference voltage V_{ref}. If V_i is smaller the V_{ref}, the output signal V_{out} will be low. The output will change to a high voltage if V_i exceeds the reference voltage. In an 8 bit flash ADC, 255 of those stages are required to define all the code transitions.

In the analog preprocessor used in a folding and interpolation ADC, a coupled differential pair is a key element. The transfer function of a coupled differential pair (CDP) can be summarized as follows. The output signal V_{out} is high if the input signal V_i is higher than V_{lo} and lower than V_{hi}. In a high speed ADC, generation of full differential signals is necessary. Two CDPs can be configured in such a way that a full differential MSB pattern is generated. For simplicity, the input range of the ADC is defined from $0V$ to $1V$. The lower CDP has an active DC region between $0V$ and $0.5V$ and can therefore be defined as $\overline{MSB}$ where the other CDP is responsible for the MSB generation.

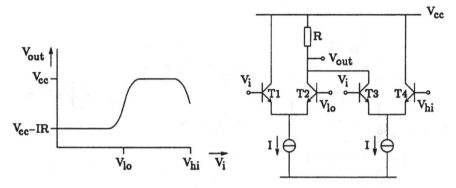

Figure 9: Coupled differential pair

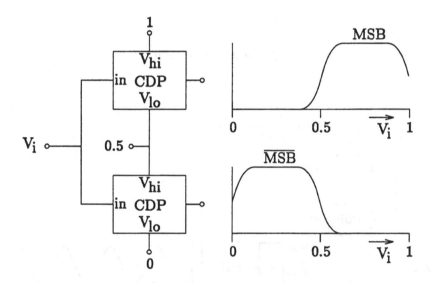

Figure 10: MSB generation

Resolution can be increased by using more CDPs in one array. For instance, to make a 2 bit ADC, 4 CDPs with reference steps of 250mV together with analog combination blocks are used to make such a device. This process of increasing resolution using more CPDs in one block can be repeated. The configuration in the present ADC has in total 8 CDPs in one block. The reference step used is decreased to 125mV. Combining electronic is also used in this configuration to make MSB, MSB-1 and MSB-2.

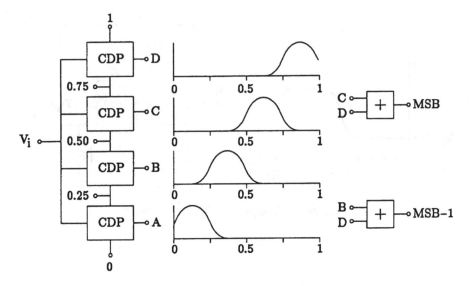

Figure 11: 2 bit ADC

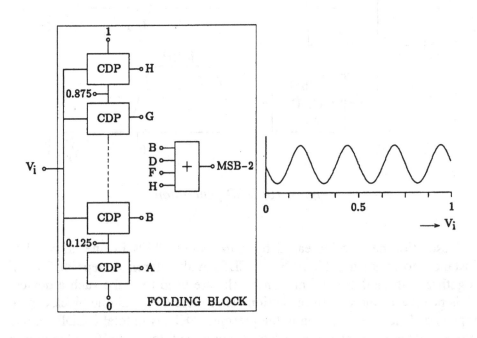

Figure 12: MSB-2 generation

For further explanation, a set of 8 CDPs, together with the combining electronics to generate MSB-2 is defined as a folding block. The folding rate of this system is defined as the number of CDPs in one folding block, in this case 8. To avoid overcrowded graphs, the differential signals are shown in there single ended version.

The designer has the freedom to increase the resolution by using for instance 16 CDP's in one folding block. There are strong practical reasons for not increasing the folding rate. At the output of a folding a signal is generated, with a frequency of approximately the frequency of the input signal multiplied with the folding rate. Converting for instance a 250 *MHz* sine wave will give an internal frequency of approximately 2 *GHz*.

Other techniques have to be used to increase the resolution without increasing the folding rate of the system. In the current folding technique, parallel use of blocks is used to increase resolution. The same technique is used in the folding and interpolation ADC. One folding block is used to generate an output wave similar to MSB-2, as explained before. The output of this folding block defines 8 code transitions.

A second folding block, with exactly the same elements is used in parallel. The only difference in the second folding block is a small offset in the reference signals used in the CDP's. A shift in reference voltages of $\frac{1}{16}V$ generates a wave pattern with code transitions exactly between the code transitions of the first folding block.

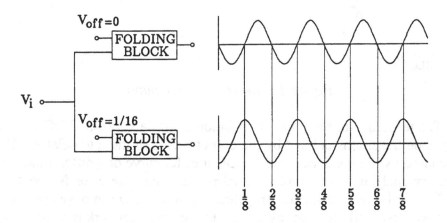

Figure 13: Parallel folding blocks

In total 4 parallel blocks are used in the most up-to-date folding and interpolation ADC to increase the resolution with 2 bits up to 5 bits. Further increasing the resolution can in theory be done by placing more blocks in parallel with of course smaller steps in the offsets of the reference voltages of the CDP's.

Increasing the resolution up to 8 bit using the parallel folding block technique would give a final design with a complexity comparable to a flash ADC because of the large amount of elements used in the analog preprocessing.

Resistive interpolation is a very cheap technique to increase the resolution without using more then the already proposed 4 folding blocks in parallel. The basic idea behind the technique is to use two outputs of consecutive folding blocks together with a set of 8 resistors. The taps of the interpolation circuit will generate the intermediate code transitions of the ADC.

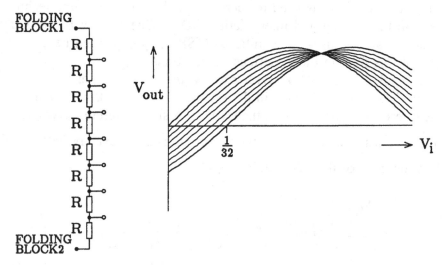

Figure 14: Resistive interpolation

The technique of resistive interpolation causes a very small DC-error in the code transitions defined by the output waves of the interpolator. This is caused by the nonlinearity in the DC transfer curve of a differential pair. Further analysis of this problem makes clear, that the error is about 0.1 LSB-units large at its maximum value. Techniques are in development to remove this small DC-error by scaling the resistive network in such a way, that it is compensating for the nonlinearity in the differential pairs. This technique is limited by the accuracy of integrating exact values of resistors and by the matching of the IC-process.

The preprocessing is now completed. Three coarse bits are generated directly by combining internal signals of a folding block. Resolution is synchronised in the bit synchronisation cell (BITSYNC) using *MSB-2* code transitions. The *MSB-2* pattern is suitable for synchronisation, because this signal is made using a folding block and so has the timing of the fine bit information.

All the explained techniques are used in the, up to so far, fastest 8 bit ADC in the world. The device was presented at the latest ISSCC. The specifications makes it clear, that fast ADC's can be made while consuming not to much power and while not coming up with a large die. The present design consumes only 850 *mW*, where the required chip area for implementing the system is a little bit more than 4 mm^2.

4.4. Design trade offs

Designing a folding and interpolation ADC places the engineer for a few trade-offs. There are in basic three sections where the resolution of the ADC is made. The first one is the folding block on its own. The number of CDPs used in one folding block, also called the folding rate, is responsible for the generation of in total ^2log(folding rate) bits by combining internal signals of a folding block and interpreting the output of the folding block also as a bit. Those bits can be defined as the coarse bits of the system.

The choice of the folding rate has to be made together with the maximum analog input frequency, because of the frequency multiplication of the internal signals in the ADC. The folding rate also expresses the reduction rate in the number of comparators.

Resolution is increased by placing folding blocks in parallel. In that part of the system ^2log(parallel blocks) bits are realised.

The last step in making the resolution of the system is done in the resistive interpolator. The number of bits made there equals ^2log(interpolation rate).

Experience in integrating the system makes it clear, that the layout has a very symmetrical structure if the number of comparators is about equal to the number of used CDPs, which equals the folding rate multiplied by the number of folding blocks used in parallel.

As an example, designing a 10 bit folding and interpolation ADC, a possible solution of the trade offs is shown in the next figure.

FOLDING RATE	16——▶ 4 bits
PARALLEL FOLDING BLOCKS	4 ——▶ 2 bits
INTERPOLATION RATE	16——▶ 4 bits
COMPARATORS	64 (1024/16)

Figure 16: 10 bit ADC proposal

5. Distortion

In this section, sampling time uncertainty, propagation delay and signal dependent delay will be described. Signal dependent delay is one of the main sources of distortion in ADC's, where the input signal V_i directly is compared with a number of reference voltages, like in a flash ADC or in a folding ADC.

5.1. Sampling time uncertainty

Sampling time uncertainty introduces additional errors when analog signals are sampled at time intervals which show a timing uncertainty. Suppose, the analog input signals $V_i = A \, sin(2\pi \, f_{in} \, t)$. During a uncertainty in the sampling clock of ΔT, the input signal will have a chance equal to $\Delta V_i = 2\pi \, f_{in} \, A \, cos(2\pi \, f_{in} \, t) \, \Delta T$. If ΔV_i may be 1 LSB, which in this case equals $\frac{2A}{2^N}$, then the maximum timing uncertainty in the system may be

$$\Delta t_{max} = \frac{2^{-n}}{\pi f_{in}} \ .$$

As an example, the sampling time uncertainty for an 8 bit ADC with an input frequency of 250 *MHz* must be smaller then 5 *ps*. For such a device, the designer must be very careful with the clockjitter in the system. Those very strong demands on the clock specifications, introduces a second problem, more related to the implementation of the system on silicon.

5.2. Propagation delay

Although the speed of a signal on silicon is very high, it still has a finite value. Because of the ε_r of silicon, which is unequal to 1, the maximum speed of a signal on silicon is less then the speed of light, let us say it will get to 2/3 of that speed. That means, that a signal will cover about 200 μm within a time of 1 *ps*. If in a system a set of comparators have to be connected to one clocksource, clockskew is introduced by interconnecting the clock in the wrong way.

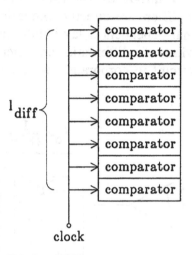

Figure 17: Clock interconnect introducing skew

In this example, the difference in length of interconnect l_{diff} of for instance 2 *mm* introduces a delay in the clock line of approximately 10 *ps*, which is to large for an 8 bit 250 *MHz* ADC. The solution to this problem is very simple by using so called binary trees to use as interconnect strategy in the sensitive parts of the system. A binary tree has the property that no skew due to difference in interconnect length is introduced.

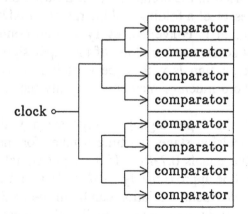

Figure 18. Binary tree

5.3. Signal dependent delay

Delay variation caused by difference in interconnect can be eliminated using binary trees. There is a second source of delay variation in another part of the system which introduces distortion. In every converter, where a set of differential amplifiers are directly comparing an input signal with a reference signal, signal dependent delay is introduced.

The nonlinear behaviour of a differential amplifier can be modelled using a amplitude limiting circuit in corporation with a first order bandlimited low pass filter.

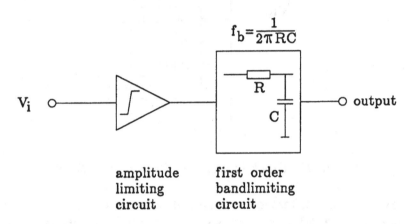

$$f_b = \frac{1}{2\pi RC}$$

amplitude limiting circuit   first order bandlimiting circuit

Figure 19: Non linear model of differential pair

The bandwidth limitation is modelled using a simple RC-network. Those sections are used in the comparators of a flash converter, but also in the analog preprocessor of a folding and interpolation ADC. If a full range sine wave input signal is converted, every section compares on another reference level and so on another slope of the input signal. The response of the RC-network on signals with slope variation is briefly described in this section. To avoid much of mathematics, only the results of the exact analysis is shown.

If the slope of the input signal is very steep with respect to the bandwidth of the decision stage, the time it takes for the output to go to half the output value equals $0.7 \cdot RC$. If the input signal has a very small slope, this delay will be approximately $1.0 \cdot RC$. In a first order bandlimited system, already $0.3 \cdot RC$ delay variation can be introduced if delay slope of the input varies from very steep to very small with respect to the bandwidth of the system.

To give an indication of this delay variation, a 200 *MHz* sine wave with an amplitude between 0*V* and 1*V* is compared on different reference levels and the delay on the output of the differential sections is plotted. The bandwidth in the nonlinear model equals 36 *ps*.

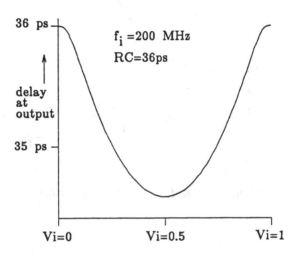

Figure 20: Delay variation example

Further analysis on the delay variation using a sine wave input signal shows that there is a direct relation between relative delay variation and bandwidth-input frequency ratio. Looking at the shape of the delay variation, only a small mismatch is made if the delay variation is modelled as a second order polynome. The relative delay variation can simply being transformed into third harmonic distortion and as a result of that a direct relation between distortion and bandwidth-input frequency ratio can be made.

This plot of the relative delay variation shows the full $0.3 \cdot RC$ delay variation at lower ratios of the bandwidth f_b divided by the signal frequency f_i. In this low ratio case, every slope outside the top or bottom of the sine wave will be interpreted as a steep slope and will give a delay of $0.7 \cdot RC$ where in the top or the bottom always a delay of RC will appear. At higher values of $\frac{f_b}{f_i}$ every slope is a small slope in respect to the bandwidth and in all parts of the input signal the delay will be approximately $1\ RC$ and no variations can be found at the outputs of the differential pairs. If the transformation from relative delay variation to third harmonic distortion is done, the final result of the signal dependent delay analysis is determined and shown in Fig.22.

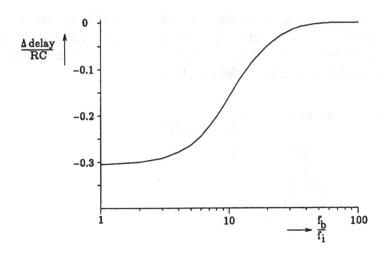

Figure 21: Relative delay variation

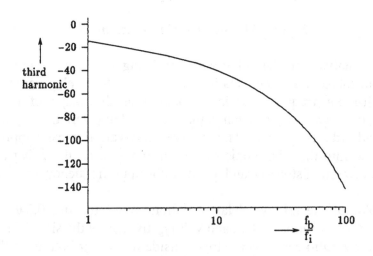

Figure 22: Third harmonic distortion

To give an indication, if a sine wave of 250 *MHz* is applied at the input of a converter, the bandwidth in the decision stages should be approximately 5 *GHz* to keep the third harmonic distortion below the -60 *dB* level. Those numbers have a mayor implication on the choice of the technology to implement such an ADC.

6. Conclusions

Folding analog to digital converters, especially those of the folding and interpolation type, are a very good option to ADC structures using flash are multi step techniques. Looking at the die area and at the power consumption, the latest folding and interpolation ADC of Philips is very competitive to those other structures. The latest device of the folding and interpolation type consumed only 850 mW and has a die size off 2.0×2.1 mm^2. A chip micrograph of this device is shown below.

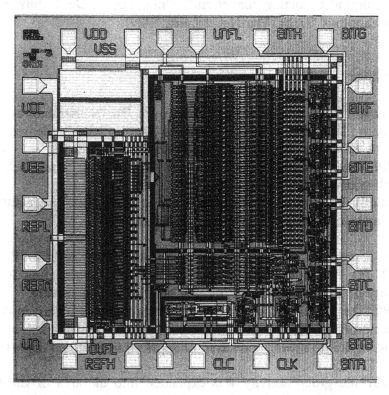

Figure 23: Chip photograph of the folding and interpolation ADC

Distortion in the system can be minimized, optimizing the system to a minimum in delay variation at the outputs of the decision stages.

The increasing frequencies in analog to digital conversion will introduce new problems implementing those systems on silicon, where even the maximum velocity of signals on silicon has an impact on the maximum performance.

References

[1] R.J. v.d. Plassche, P. Baltus, "An 8 bit 100 MHz Full Nyquist ADC", IEEE Journal of Solid State Circuits, Vol.23, No.6, December 1988, pp.1334-1344.

[2] R. v.d. Grift et al., "An 8 bit video ADC incorporating folding and interpolation techniques", IEEE Journal of Solid State Circuits, Vol.22, No.6, December 1987, pp.944-953.

[3] R.J. v.d. Plassche, "High speed and high resolution analog to digital and digital to analog converters", Ph.D. Thesis, Delft University of Technology, Delft, The Netherlands, 1989.

Biographies

Johan van Valburg was born in Wadenoyen, The Netherlands on the 3rd of January 1963. In 1986 he received the bachelor's degree at the University of Technology in Enschede, The Netherlands

In 1987 he joined the Philips Research Laboratories, where he became involved with adaptive digital signal processing

Currently he is working on high speed digital signal processing with special interest on high speed analog to digital conversion.

Rudy J. van de Plassche was born in the Netherlands in 1941. He received the M.S and Ph.D. degrees in electrical engineering from the university of Technology Delft in 1964 and 1989, respectively.

In 1965 he joined Philips Research Laboratories in Eindhoven working in the area of instrumentation and control systems. His first subject was a magnetic digital-to-analog converter for applications in direct digital control systems (DDC).

In 1967 the group activity was expanded into the area of Electronic Instrumentation where he started research into analog and combined analog and digital circuits. During this time IC technology matured and a number of new circuits were designed. To mention are: operational amplifiers, instrumentation amplifiers, analog multipliers, a chipset to construct an auto ranging digital multimeter (PM 2517) and analog-to-digital and digital-to-analog converters for applications in digital audio and digital video systems.

In 1983 he joined Philips Research Labs in Sunnyvale where he became a group manager for the advanced design group.

In 1986 he returned to Philips Research Labs in Eindhoven where he got involved in high-speed circuit design for optical communications using high-speed bipolar and GaAs technologies.

In 1988 he joined the Radio and Data Transmission group where he is involved in digital receiver architectures.

His research resulted in at least 40 publications and 50 issued US patents. He is a Fellow of IEEE, a member of the ISSCC technical program committee and a member of the ESSCIRC technical program committee. He has been a number of times guest editor for the Journal of Solid-State circuits and an associate editor from 1986-1989.

Since 1989 he is a part time professor at the University of Technology Eindhoven in the area of IC's for Telecommunication. He received the "1988 Veder" award from the dutch Stichting Wetenschappelijk Radiofonds Veder.

Currently he is working on a graduate textbook about Integrated A/D and D/A conversion.

Oversampled Analog-to-Digital Converters [1]

Shujaat Nadeem and Charles G. Sodini

Department of Electrical Engineering and Computer Science
Massachusetts Institute of Technology
Cambridge
U.S.A.

Abstract

Oversampled Analog-to-Digital conversion has been demonstrated to be an effective technique for high resolution analog-to-digital (A/D) conversion that is tolerant of process imperfections. The design of analog modulators for oversampled analog-to-digital converters can be divided into two main categories, the multistage or cascaded modulator (MASH) and the single loop modulators. This paper gives an overview of the oversampled Analog-to-Digital modulator architectures and their tradeoffs. The design and analysis of a stable N-th order single loop modulator and of multistage modulator topologies is covered. A comparative analysis of these two dominant architectures is also presented, followed by a brief look at the future directions of oversampled converters.

1. Introduction

The advent of VLSI digital IC technologies has made it attractive to perform many signal processing functions in the digital domain placing important emphasis on analog-to-digital conversion [1]. For high

---

[1]This work was supported with funding from Analog Devices, GE, and IBM.

resolution, band-limited signal conversion applications, oversampled A/D converters have become a dominant architecture. This technique has been shown to reliably provide high resolution without trimming or high precision components.

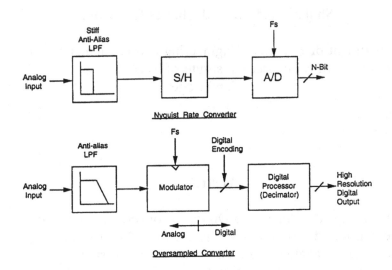

Figure 1: Block diagrams of Oversampled A/D and Nyquist A/D

A generalized oversampled A/D converter system and a conventional Nyquist rate A/D converter block diagram are shown in Fig.1. The block diagram of the oversampled converter shows three main system blocks: the anti-alias filter, the analog modulator and the digital decimator. The modulator samples the analog signal at very high clock rates and pushes the quantization noise out of the baseband to higher frequencies. The digital decimator is a very stiff low pass filter in the digital domain which takes the low resolution modulator output and produces the high resolution digital word at Nyquist rate by removing the out of band quantization noise [2]. The Nyquist rate converters require stiff anti-alias filters and accurate sample and hold circuits in front of the analog-to-digital converter block.

The major advantage of oversampled A/D system is that the analog circuit complexity can be greatly reduced if the encoding is selected such that the modulator only needs to resolve a coarse quantization (frequently a single bit) [3]. Also, if oversampling rates are high, the baseband is a small portion of the sampling frequency. Consequently, constraints on the analog anti-aliasing filter can be relaxed, permitting gradual roll-off, linear phase and easy construction with passive components. The precision filtering requirement is now relegated to the digital domain, where a "brick-wall" anti-aliasing filter is needed to decimate the digital signal down to the Nyquist rate. Additional benefits can be gained with the digital processor which can also provide on-chip functions such as equalization, echo cancellation, etc. Thus, this system can provide integrated analog and digital functions and is compatible with digital VLSI technologies [4].

In this chapter we will discuss the design and implementation issues for oversampled modulators and compare some of the major oversampled modulator architectures proposed in recent years.

1.1. Delta-Sigma Modulators

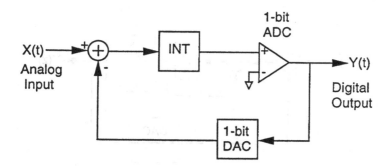

Figure 2: Block diagram of delta-sigma modulator

Fig.2 illustrates the simplest form of an oversampled interpolative modulator, which features an integrator, a 1-bit ADC and DAC, and a summer. This topology, known as the delta-sigma [5], uses feedback to lock onto a band-limited input $X(t)$. Unless the input $X(t)$ exactly equals one of the discrete DAC output levels, a tracking error results. The integrator accumulates the tracking error over time and the in-loop ADC feeds back a value that will minimize the accumulated tracking error. Thus, the DAC output toggles about the input $X(t)$ so that the average DAC output is approximately equal to the average of the input.

The operation of the delta-sigma modulator can be analyzed quantitatively by modelling the integrator with its discrete-time equivalent and the quantization process by an additive noise source $E(z)$ as illustrated in Fig.3. $E(z)$ is assumed to be a white and statistically uncorrelated noise source [6-8]. With this linearized model of the delta-sigma modulator, it can be shown that

$$Y(z) = X(z)z^{-1} + E(z)(1-z^{-1}) \qquad (1)$$

Interpolative modulators are also called "noise-shaping" coders because of their effect on the quantization noise $E(z)$ as seen at the output $Y(z)$. A plot of the quantization noise response $|H_e(z)| = |1-z^{-1}|$ is shown in Fig.4. When the modulator is sampling much higher than the Nyquist rate, the baseband is in a region where the quantization noise will be greatly attenuated. Although only a coarse quantization is made by the modulator, the bulk of the quantization noise has been pushed to higher frequencies which can be removed by the subsequent digital processing stage. Thus, the final output is a high-resolution digital representation of the input.

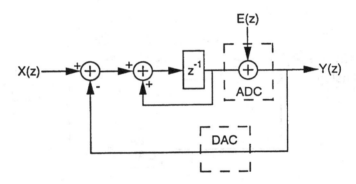

Figure 3: Discrete time equivalent of delta-sigma loop

The single-bit encoding scheme used by the delta-sigma modulator and similar interpolative modulators has a number of additional advantages: (1) the format is compatible for serial data transmission and storage systems, (2) subsequent digital processing can be simplified because multiplications and additions reduce to simple logic operations, and (3) highly linear converters are possible due to the inherent linearity of the in-loop 1-bit DAC.

The last point above deserves further explanation. The integral nonlinearity of the in-loop DAC often limits the harmonic distortion performance of many oversampling A/D converters [9]. A multi-bit DAC has many discrete output levels that must be precisely defined to prevent linearity error. With only two discrete values, a one-bit DAC always defines a linear transformation between the analog and digital domains. Gain and offset error can still exist in a one-bit DAC, but linearity error is avoided without precision trimming of output levels.

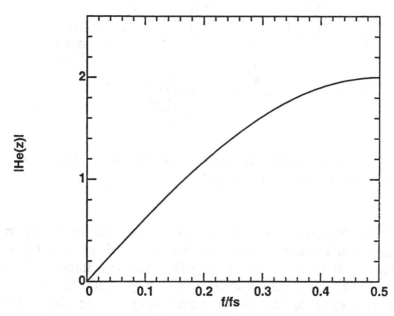

Figure 4: Calculated quantization noise magnitude response $|H_E(z)|$ *for delta-sigma loop*

Many problems arise when implementing an A/D converter using a delta-sigma modulator. Most prominent is that the quantization noise actually is signal dependent [10] and not statistically uncorrelated as is usually assumed. This is related to the number of state variables in a system. For a delta-sigma modulator , the state of the system is determined by the integrator output value along with the input value. With only one state variable, the loop can lock itself into a mode where the output bit stream repeats in a pattern. Consequently, the spectrum of the output may contain substantial noise energy concentrated at multiples of the repetition frequency. To counter this effect, dithering has been used with delta-sigma modulators to randomize the input so that repeating bit patterns will not form. However, this is not an attractive technique since the input dynamic range is lowered by the additional dither noise.

1.2. Second-Order Delta-Sigma Modulators

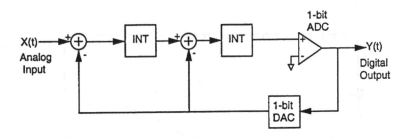

Figure 5: The second order delta-sigma modulator. This is a delta-sigma within a delta-sigma loop.

Candy [11] has done substantial work on a second-order interpolative modulator (the *double-loop*) based on the idea of embedding a delta-sigma loop within the main loop (Fig.5). With two integrators in the loop, the quantization noise response rises as a quadratic function of frequency as compared to the linear relationship of the first-order topology. The higher order causes quantization noise to be further suppressed in the low frequency baseband and to rise more sharply in the higher frequencies. The net effect is that the total power of the quantization noise in baseband is further reduced, and therefore, a higher effective resolution can be achieved for the same oversampling ratio. Also, due to the two integrators in the loop, repeating bit patterns are less likely to occur and, as a result, the quantization noise tends to be less signal dependent. Thus, the double-loop increases the SNR possible for a given oversampling ratio while randomizing the output bit pattern so that noise spikes in the output spectra appear less frequently.

The double-loop has been successfully demonstrated in a number of implementations [11,12]. An example of the performance that can be achieved with a second order interpolative modulator design is found in [13]. For an oversampling ratio of 256, a dynamic range of approximately 90 *dB* was achieved for a sampling frequency of 3 *MHz*. The converter was fabricated in a 2 μm CMOS technology and used less than 15 *mW* with a single 5 *Volt* supply.

2. Higher Order Single Loop Modulators

In order to improve the signal-to-noise ratio without raising the oversampling ratio, higher order loops have been explored. Extending the idea of multiple integration loops to achieve higher order modulators has proven quite difficult to stabilize [14,15]. When more than one integrator is in the loop, it is possible for the system to be excited in a mode where a large amplitude, low frequency oscillation persists in the integrators. Approaches to stabilize the loops have fallen into two categories: (1) increasing the size (number of bits) of the in-loop ADC-DAC to allow a greater dynamic range of signals in the loop [16], and (2) cascading multiple stages of first order delta-sigma modulators [15,17]. In the first case nonlinearities in the in-loop multi-bit ADC-DAC become the limiting factor for overall resolution while component mismatches in multi-stage modulators determine the resolution for the second case. In this section we will analyze single bit higher order single loop modulator architectures.

2.1. N-th Order Modulator Topology

The N-th order interpolative modulator to be described is shown in Fig.6. This architecture was the solution to design higher order stable single loop modulators by controlling the pole locations of the linearized loop [18] [19]. This N-th order modulator architecture can be treated as a feedback loop with the loop filter H_z in the feedforward path only. It is important to note that linear analysis is used only to determine a starting point for modulator design. Further design refinements and verification are accomplished through software simulations which will be described for this architecture in the next section.

This section concentrates on the results of a fourth-order implementation, the analysis and design techniques discussed here apply equally well to other order topologies. The salient features of this topology are the feedforward structure with loop coefficients $A_0, \ldots, A_N$ and a feedback structure with coefficients $B_0, \ldots, B_N$. The in-loop quantizer[2] is reduced to a single bit for previously cited reasons. As will be demonstrated, the large number of loop coefficients not only allows one to stabilize loops of any order, but also allows optimizing the quantization noise shape to improve performance.

---

[2]The term *quantizer* refers to a cascaded ADC--DAC combination. The input and output are both analog signals which differ by the error due to the quantization process.

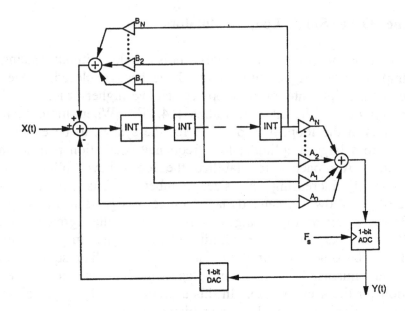

Figure 6: N-th order loop topology with feedforward and feedback coefficients for stabilizing the modulator

2.1.1. System Function

The system function for the topology of Fig.6 is derived by replacing the integrators with their discrete-time equivalents and by modelling the in-loop ADC as an additive noise source $E(z)$ [11]. A linear model is used to facilitate design of the loop coefficients, which will be subsequently described. The DAC is assumed to be ideal and delay-less, but a z^{-1} delay is associated with the ADC. The resulting system function can be expressed as a combined response to the input $X(z)$ and the quantization noise $E(z)$,

$$Y(z) = H_X(z)X(z) + H_E(z)E(z) , \qquad (2)$$

where

$$H_X(z) = \frac{\sum_{i=0}^{N} A_i(z-1)^{N-i}}{z\left((z-1)^N - \sum_{i=1}^{N} B_i(z-1)^{N-i}\right) + \sum_{i=0}^{N} A_i(z-1)^{N-i}} \qquad (3)$$

$$H_E(z) = \frac{(z-1)^N - \displaystyle\sum_{i=1}^{N} B_i(z-1)^{N-i}}{z\left((z-1)^N - \displaystyle\sum_{i=1}^{N} B_i(z-1)^{N-i}\right) + \displaystyle\sum_{i=0}^{N} A_i(z-1)^{N-i}} \qquad (4)$$

To gain an intuitive understanding of the system function, it is helpful to consider the case where the feedback coefficients $B_1, \dots, B_N$ are set to zero. With this assumption, equation (2) can be simplified for the case of low frequency baseband signals by noting that $(z-1) \approx j\Omega$ and $|j\Omega| \ll 1$ where $\Omega = 2\pi(f/f_s)$, f_s is the sampling frequency, and f is the frequency of interest. The system response reduces to

$$Y(z) \approx X(z) + \frac{E(z)(j\Omega)^N}{A_N} \qquad (5)$$

which shows that when N is sufficiently greater than 1 and the oversampling ratio is large, i.e. $|j\Omega| \ll 1$, the quantization noise $E(z)$ is greatly attenuated. This demonstrates the accurate tracking aspect of the modulator for low frequency signals.

2.1.2. Design of Loop Coefficients

Treating the loop as a quantization noise filter allows linear filter design techniques to be used. The primary design criteria is to minimize the quantization noise energy in baseband. However, suppressing quantization noise in the low frequency baseband produces the adverse effect of increasing the quantization noise level at higher frequencies. In order to control the shape of the quantization noise we have to design the loop poles to guarantee stability and choose the placement of zeros to suppress the quantization noise in the baseband. The zeros of the quantization noise transfer function $H_E(z)$ are controlled by the B-coefficients and the poles of the system are a function of both A and B coefficients.

Many methodologies can be used to determine the z-domain pole-zero values for $H_E(z)$. Once the pole-zero locations are known, it is a simple matter to determine the loop coefficients. We have chosen to use a Butterworth filter design because of its flat characteristics and its relative insensitivity to coefficient errors.

A$_i$ Coefficients: Again, it is useful to consider the case where the B_i coefficients are set to zero and to first concentrate on the A_i coefficients. The process involves using a discrete filter design programme to determine the z-domain pole and zero placement for the required baseband, clock frequency and transition width. Once the desired values of the z-domain poles are known, values for A_0, ... , A_N can be determined as follows. Let $H_D(z)$ be the desired N-th order response defined as

$$H_D(z) \equiv \frac{K(z-1)^N}{(z-p_1)(z-p_2)\cdots(z-p_N)} \tag{6}$$

where $p_1, p_2, ... , p_N$ are the z-domain poles. Equation (6) can be related to $H_E(z)$ by the simple relationship,

$$H_D(z) = zH_E(z) \tag{7}$$

Expansion of the denominator terms gives

$$\frac{K(z-1)^N}{z^N + z^{N-1}C_{N-1} + \cdots + C_0} = \frac{z(z-1)^N}{z^{N+1} + z^N D_N + \cdots + D_0} \tag{8}$$

where C_i and D_i are coefficients resulting from the expansion. Equating similar denominator terms produces a set of linear equations which determine the A_i coefficients. For the case $N= 4$, $K= 1$, $f_s= 2.1\ MHz$, $f_b= 20\ kHz$, and s-domain poles at -1.8074 $\pm$ 0.7486j , -0.7486 $\pm$ 1.8074j radians, the resulting A_i coefficients are

$$
\begin{aligned}
A_0 &= 0.8653 \\
A_1 &= 1.1920 \\
A_2 &= 0.3906 \\
A_3 &= 0.06926 \\
A_4 &= 0.005395
\end{aligned}
$$

B$_i$ Coefficients: In the section above, all the zeros of the quantization noise response $H_E(z)$ reside at DC ($z=1$). As a consequence, the noise response rises monotonically out of baseband as an N-th order function. One finds that a small portion of the noise spectrum at the upper edge of baseband will dominate the total in-band noise energy. It is apparent that further shaping of the quantization noise will improve performance and this can be accomplished by using the B_i coefficients to move the zeros away from DC.

The methodology developed here is to use the equal-ripple characteristics of the Chebyshev polynomials. These polynomials are defined by

$$T_0(x) = 1$$
$$T_1(x) = x$$
$$T_N(x) = 2xT_{N-1}(x) - T_{N-2}(x)$$

What makes them special is that $|T_N(x)| \leq 1$ for $|x| \leq 1$, and $|T_N(x)| > 1$ for $|x| > 1$. This property of the Chebyshev polynomials can be used to define a desired baseband response through proper scaling. For the case of large oversampling rates, i.e. $f_b \ll f_s$, where f_b is the baseband frequency and f_s is the sampling frequency, the desired z-domain zeros z_i are located at

$$z_i = e^{j2\pi \frac{f_b}{f_s}X_i}, \tag{9}$$

where X_i are the roots of T_N. The B_i coefficients can then be determined using the method outlined above for determining the A_i coefficients. For $f_s = 2.1\ MHz$ and $f_b = 20\ kHz$, the resulting B_i coefficients are

$$B_1 = -3.540 \times 10^{-3}$$
$$B_2 = -3.542 \times 10^{-3}$$
$$B_3 = -3.134 \times 10^{-6}$$
$$B_4 = -1.567 \times 10^{-6}$$

A plot of the quantization noise response for a $N = 4$ loop implementing the above coefficients is shown in Fig.7.

From above and from Equation (4), it is apparent that the zeros only depend on the B_i coefficients. The zeros will usually be located near $z=1$ because of high oversampling ratios, ensuring that the B coefficient values will be small. The effect of B coefficients in determining the A coefficients is then negligible. Thus, for large oversampling ratios, the B coefficients determine the zeros and the A coefficients determine the poles of the quantization noise response. It is important to realize that if the B coefficients are not negligible they could affect the pole locations of the loop and thus move one or more poles out of the unit circle and result in instability. In cases where the zeros are not close to DC, as in the case of bandpass modulators, both A and B coefficients determine the pole and zero locations.

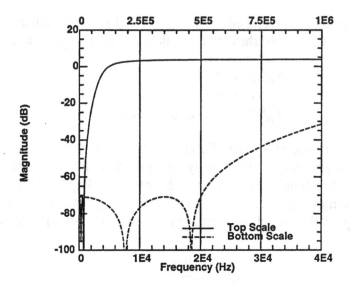

Figure 7: Calculated quantization noise spectrum for a fourth-order loop. Top axis is from 0 to 1 MHz; bottom axis shows details from 0 to 40 kHz.

2.1.3. Quantization Noise

The effective resolution of the modulator can be calculated by evaluating P^2, the total quantization noise power in baseband, which is given by

$$p^2 = \int_0^{f_b} |H_E(f)\, E(f)|^2 \, df \,, \tag{10}$$

where

$$|H_E(f)| \approx \left| \left(\frac{\pi f_b}{f_s}\right)^N \left(\frac{2}{A_N}\right) T_N\left(\frac{f}{f_b}\right) \right| \qquad for \ \ f \le f_b \ll f_s \tag{11}$$

It has been shown that for delta-sigma modulators $E(f)$ can be modelled as an additive white noise source of magnitude [6-8]

$$|E(f)| = e_0 \sqrt{\frac{2}{f_s}} \qquad (12)$$

where e_0 is the average quantization noise and is related to the quantizer step size σ by

$$e_0 = \frac{\sigma}{\sqrt{12}} . \qquad (13)$$

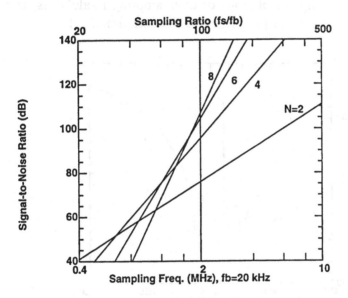

Figure 8: Calculated effective resolution as a function of oversampling ratio and modulator order for various order loops

Equation (10) has been evaluated and is graphically presented in Fig.8. As the modulator order increases, the lines become steeper, implying a greater payoff from oversampling. For 96 *dB* resolution across the 20 *kHz* audio range, Fig.8 shows that a second-order loop requires a sampling rate of 5 *MHz*, whereas a fourth order modulator requires only 2 *MHz*. In this case, the additional analog circuit complexity of two integrators can reduce the sampling rate by a factor of 2½.

2.1.4. Simulation Results

The oversampled modulator is inherently a nonlinear system. The linear analysis of the loop allows us to design the loop coefficients and

198

gain some intuitive understanding of the system. In order to verify stable system operation and understand the effect of system non-idealities a software simulator is required. The simulator mimics circuit behavior in the discrete-time domain [4] with modules that model block-level circuit functions. Each model contains associated non-idealities so that the effect of each non-ideality can be examined. The modules can then be connected, via software, to simulate almost any topology for oversampling modulators. This simulator is especially useful for CMOS implementations of oversampling modulators, in which many analog functions are performed in discrete-time [20].

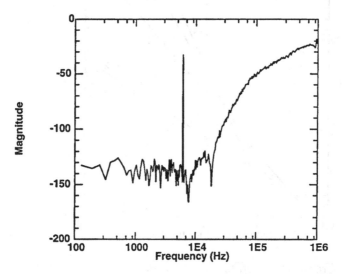

Figure 9: Typical simulation spectrum of an ideal fourth-order loop. Sinusoidal input with an amplitude of 0.1(normalized to DAC output), a DC-offset of 0.02, and a frequency of 6.1 kHz. The sampling frequency is 2.1 MHz. Note the notches in the baseband noise floor due to the optimized zeros.

A typical spectrum of the modulator's output bit stream is shown in Fig.9. For all simulations, the FFT's were performed on over 32,000 data points using a raised-cosine window (von Hann). Various simulation results are presented in the following sections for the fourth-order modulator previously described.

Finite Op-amp Gain: Finite op-amp gain causes the inverting op-amp terminal to reflect the output voltage rather than behave as a virtual ground. Consequently, not all of the charge on the sampling capacitor will be transferred to the integrating capacitor, resulting in a "leaky" integration. This effect is still a linear process and can be incorporated into the linear SC integrator model by the addition of leakage and integration gain coefficients. In the system function, integration gain errors are manifested as errors in the loop coefficients.

The effect of finite op-amp gain on fourth-order modulator performance can be seen in the signal-to-noise ratio curves of Fig.10. For op-amp gains less than 200, the performance has degraded by about 2 *dB*. However, with gains of 1000 or more, the performance is nearly ideal.

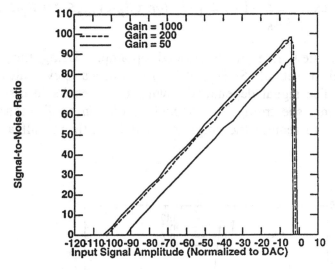

Figure 10: Simulated SNR versus relative input amplitude for fourth-order loop showing effects of finite op-amp gain. Sinusoidal input with a DC-offset of 0.02, a frequency of 6.1 kHz, a clock frequency of 2.1 MHz, and an oversampling ratio of 48.

Integrator Settling Time: Typically, oversampled modulators are operated with clock frequencies over 1 *MHz*, thus settling becomes an issue in determining performance. This is especially true for switched-capacitor implementations, where each integrator can be expected to step to a new output level during every clock cycle. If the integrator is modeled as a single-pole system, then step responses will be exponential in nature. The fractional error introduced by this response is

200

$$\varepsilon \equiv \frac{V_{actual} - V_{step}}{V_{step}} = -\exp(-T_s/\tau) \qquad for \ V_{step} \leq \tau S_R , \qquad (14)$$

where T_s is sampling period, τ the time constant of the exponential response, and V_{step} the output step size. Thus, for a given T_s, ε remains constant. However, if the step size is sufficiently large, then slewing occurs and the integrator will settle with a combination of slew and exponential characteristics. The fractional error in this case is

$$\varepsilon = -\frac{\tau S_R}{V_{step}} \exp\left(\frac{V_{step}}{\tau S_R} - 1 - \frac{T_s}{\tau}\right) \qquad for \ \tau S_R < V_{step} \leq (\tau + T_S)S_R \quad (15)$$

where S_R is the slew rate of the loaded op-amp. Using the above described model a set of SNR curves for various slew rates was produced [19]. The signal dependent slewing error results in distortion of the output signal spectrum, thus for high resolution applications it is preferred practice to require the amplifier to fully settle in half a clock cycle.

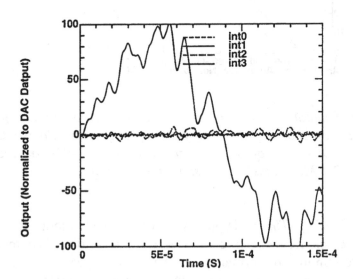

Figure 11: Simulated integrator outputs for a fourth-order loop. Sinusoidal input signal with an amplitude of 0.1, a DC-offset of 0.02, a frequency of 6.1 kHz, and a clock of 2.1 MHz. The integrator gain is 1.

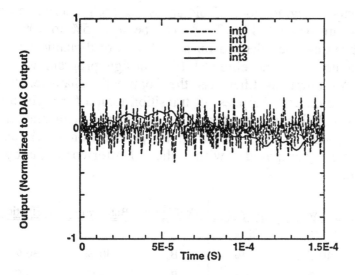

Figure 12: Simulated integrator outputs for a fourth-order loop. Sinusoidal input signal with an amplitude of 0.1, a DC-offset of 0.02, a frequency of 6.1 kHz, and clock of 2.1 MHz. The integrator gain is 0.2

Limited Op-amp Swing: The effect of limited output swing has already been documented [4,13]. Because the integrators contain state information, clipping of output levels will cause loss of state information, thus performance will be degraded. For higher-order loops, this is a serious problem because the last integrator has very large signal swings. A solution is to design the discrete-time integrators with a time-constant less than unity, thereby attenuating the signal levels at the integrator outputs. This attenuation is then compensated by multiplying the values of the A_i and B_i coefficients so that the overall loop transmission remains the same. Fig.11 shows superimposed images of the four integrator outputs without the "gain compensation", and Fig.12 shows the same outputs with the "gain compensation''. Clearly, the last integrator now has signal levels well within the maximum swing bounds. Thus with proper system design, output swing is not a limitation for this topology.

A_i and B_i Coefficient Errors: As mentioned in the preceding section, the A_i and B_i coefficients determine the pole and zero locations of the system. Errors from process mismatch change the coefficient values and

cause performance degradation. Tolerances to each coefficient were determined and are listed in Fig.13. As can be seen, 5% to 30% errors in each coefficient can be tolerated for a 3 *dB* loss of dynamic range. A set of coefficients using the elliptical filter design program were also determined. An elliptical filter has the "optimal" characteristics of maximum attenuation and sharpness of transition region for a given filter order. From simulation results, the elliptical filter implementation was more susceptible to instability for a given perturbation of A_i and B_i coefficients as compared with the Butterworth-Chebyshev implementation.

Coefficient	Butterworth	Elliptic	Coefficient	Butterworth	Elliptic
A_0	5 %	5 %			
A_1	10 %	5 %	B_1	30 %	30 %
A_2	20 %	10 %	B_2	10 %	10 %
A_3	30 %	20 %	B_3	30 %	20 %
A_4	40 %	20 %	B_4	30 %	30 %

Figure 13: Maximum error tolerance of A_i and B_i coefficients which will cause less than 3 dB loss of dynamic range

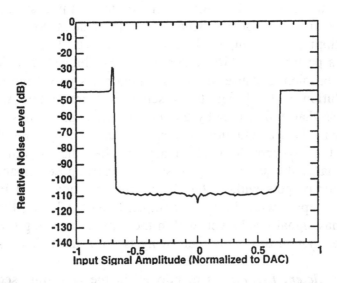

Figure 14: Simulated fourth-order loop dc idle-channel noise. The lack of noise spikes within the valid input region indicates signal-independent noise.

DC Idle-channel Noise: One major problem with first- and second-order loops is that their quantization noise is highly DC-bias dependent [6,21]. The first-order system contains only one state variable, thus it is very likely that repeating bit patterns will be produced. In a second-order system, somewhat better performance is possible because the states are more random, hence quantization noise is less signal dependent. Presumably, for a fourth-order system, even better randomization should be achieved. This is indeed the case and is reflected in the flatness and lack of noise spikes with DC inputs as shown in Fig.14. Thus, no dither signal is necessary.

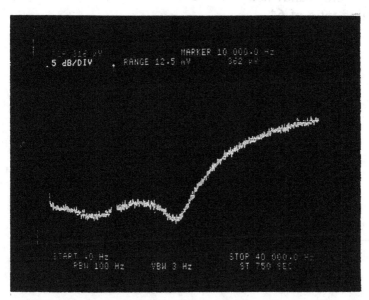

Figure 15: Output spectrum of breadboard fourth-order loop showing quantization noise around baseband region (0 - 40 kHz). The clock rate is 2.1 MHz. The input is a 10 kHz sinusoid with an amplitude of -70 dB full scale.

From a system viewpoint, the fourth-order loop is quite immune to many nonidealities and does not suffer from the DC bias problems associated with lower-order systems. These simulation results have been used to determine the design specifications for a CMOS implementation of a fourth-order modulator for digital audio applications [22]. Fig.15 is an output spectrum of the breadboard version of the fourth order modulator described in this section [18]. Notice the large amount of quantization noise, except in baseband where it is greatly suppressed by the response of the loop. A 10 *kHz* input signal component is also shown with no harmonic distortion visible, revealing the extreme linearity of the modulation process. As mentioned in the introduction, the excellent linearity is aided by the use of a one-bit DAC which is inherently linear because it has only two discrete output levels. Measurements indicated a dynamic range of 90 *dB* with less than 0.1% total harmonic distortion (THD) for inputs smaller than -3 *dB* and less than 0.01% THD for inputs smaller than -21 *dB*. Subsequent designs using higher order single loop topologies have shown impressive results. Welland [24] has achieved 92 *dB* signal-to-noise plus distortion ratio over the audio bandwidth with a fourth order architecture and an oversampling ratio of 64.

2.2. *N*-th Order Modulator without Active Summers

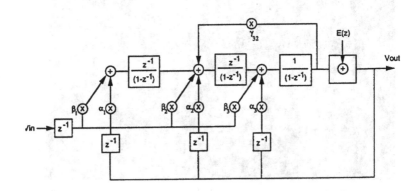

Figure 16: Third order example of N-th order modulator without active summers

Ferguson et al. [25] reported a single-bit multi-order loop architecture that does not require coefficient summers at the input and the output as in [19] and [24]. The block diagram of this variation on the original *N*-th order topology described in the previous section is shown in Fig.16. In this architecture the feedback path also has a shaping function whereas the *N*-th order architecture described in the previous sections has a loop filter in the feedforward path only. The distributed nature of the quantized feedback signal in this realization enhances the stability characteristics even under conditions where one or more of the integrators are in saturation. The ideal input signal transfer function for an N-th order loop is given by:

$$\frac{Y(z)}{X(z)} = \frac{\sum\limits_{i=1}^{n} \beta_i(z) \prod\limits_{j=1}^{n} G_i(z)}{1 + \sum\limits_{i=1}^{n} \alpha_i(z) \prod\limits_{j=1}^{n} G_i(z)} \tag{16}$$

And the transfer function of the quantization noise is given by:

$$\frac{Y(z)}{Q(z)} = \frac{1}{1 + \sum\limits_{i=1}^{n} \alpha_i(z) \prod\limits_{j=1}^{n} G_i(z)} \tag{17}$$

The feedforward block $G(z)$ is designed to be a series connection of integrators and resonators and feedforward coefficients $\beta(z)$ can also be added to $G(z)$ to reduce the amount of signal seen at the output of intermediate opamps and to improve the transient response of the loop. In practice $\alpha_i(z)$ and $\beta_i(z)$ are simply constant terms and the various $G_i(z)$ are composed of integrators and resonators whose transfer functions are given below.

$$G_{int}(z) = \frac{z^{-1}}{1 - z^{-1}} \tag{18}$$

$$G_{res}(z) = \frac{z^{-2}}{1 - 2\cos\theta\, z^{-1} + z^{-2}} \tag{19}$$

The figure shows the third order version of this architecture. The associated input and quantization transfer function for this system are:

$$\frac{Y(z)}{X(z)} = \frac{\beta_1 z^{-3} + \beta_2 z^{-2}(1 - z^{-1}) + \beta_3 z^{-1}(1 - z^{-1})^2}{(1 - z^{-1})^3 + \alpha_1 z^{-3} + \alpha_2 z^{-2}(1 - z^{-1}) + \alpha_3 z^{-1}(1 - z^{-1})^2 + \gamma_{32} z^{-1}(1 - z^{-1})}$$

$$\frac{Y(z)}{Q(z)} = \frac{(1 - z^{-1})^3 + \gamma_{32} z^{-1}(1 - z^{-1})}{(1 - z^{-1})^3 + \alpha_1 z^{-3} + \alpha_2 z^{-2}(1 - z^{-1}) + \alpha_3 z^{-1}(1 - z^{-1})^2 + \gamma_{32} z^{-1}(1 - z^{-1})}$$

Once the desired values of the z-domain poles and zeros are known values for the loop coefficients can once again be determined by using the coefficient design technique described for the general N-th order topology. Ferguson et al. used a fifth order version of this architecture to achieve a signal to noise plus distortion ratio of 105 dB over the audio bandwidth [26]. This result corresponds to the 18-bit analog-to-digital conversion.

2.3. Stability of Single Loop Modulators

We use a linear additive quantization noise source model for the one bit quantizer in the single loop modulators. The design methodology for the single loop modulators uses a white additive noise source model for the quantizer to design the loop coefficients. It is very important to realize that the stability of the linearized modulator loop does not guarantee the stable behaviour of the nonlinear modulator. The loop is said to be unstable when the signal-to-noise ratio of the output falls drastically and the mechanisms that can cause this are:

- Pole locations on the z-plane of the linearized system.

- Loss of state information by integrator output clipping.

- Initial state of integrators required to be zero.

- Tolerance of loop coefficients.

- Quantizer overload.

The design of a single loop higher order modulator uses the linearized system function of the feedback loop to design the pole locations to be inside the unit circle on the z-plane. Unlike the linear systems this criteria alone cannot guarantee the stability of the inherently nonlinear loop of the modulator.

The finite output swing of a non-ideal amplifier could also result in loss of state information in the integrator. Software simulations are used to determine the maximum integrator amplitudes and if they exceed the linear range of the amplifiers, signal attenuation by scaling the integrator time constant is used. This integrator time constant scaling is compensated by appropriately changing the loop coefficient values to preserve the loop system function. The initial integrator states have to be set to zero for maximum input range because if the initial integrator state is not zero the input dynamic range of the loop is further lowered.

The absolute value of the loop coefficients is also a parameter which is bounded by an upper and lower limit. For switched capacitor CMOS designs the loop coefficients are implemented as ratio of capacitors. Thus very large and very small valued coefficients could become impractical to implement. As we already discussed, the closed loop poles and zeros are controlled by the loop coefficients. Thus the coefficient implementation has to take into account that the value of the coefficients does not vary more than the allowed tolerance for a particular design.

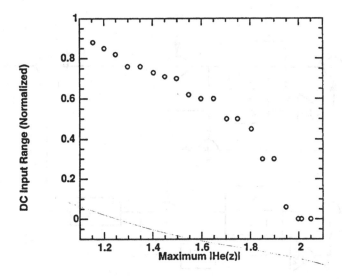

Figure 17: Simulated DC input range of a fourth-order loop for various values of $|H_E(z)|$. With zero input, the loop becomes unstable when $|H_E(z)| > 2$. The vertical axis is normalized to quantizer step-size.

Another mechanism for instability is due to the limited input range of the quantizer [23] which places further constraints on the design of the loop filter [27]. A signal at the input of the quantizer which exceeds the quantizer limits is reflected by an increase in the amount of quantization noise $|E(z)|$. This excess noise is circulated through the loop and can cause an even larger signal to appear at the quantizer input, eventually causing instability. The high frequency gain of the quantization noise response $|H_E(z)|$ for $|z|=1$ is inversely related to the DC input range of the loop as shown in Fig.17. An upper limit on $|H_E(z)|$ can be determined beyond which the modulator loop is unstable in spite of stable pole locations on the z-plane. This criteria can only be used as a rule of thumb to aid in the design of linear loop coefficients. The rigorous mathematical characterization of loop stability for single loop modulators is still an active area of research [10,28,29].

3. Multi-loop Cascaded Modulators

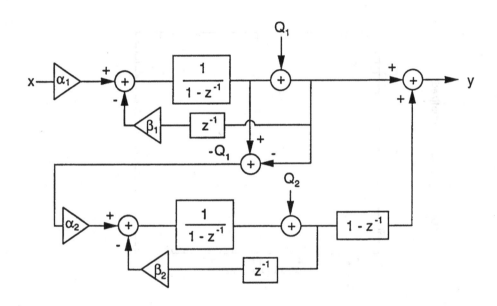

Figure 18: Second Order Multistage Architecture

An alternative approach to achieving higher-order noise-shaping functions is to cascade first or second-order delta-sigma modulators. This concept was first proposed by Hayashi et al. where two first-order modulators are cascaded as shown in Fig.18 [17]. The principal advantage of cascade modulators is that higher-order oversampling can be achieved without the characteristic conditional stability of the high-order single loop modulators. The second order loop is built out of the cascade of two first order delta-sigma modulators. The single bit quantizers in the loop are modelled as the additive white quantization noise sources [6] and the switched capacitor integrators have been modelled with the z-transform of the integrator transfer function. Using the linear model and assuming the loop coefficients α_i and β_i are ideally equal to one, DAC is ideal and delay free, and the ADC has a unit delay the resulting system function is

$$Y = X(z) + (1-z^{-1})^2 Q_2 \tag{22}$$

We see that for the ideal system the first stage quantization noise is exactly cancelled and the output consists of the input signal plus the second order difference of the quantization error from the second stage. For an N-stage ideal multistage modulator the output consists of the input signal and the N-th order difference of the quantization error. The quantization noise from all the stages except the last is cancelled.

3.1. Effect of Capacitor Ratio Mismatch

The block diagram of the second order multistage modulator models the integrator gain nonidealities as the gain coefficients α_i in the input signal path and β_i in the feedback signal path. The subscript i stands for the i-th stage of the modulator. We can see that if the path gains of the two stages do not match or are not exactly equal to one we do not get exact cancellation of the quantization noise of the first stage. This is best illustrated if we derive the transfer function of the two stage cascaded modulator architecture shown in Fig.18.

$$V_0(z) = V_i(z) \frac{(\alpha_1 z^3 + (\beta_2 - 1)\alpha_1 z^2)}{z^3 + (\beta_2 + \beta_1 - 2)z^2 + ((\beta_1 - 1)\beta_2 - \beta_1 + 1)z}$$

$$+ Q_1 \frac{(1-\alpha_2)z^3 + ((2-\beta_1)\alpha_2 + \beta_2 - 2)z^2 + ((\beta_1 - 1)\alpha_2 - \beta_2 + 1)z}{z^3 + (\beta_2 + \beta_1 - 2)z^2 + ((\beta_1 - 1)\beta_2 - \beta_1 + 1)z}$$

$$+ Q_2 \frac{(z^3 + (\beta_1 - 3)z^2 + (3 - 2\beta_1)z + \beta_1 - 1)}{z^3 + (\beta_2 + \beta_1 - 2)z^2 + ((\beta_1 - 1)\beta_2 - \beta_1 + 1)z} \tag{23}$$

The above transfer function shows that if the alpha and beta coefficients are not exactly matched and equal to one the first stage quantization noise will leak into the output. Thus the upper limit on the resolution of the system will be determined by the leakage of the first stage quantization noise to the output. This resolution dependence on the coefficient matching, namely the capacitor matching for switched capacitor implementation, is the biggest drawback of the multistage architecture.

3.2. Effect of Finite Gain and Incomplete Settling

The other important non-ideality to be taken into account for an accurate system block diagram is the finite opamp gain and settling time. The transfer function of an integrator with finite gain A is:

$$\frac{V_0(z)}{V_i(z)} = \frac{K_A z^{-1}}{1 - K_B z^{-1}} \tag{24}$$

The non-ideal effect of incomplete settling can be modelled as a coefficient G_s scaling the integrator transfer function and is a function of the opamp settling speed and the clock frequency of the modulator. This model assumes that the integrators are not slew limited and the settling error is determined by linear settling of amplifiers. The integrator transfer function which includes both the finite gain and the incomplete settling effects can be written as:

$$\frac{V_0(z)}{V_i(z)} = \frac{G_s K_A z^{-1}}{1 - K_B z^{-1}} \tag{25}$$

The gain error K_A and the pole error K_B are a function of the open loop gain and circuit elements of each respective path like the sampling and the integrating capacitor in a switched capacitor implementation, where the sampling capacitor is different for the input and feedback path. Thus the input and the feedback signal paths in the first order modulator will have different expressions for the gain and pole errors. Fig.19 shows the block diagram of the second order multistage modulator which includes the finite gain and incomplete settling nonidealities. This block diagram uses the same settling error coefficient G_s for each of the stages assuming the same amplifier is being used for every stage.

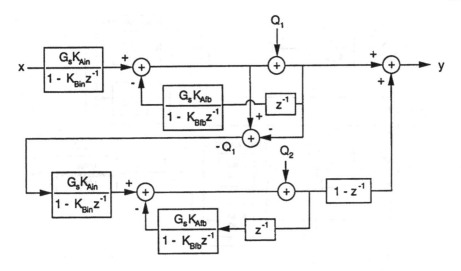

Figure 19: Block diagram of multistage architecture including the finite gain and incomplete settling nonidealities

3.3. Cascaded Modulator Implementations

Several implementations of cascaded modulators using combinations of first and second-order delta-sigma modulators have been proposed [30,15]. Matsuya et al. achieved 16 bit resolution over the audio range with a triple noise shaping architecture [15].

Recently cascaded multi-bit modulators have been employed to achieve high signal-to-noise ratios for higher baseband frequency signals. An example of such a modulator is shown in Fig.20. The first loop is a second-order delta-sigma modulator which is cascaded with a first-order delta-sigma modulator with a multi-bit quantizer. Multi-bit quantizers have non-linearities due to imperfections in the DAC. However, this configuration is constructed such that the non-linearity from the multi-bit DAC is noise shaped by the second-order delta-sigma modulator. The multi-bit quantizer allows for a smaller oversampling ratio and hence, higher-frequency baseband signals. An implementation of this topology by Brandt and Wooley achieved a dynamic range of 74 *dB* for a signal bandwidth of 1 *MHz* fabricated in a 1 μm CMOS technology [31].

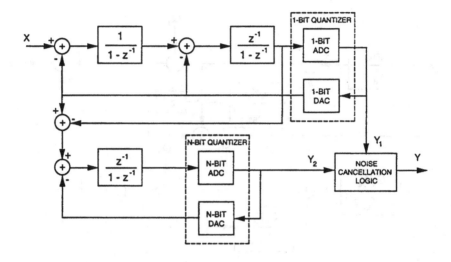

Figure 20: A discrete time equivalent of a 2 stage (2-1) cascade of a 2-nd order delta-sigma modulator and a first order delta-sigma modulator with an N-bit quantizer

4. Comparison of Multistage versus Single Loop Modulators

The single loop higher order and the multistage topologies have emerged as the dominant oversampled modulator architectures for high resolution applications. In this section we compare the two architectures to bring out the advantages and disadvantages of both in a concise manner.

4.1. Stability

The stability of the single loop higher order modulators has been discussed in a previous section. The single loop modulators are characterized under the conditionally stable class of systems. The reason for this characterization is that if any one of the requirements of stability are not satisfied the system becomes unstable. The stability requirements for the single loop modulator are:

- Poles of the linearized transfer function inside the unit circle

- Integrator amplifiers should not saturate

- Reset mechanism for integrators to guarantee zero initial state

- $|H_E(z)|$ for $|z|=1$ has to be chosen to avoid quantizer overload

The higher order cascaded modulator architecture does not suffer from any inherent stability problems as long as each of the stage is a first order delta-sigma loop. If higher order modulators are used for cascading then even the multistage architecture cannot have guaranteed stability. This comparison clearly shows that a lot more design work is required to make an N-th order single loop stable than the first order cascaded architecture which is inherently stable.

4.2. Quantization Noise Shaping Zeros - Bandpass

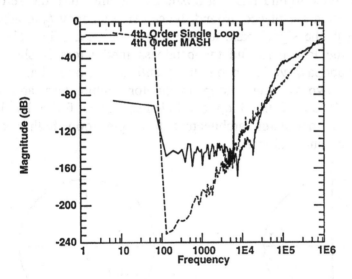

Figure 21: Software simulation of fourth order single loop modulator and the fourth order multistage modulator with ideally matched coefficients.

The coefficient design methodology for the single loop higher order modulators, as described in a previous section, shows the benefit of placing the zeros of the quantization noise transfer function in the baseband instead of DC. In the single loop architecture the loop coefficients control the location of the zeros whereas the multistage architecture has no means of moving the zeros away from DC when using first order cascades. The software simulation of the output of a fourth order single loop modulator and the fourth order multistage modulator with ideally matched integrator gains is shown in Fig.21. As can be seen in the simulation the total integrated noise power in the baseband can be better optimized by moving the zeros out of DC. The advantage of optimized zero placement is directly proportional to the signal bandwidth required from the modulator.

The availability of the zero location control in the single loop modulators also allows the design of bandpass delta-sigma modulators as opposed to the traditional lowpass designs [32,33]. The conventional design of the analog modulators for the oversampled analog-to-digital converters suppresses the quantization noise in the low frequency region by placing the zeros around DC. In a bandpass architecture the zeros of the transfer function are placed around the center frequency f_0 to achieve high resolution in the signal band around that frequency. The pole zero placement for the low pass and the bandpass modulators is shown in Fig.22. The bandpass variation on the quantization noise shaping can use the same lowpass single loop modulator architecture and only requires recalculation of the loop coefficients. Thus for a bandpass variation on the modulator architecture a single loop higher order architecture is the only choice.

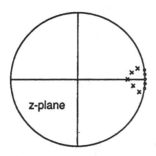

Pole-Zero Placement
for Lowpass Modulator

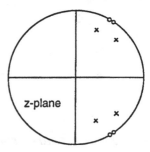

Pole-Zero Placement
for Bandpass Modulator

Figure 22: Pole-Zero placement to implement bandpass modulators

4.3. Coefficient Mismatch

The section on the multistage modulator architecture derives the transfer function for a second order cascaded modulator. It was shown that the first stage quantization noise is only cancelled to within the matching accuracy of the α and β coefficients between the first and the second stage. The coefficient mismatch puts an upper limit on the resolution achievable by a cascade of N-stages. The ultimate resolution is limited by leakage of the first stage quantization noise to the output. Matsuya et. al. achieved 16-bit performance over audio bandwidth of 20 kHz using a third order cascaded modulator architecture, and required a capacitance tolerance (3σ) of better than one percent [15].

The other schemes suggested in the literature for improving the coefficient mismatch in this topology rely on increasing the order of the first stage of the multistage modulator [34,35]. That approach is based on the philosophy to shape the quantization noise of the first stage with a second order difference so that it is the second order shaped noise that leaks to the output. The noise leakage due to component mismatch is reduced with this technique at the expense of conditional stability.

The single loop modulator performance is not limited by the coefficient mismatch. The only design criteria that needs to be simulated during the design of the single loop modulator is the coefficient tolerance in the loop. The variability of the coefficient values effects the exact location of the poles and zeros of the system function. If the system coefficient values change drastically it can move the poles of the system out of the unit circle and thus make the system unstable. Typical coefficient tolerance for the single loop designs exceed $\pm$ 10%, thus making it very robust to any capacitance mismatch or parasitic errors. Thus we can conclude that the SNR achievable from a multistage architecture is a strong function of the capacitance tolerance (1%) whereas the single loop modulator performance is independent of component mismatch.

4.4. Multi-bit Output Decimator Complexity

The output of the multistage modulator is a multi-bit word as opposed to the single bit output from the single loop modulator. The complexity of the decimator architecture which performs the lowpass filtering on the output bit stream of the modulator is a function of the number of bits at the output of the modulator. The FIR filtering process for a single bit output becomes very simple since the coefficient multiplications become simple additions and no multipliers are needed. The multi-bit output requires the multipliers and thus the decimation architecture for a multi-bit output system becomes relatively more complex. The payoff for using a single loop modulator increases if the decimation architecture is a single stage decimation process.

4.5. Finite Gain and Incomplete Settling of the Amplifier

We have previously derived the transfer function for the two stage modulator which included coefficient mismatch and ideal amplifier. Now we would like to assume perfect coefficient matching and non-ideal amplifier behaviour for the derivation of the system function of a second order multistage modulator. This would allow us to see the effect of finite gain and incomplete settling on the multistage modulator performance for ideally matched coefficients. The block diagram of a second order multistage modulator modelling the finite gain and incomplete settling is shown in Fig.19. The transfer function of the non-ideal integrator is given by:

$$\frac{V_0(z)}{V_i(z)} = \frac{G_s K_A z^{-1}}{1 - K_B z^{-1}} \tag{26}$$

If we assume ideally matched components the gain and pole errors become:

$$K_A = \frac{A}{1 + (1 + A)} \tag{27}$$

$$K_B = \frac{(1 + A)}{2 + A} \tag{28}$$

If we use the non-ideal transfer function for the integrator in the second order multistage modulator shown in Fig.19 we get the transfer function:

$$V_0(z)\Big[z^3 + (G_{s1}K_{A1} + G_{s2}K_{A2} - K_{B1} - K_{B2})z^2$$
$$+ ((K_{B1} - G_{s1}K_{A1})K_{B2} - G_{s2}K_{A2}K_{B1} + G_{s1}G_{s2}K_{A1}K_{A2})z\Big]$$

$$= V_i(z)\Big[G_{s1}K_{A1}z^3 + (G_{s1}G_{s2}K_{A1}K_{A2} - G_{s1}K_{A1}K_{B2})z^2\Big]$$
$$+ Q_1(z)\Big[(1 - G_{s2}K_{A2})z^3$$
$$+ (-K_{B2} + (G_{s2}K_{A2} - 1)K_{B1} + (2G_{s2} - G_{s1}G_{s2}K_{A1})K_{A2})z^2$$
$$+ (K_{B1}K_{B2} - 2G_{s2}K_{A2}K_{B1} + G_{s1}G_{s2}K_{A1}K_{A2})z\Big]$$
$$+ Q_2(z)\Big[z^3 + (-K_{B2} - K_{B1} + G_{s1}K_{A1} - 1)z^2$$
$$+ ((K_{B1} - G_{s1}K_{A1} + 1)K_{B2} + K_{B1} - G_{s1}K_{A1})z$$
$$+ (G_{s1}K_{A1} - K_{B1})K_{B2}\Big]$$

$$(29)$$

This derivation of the system function shows that even with the ideal matching of the coefficients the finite gain and incomplete settling of the amplifier will result in the leakage of the first order quantization noise to the output. The required SNR specification would dictate the minimum value on the open loop gain and the incomplete settling coefficient of the amplifier of the integrator. For a 16-bit performance from a third order architecture the amplifier is typically required to settle to within 0.001% of the final value and the amplifier gain has to be greater than 75 dB [36]. The high gain-bandwidth product requirements put an upper limit on the fastest sampling rate for the multistage converters. In order to overcome the high open loop gain requirement an auto-zeroed integrator can be used to cancel the finite gain errors and thus reduce the open loop gain requirements on the amplifier by increasing the design complexity of the integrator block [36].

The effect of finite gain in the single loop modulators is analogous to the loop coefficient variation. The coefficient tolerance on the single loop coefficient is relaxed enough that the amplifier gain only needs to be greater than the oversampling ratio of the modulator to achieve ideal performance [37,19]. Thus we see that the integrator design requirements for the single loop modulator are much more relaxed than the multistage modulator architecture.

4.6. Input Signal Transfer Function

The last issue to consider is a comparison of the input transfer function for the two architectures. The general transfer function for an analog modulator for oversampled analog-to-digital converter can be written as:

$$Y(z) = H_X(z)X(z) + H_E(z)E(z) \qquad (30)$$

Until now we have been discussing $H_E(z)$ performance for the two architectures but it is important to compare the behaviour of the signal transfer function $H_X(z)$. The signal transfer functions for the two architectures have been mentioned earlier and are repeated here for comparison:

$$Single\ Loop:\ H_X(z) = \frac{\displaystyle\sum_{i=0}^{N} A_i(z-1)^{N-i}}{z\left[(z-1)^N - \displaystyle\sum_{i=0}^{N} B_i(z-1)^{N-i}\right] + \displaystyle\sum_{i=0}^{N} A_i(z-1)^{N-i}} \qquad (31)$$

$$Mash:\ H_X(z) = 1 \qquad (32)$$

The signal transfer function for the single loop modulator is a function of the frequency. For low frequency signals $(z - 1) \approx j\Omega$ where $|j\Omega| < 1$ and $\Omega = 2\pi(f/f_s)$; f_s is the sampling frequency and f is the frequency of interest. Thus for the baseband signals $H_X(z) \approx 1$ for the single loop modulator whereas the MASH architecture has exactly one for the signal transfer function. A transformation of the single loop N-th order architecture under discussion gives us the architecture shown in Fig.16. This architecture as discussed earlier does not require any active summers. In this architecture the digital feedback signal is fed back at all the integrator inputs instead of just at the first integrator [26]. The poles of this system are controlled by the α_i and γ_i coefficients and the zeros of the quantization noise transfer function are controlled by the γ_i coefficients. The input signal transfer function also has zeros which are a function of the β_i feedforward input coefficients shown in the figure. The availability of the β-coefficients allows to shape the signal tranfer transfern and compensates for any frequency dependent attenuation. Whereas in the original N-th order loop the signal transfer function was determined by the A-coefficients which were dictated by the pole locations.

The availability of the β-coefficients in the transformed N-th order architecture can be used to place zeros to cancel out the poles in the signal transfer function. We must realize that in actuality the mismatch of the zeros and pole placement in the signal transfer function will result in a slow timeconstant tail for a step input. This is in contrast with the exact value of one for the signal transfer function in the multistage topology. The importance of having the exact value of one for the signal transfer function starts to play a role when a single modulator needs to be multiplexed between more than one input. In applications of multichannel data acquisition, multistage topology offers a significant inherent advantage of unity gain signal transfer function over the single loop modulator.

5. Future Directions of Oversampled Converters

In the last few years oversampled converters have become the preferred architecture for low frequency high resolution applications. Two main architectures of oversampled modulators have been discussed in the context of low frequency signals. The area of radio communication will see increased applications of oversampled analog-to-digital converters in the near future. In digital radios bandpass oversampled modulators can be used to perform the analog-to-digital conversion function on the signal band around the carrier frequency [32,33]. The bandpass oversampled modulator falls under the class of a single loop interpolative modulator architecture. Bandpass modulators use the zeros of the transfer function to suppress the quantization noise in a narrow band around Ω_0. Thus the single bit stream gives an accurate representation of the narrow band signal around the carrier frequency. This single bit digital stream can be filtered in the digital domain to reproduce the analog signal in high resolution digital format for further digital processing. This architecture provides the analog-to-digital conversion and accurate digital filtering in a single block.

Oversampled converters have always been analyzed and understood as signal acquisition systems. The single bit output of the modulator is fed into the digital decimator which uses linear decoding scheme like low pass filtering to produce a high accuracy digital representation for the baseband signal. The application of oversampled converters for data acquisition applications where fast transients also need to be captured sounds counter intuitive at first sight, but it is possible to produce high resolution data acquisition converters by using nonlinear decoding schemes [38,39]. This decade could see the practical implementations of data acquisition converters using nonlinear decoding algorithms or other techniques.

First-order and second-order delta-sigma modulators will remain popular for medium resolution signals (12-14 bits) with baseband frequencies less than 50 kHz. As the number of analog signal inputs increases the area and power efficient oversampled architectures will become useful in applications which require multi-channel analog-to-digital conversions. The requirements of the digital decimator are significantly relaxed for the first and second-order delta-sigma converters. Thus with the decreasing channel lengths of the CMOS process very compact and area efficient multichannel oversampled architectures could see increased applications.

Oversampled architectures will continue to push up into higher baseband frequencies. At the present time it is not clear where the boundary lies between oversampling architectures and more conventional Nyquist converters as a function of baseband frequency. It is clear, however, that oversampled converters will be a major force in band-limited signal-acquisition systems in both high-resolution, small number of channel systems with frequencies under 100 kHz and for medium-resolution large multi-channel systems.

References

[1] D.A. Hodges, P.R. Gray, and R.W. Brodersen, "Potential of MOS Technologies for Analog Integrated Circuits", *IEEE J. of Solid-State Circuits*, vol. SC-13, pp.285-294, June 1978.

[2] J.C. Candy, "Decimation for sigma delta modulation", *IEEE Transactions on Communications*, vol. COM-34, pp.72-76, Jan. 1986.

[3] J.R. Fox and G.J. Garrison, "Analog to Digital Conversion using Sigma-Delta Modulation and Digital Signal Processing", in *Proceedings MIT VLSI Conference*, pp.101-112, Jan. 1982.

[4] M.W. Hauser and R.W. Brodersen, "Circuit and Technology Considerations for MOS Delta-Sigma A/D Converters",' in *Proceedings 1986 International Symposium on Circuits and Systems*, pp.1310-1315, May 1986.

[5] H. Inose, Y. Yasuda, and J. Murakami, "A Telemetering System by Code Modulation $\Delta\Sigma$-Modulation", *IRE Trans. Space Electronics and Telemetry*, vol.SET-8, pp.204-209, Sept. 1962.

[6] J.C. Candy, "The Structure of Quantization Noise from Sigma-Delta Modulation", *IEEE Transactions on Communications*, vol.COM-29, pp.1316-1323, Sept. 1981.

[7] W.R. Bennett, "Spectra of Quantized Signals", *Bell System Technical Journal*, Vol.27, pp.446-472, July 1948.

[8] D.J. Goodman, "Delta Modulation Granular Quantizing Noise", *Bell System Technical Journal*, pp.1197-1218, May 1969.

[9] J.W. Scott, W. L. Lee, C.H. Giancarlo, and C.G. Sodini, "CMOS Implementation of an Immediately Adaptive Delta Modulator", *IEEE Journal of Solid-State Circuits*, vol. SC-21, pp.1088-1095, Dec. 1986.

[10] R.M. Gray, "Oversampled Sigma-Delta Modulation", *IEEE Transactions on Communications*, vol. COM-35, pp. 481-489, May 1987.

[11] J.C. Candy, "A Use of Double Integration in Sigma Delta Modulation", *IEEE Transactions on Communications*, vol.COM-33, pp.249-258, March 1985.

[12] R. Koch and B. Heise, "A 120 kHz Sigma-Delta A/D Converter", in *ISSCC Digest of Technical Papers*, pp.138-141, Feb. 1986.

[13] B.E. Boser and B.A. Wooley, "Design of a CMOS Second-Order Sigma-Delta Modulator", in *ISSCC Digest of Technical Papers*, pp. 258-259, Feb. 1988.

[14] S.H. Ardalan and J.J. Paulos, "Stability analysis of high-order sigma-delta modulators", in *Proceedings 1986 International Symposium on Circuits and Systems*, pp.715-719, May 1986.

[15] Y. Matsuya, K. Uchimura, A. Iwata, T. Kobayashi, M. Ishikawa, and T. Yoshitome, "A 16-bit Oversampling A-to-D Conversion Technology using Triple Integration Noise Shaping", *IEEE Journal of Solid-State Circuits*, vol.SC-22, pp.921--929, Dec. 1987.

[16] M.J. Hawksford, "N-th order recursive sigma-ADC machinery at the analogue-digital gateway", in *Audio Engineering Society*

Convention, May 1985.

[17] T. Hayashi, Y. Inabe, K. Uchimura, and T. Kimura, "A Multistage Delta-Sigma Modulator without Double Integration Loop", in *ISSCC Digest of Technical Papers*, pp.182-183, Feb. 1986.

[18] W.L. Lee and C.G. Sodini, "A topology for higher order interpolative coders", in *Proceedings 1987 International Symposium on Circuits and Systems*, pp.459-462, May 1987.

[19] K.H. Chao, S. Nadeem, W.L. Lee, and C.G. Sodini, "A High Order Topology for Interpolative Modulators for Oversampling A/D Converters", *IEEE Transactions on Circuits and Systems*, Vol.37, pp.309-318, March 1990.

[20] K.H. Chao, *Limitations of Interpolative Modulators for Oversampling A/D Converters*. Master's thesis, Massachusetts Institute of Technology, Cambridge, MA, 1988.

[21] B.E. Boser and B.A. Wooley, "Quantization error spectrum of sigma-delta modulators", in *Proceedings 1988 International Symposium on Circuits and Systems*, June 1988.

[22] S. Nadeem, *Design and Implementation of Fourth Order Modulator For 16-bit Oversampled A/D Converter*. Master's thesis, Massachusetts Institute of Technology, Cambridge, MA, 1989.

[23] W.L. Lee, *A Novel Higher Order Interpolative Modulator Topology for High Resolution Oversampling A/D* Converters. Master's thesis, Massachusetts Institute of Technology, Cambridge, MA, 1987.

[24] D. Welland, B.P.D. Signore, and E.J. Swanson, "A Stereo 16-Bit Delta-Sigma A/D Converter for Digital Audio", in *85-th Convention of the Audio Engineering Society*, pp.0, Nov. 1988.

[25] P. Ferguson, A. Ganesan, and R. Adams, "One Bit Higher Order Sigma-Delta A/D Converters", in *IEEE Proceedings of the International Symposium on Circuits and Systems*, pp.890-893, 1990.

[26] P. Ferguson, A. Ganesan, R. Adams, S. Vincelette, R. Libert, A. Volpe, D. Andreas, A. Charpentier, and J. Dattorro, "An 18-bit 20 kHz Dual Sigma-Delta A/D Converter", in *IEEE International Solid-State Circuits Conference*, pp.890-893, 1991.

[27] S.H. Ardalan and J.J. Paulos, "An analysis of nonlinear behavior in delta-sigma modulators", *IEEE Transactions on Circuits and Systems*, vol.CAS-34, pp.593-603, June 1987.

[28] S. Hein and A. Zakhor, "Lower Bounds on the MSE of the Single and Double Loop Sigma Delta Modulators", in *IEEE International Symposium on Circuits and Systems*, pp.1751-1755, May 1990.

[29] S. Hein and A. Zakhor, "On the Stability of Interpolative Sigma Delta Modulators", in *IEEE International Symposium on Circuits and Systems*, pp.1621-1624, June 1991.

[30] L. Longo and M. Copeland, "A 13 bit ISDN-band Oversampled ADC Using Two-Stage Third Order Noise Shaping", *IEEE Custom IC Conference*, pp.21.2.1-21.2.4, Jan. 1988.

[31] B.P. Brandt and B.A. Wooley, "A CMOS Oversampling A/D Converter With 12-Bit Resolution at Conversion Rates Above 1 MHz", *IEEE Solid State Circuits Conference*, Feb. 1991.

[32] R. Schreier and W.M. Snelgrove, "Bandpass Sigma-Delta Modulation", in *Electronic Letters*, pp.1560-1561, Nov. 1989.

[33] R. Schreier and W.M. Snelgrove, "Decimation for Bandpass Sigma-Delta Analog-to-Digital Conversion", in *IEEE International Symposium on Circuits and Systems*, pp.1801-1804, May 1990.

[34] L.A. Williams III and B.A. Wooley, "Third-Order Cascaded Sigma-Delta Modulators", *IEEE Transactions on Circuits and Systems*, pp.489-497, May 1991.

[35] D.B. Ribner, R.D. Baertsch, S.L. Garverick, D.T. McGrath, J.E. Krisciunas, and T. Fuji, "16b Third-Order Sigma Delta Modulator with Reduced Sensitivity to Nonidealities", in *IEEE Solid-State Circuits Conference*, pp.66-67, Feb. 1991.

[36] M. Rebeschini, N.R.V. Bavel, P. Rakers, R. Greene, J. Caldwell, and J.R. Haug, "A 16-b 160-kHz CMOS A/D Converter Using Sigma-Delta Modulation", in *IEEE Journal of Solid-State Circuits*, pp.431-440, Apr. 1990.

[37] B.E. Boser and B.A. Wooley, "The Design of Sigma-Delta Modulation Analog-to-Digital Converters", in *IEEE Journal of Solid-State Circuits*, Dec. 1988.

[38] S. Hein and A. Zakhor, "Optimal decoding for data acquisition

applications of Sigma Delta modulators", Apr. 1991. U.S. Patent Application No. 07/694,294.

[39] S. Hein and A. Zakhor, "Optimal decoding for data acquisition applications of Sigma Delta modulators", Feb. 1993. Accepted for publication in *IEEE Transactions on Signal Processing*.

Biographies

Shujaat Nadeem was born in Multan, Pakistan, in 1965. He received his B.S. and M.S. degrees in electrical engineering from Massachusetts Institute of Technology in 1989. He is currently working towards his Ph.D. degree in electrical engineering at Massachusetts Institute of Technology. His research interests are in the area of high resolution analog-to-digital convertors and analog circuit applications to image and signal processing.

Mr. Nadeem is a member of Tau Beta Pi, Eta Kappa Nu and Sigma Xi. He is also the recipient of the IEEE Circuits and Systems Darlington Award for the best paper of 1990.

Charles G. Sodini was born in Pittsburgh, PA, in 1952. He received the B.S.E.E. from Purdue University, Lafayette, IN, in 1974, and the M.S.E.E. and Ph.D. degrees from the University of California, Berkeley, in 1981 and 1982 respectively.

He was a member of the Technical staff at Hewlett-Packard Laboratories from 1974 to 1982, where he worked on the design of MOS memory and later on the development of MOS devices with very thin gate dielectrics. He joined the faculty of the Massachusetts Institute of Technology, Cambridge, MA, in 1983 where he is currently an Associate Professor in the Department of Electrical Engineering and Computer Science. His research interests are focused on IC fabrication, device modelling, and device level circuit design, with emphasis on analog and memory circuits.

Dr. Sodini held the Analog Devices Career Development Professorship of MIT's Department of Electrical Engineering and Computer Science and was awarded the IBM Faculty Development Award from 1985 to 1987. He served on a variety of IEEE Conference Committees including the International Electron Device Meeting where he was the 1989 General Chairman. He is currently the Technical Program Co-Chairman for the 1992 Symposium on VLSI Circuits.

High Speed 1-bit Sigma Delta Modulators

Tapani Ritoniemi

Tampere University of Technology
P.O.Box 553
Tampere, SF-33101
Finland

Abstract

High speed 1-bit quantizer sigma delta ($\Sigma\Delta$-) modulator topologies and some of their applications are discussed. It is shown that a 14-bit dynamic range can be achieved with the oversampling ratio of 16 by using cascaded high-order 1-bit $\Sigma\Delta$-modulators whose noise transfer function can be modified by placing some of the zeros in the baseband. Compared to cascaded low-order $\Sigma\Delta$-modulator structures, these modulators are less sensitive to circuit imperfections. The moderate requirements allow us to use simplified high speed circuit structures. The operational transconductance amplifier (OTA) structure realization based on the use of a minimum number of active transistors in the signal path is discussed. Also an example of $\Sigma\Delta$-modulation in digital data transmission at the intermediate frequency is shown. This arrangement has some advantages over conventional systems including the fast recovery from the power-down mode, which is a crucial problem in the portable digital telecommunication applications.

1. Introduction

$\Sigma\Delta$-modulation has proved to be an attractive method for performing analog-to-digital (A/D) conversion for relatively low bandwidth signals. The theoretical upper limit for the baseband signal to quantization noise ratio (S/N_q) is given by

$$S/N_q > (6 \times N + 3) \times \log_2(M) \quad dB, \tag{1}$$

where M is the oversampling ratio and N is the modulator order. The S/N_q depends also on the modulator structure and the location of the zeros of the noise transfer function (NTF) [2]. Fig.1 shows some simulated S/N_q's as a function of the oversampling ratio for various 1-bit modulator structures.

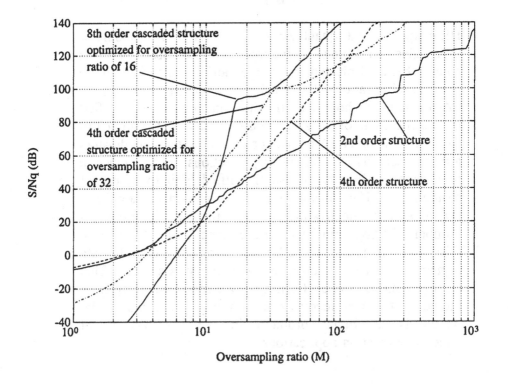

Figure 1: Simulated S/N_q's for various types of $\Sigma\Delta$-modulators as a function of the oversampling ratio.

The noise shaping in the $\Sigma\Delta$-modulation enables us to use coarse 1-bit quantization which can be realized accurately. The modulator provides a lowpass type transfer function for the input signal and a highpass type (noise shaping) transfer function for the quantization noise. This highpass filter attenuates the uniformly distributed quantization noise in the baseband. The digital data containing shaped quantization noise lying out-of-baseband can be filtered using a digital lowpass filter. The output of the filter is the desired A/D converted signal. All converters based on the use of $\Sigma\Delta$-modulation have well-known advantages over conventional A/D-converters due to their internal structure, such as linearity, monotony, and an embedded digital anti-aliasing filter.

Wide bandwidth $\Sigma\Delta$-converters are made possible by the evolution of CMOS technology which enables the fabrication of 50 MHz sampling rate switched-capacitor integrated circuits [3,4]. The modulator architecture development towards higher-order modulator structures [5,9], which provide the same S/N_q with lower oversampling ratios, can be seen in Fig.1. This has allowed realizing 500 kHz sampling rate A/D-converters with a 14-bit accuracy [5].

The unconditional stability of the modulator operation at first seemed to prevent the wide use of the higher-order modulators. Examples of stable $\Sigma\Delta$-converters are given in [5,7,8,9].

2. Modulator Structures for Low Oversampling Ratios

The main advantage of using cascaded first-order modulators [10,11] is the unconditional stability. The NTF of this structure is not optimal since all the zeros are restricted to lie at 0 Hz (DC). Also the component matching and integrator settling accuracy requirements are difficult to be achieved. However, with an oversampling ratio of 64, a 16-bit S/N_q can be achieved [10].

The use of second-order modulator blocks in a cascaded structure lowers the requirements of the component matching and integrator settling accuracy [12]. The second-order modulator blocks allow us to optimize the location of one complex NTF zero pair. This decreases the amount of the in-band quantization noise considerably. The block diagram of a fourth-order $\Sigma\Delta$-modulator [4] is shown in Fig.2. The optimized fourth-order modulator gives over a 100 dB S/N_q with the oversampling ratio of 32 and with modest component requirements [4].

228

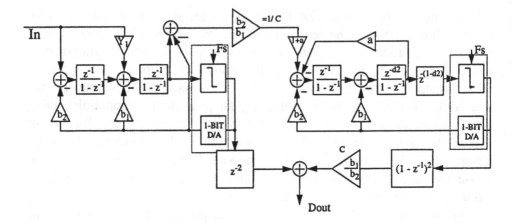

Figure 2: Block diagram of a fourth-order cascaded ΣΔ-modulator with one complex conjugate NTF zero pair.

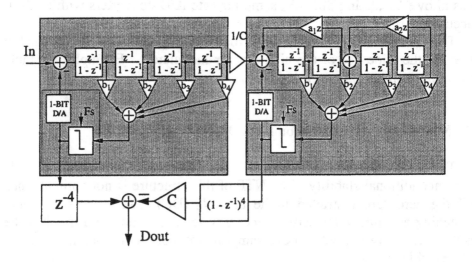

Figure 3: Block diagram of a 8th-order cascaded ΣΔ-modulator with two complex conjugate NTF zero pairs.

The next interesting modulator structure is the 8th-order cascaded structure which uses fourth-order modulator blocks as shown in Fig.2. This architecture allows us to optimize the location of two complex NTF zero pairs, resulting in a significant reduction in the quantization noise level in the baseband. Theoretically, over a 90 *dB* S/N_q can be achieved with an oversampling ratio of 16. With a 1 % mismatching between the modulator blocks, the S/N_q decreases to 82 dB. With this structure, it is possible to realize a 14-bit converter using an oversampling ratio as low as 16. The main drawback of using fourth-order modulator blocks is the conditional stability. The stability of the modulator can be guaranteed for a limited input signal, which means that the normal stable operation is achieved by scaling integrator voltages in a proper way. When this modulator is unstable, the operation can be turned back to stable operation by resetting all the integrators [2]. Another drawback of 8th-order modulators is a huge out-of-band quantization noise and, therefore, an extremely selective digital filter is needed. The quantization noise power of the modulator is about 50 dB over the signal power. This means that the required stopband attenuation for the digital filter is about 150 dB.

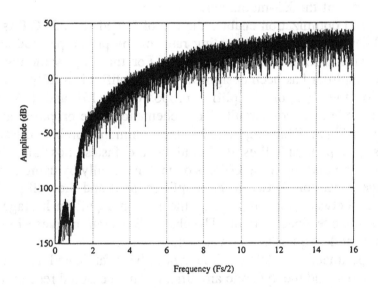

Figure 4: Spectrum of a 8th-order modulator output data, 2^{16}-point FFT, 2^{15} are shown.

The cascaded 8th-order $\Sigma\Delta$-modulator seems to be a good candidate for implementing a 14-bit 5 MHz A/D-converter with an input sampling frequency of 80 MHz. The hard problem with these structures is the implementation of the digital decimation filter. In this case, both the sampling rate and required attenuation are very high, making the realization of these filters a very challenging and demanding problem.

3. OTA Structures for High Sampling Rate

The maximum sampling frequency of the $\Sigma\Delta$-modulator is limited by several time constants in modulator components, and finally by the MOS transistors. The time-critical path for the signal is the loop consisting of the comparator, the 1-bit D/A, and the SC-integrator. The signal goes around this loop once in each output sample. The SC-technology requires at least 2-phase non-overlapping clock signals. During clock phase 1, the capacitors of the 1-bit D/A are charged and the comparator is settled. Clock phase 2 is used for integrating the charge of the previous integrator and the charge of the 1-bit D/A convertor. The resolution requirements for the comparator are so coarse that accuracy of the fast regenerative latch is adequate. Since the delay of the comparator is much smaller than that of the integrating amplifiers, the speed of the integrator is limiting the sampling rate of the $\Sigma\Delta$-modulator.

The speed optimization enables the use of folded cascode OTAs for the $\Sigma\Delta$-modulation at 50 *MHz* sampling rate for the prototype modulator [4] realized with a 1.2 μm CMOS process. For this high sampling rate, a dynamic amplifier has been designed. This amplifier is a fully differential version of the dynamic amplifier proposed in [15] and [14] with a common-mode feedback circuit. The schematic of the proposed amplifier is shown in Fig.5. This amplifier has the minimum number of transistors in the signal path and thus it should be the fastest possible amplifier structure. The biasing of the OTA is done dynamically by using capacitors as voltage generators between the amplifier input and the pull-push stage as well as between the output and the mirrored pull-push stage in the common-mode feedback circuit. The class-AB operation gives a high slew-rate and a modest quiescent power consumption.

The experimental modulator circuits have been fabricated using both the folded cascode and the dynamic amplifiers. The measured results from the first silicon are shown in Fig.6. Considerable improvements are expected at the next processing iteration. The proposed dynamic amplifier reaches the same performance as the folded cascode amplifier but with a lower power consumption.

The more advanced CMOS and BiCMOS technologies enable the use of higher sampling rates up to 200 *MHz* [16]. Using BiCMOS amplifiers of which the gain is known accurately beforehand, the changes in the transfer functions caused by this finite gain can be compensated for. This enables the designer to match the modulator blocks to each other precisely and S/N_q is reduced only slightly from the theoretical value.

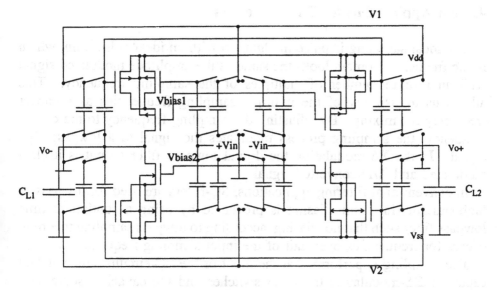

Figure 5: Circuit diagram of the dynamically biased OTA.

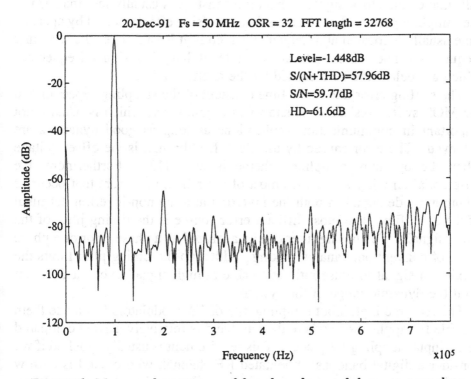

20-Dec-91 Fs = 50 MHz OSR = 32 FFT length = 32768

Level=-1.448dB
S/(N+THD)=57.96dB
S/N=59.77dB
HD=61.6dB

Figure 6: Measured spectrum of fourth-order modulator output data sampled at 50 MHz [3].

4. New Applications for $\Sigma\Delta$ -Converters

The ideal sampling is pulse modulation with an ideal pulse train, which in the frequency domain looks the same as the amplitude modulated signal with an infinite number of multiples of the sampling frequency. This allows us to use any of the multiple sampling frequencies as a carrier frequency for mixing. By adjusting the sampling frequency to the carrier frequency, the sampling process modulates the signal to a lowpass type signal. In the $\Sigma\Delta$-modulation the digital lowpass filter output is itself a modulated and A/D-converted signal.

In conventional filtering applications, SC-filters are used for lowpass or highpass filtering, and aliasing is prevented by using a continuous-time lowpass filter such that no aliasing, according to lowpass sampling theorem, occurs for frequencies over half of the input sampling frequency.

The sampling is performed at the input of the SC-circuits. In switched capacitor $\Sigma\Delta$-modulators the input switches and the capacitor serve as a very fast sampling circuit. The sampling aperture time (modulating pulse width) is the turn-off time of a MOS transistor or roughly the rise- or the fall-time of the clock signal. This time constant is usually less than 1/5 of the sampling period. The transmission zero (sinc effect) caused by aperture time usually is located at a frequency which is at least 5 times the sampling frequency. The SC-circuits are capable of handling signal frequencies which are below the zero caused by the aperture time.

The settling error due to the time constant of the sampling capacitor and the MOS switch resistance is shown as a gain error, which is usually not important in communication applications as long as good dynamics are achieved. The error caused by the clock feedthrough is the offset voltage when the optimum sampling scheme is used [17]. Furthermore, the differential circuitry suppresses most of the remaining errors that occur as common-mode signal in both the inverting and the non-inverting terminal of the amplifier. The most difficult error source is the timing jitter of the clock signals. The timing jitter can be analyzed as a phase error or a phase noise of conventional analog mixing. Although the timing jitter limits the maximum signal-to-noise ratio attainable at high signal levels, it does not limit the dynamic range of the system.

The required high input sampling rate of $\Sigma\Delta$-modulators has made them suitable for applications where the baseband is relatively narrow compared to the input sampling frequency. This requirement is usually satisfied if we consider a digital bandpass modulated RF-channel, whose band is narrow and situated at relatively high frequencies. Conventionally, we must first multiply the signal by the channel carrier transferring the channel signal to low frequencies. This signal is then converted to a digital form, and finally the modulated symbols are detected.

According to the bandpass sampling theorem [20], the bandpass signal with a bandwidth less than the sampling frequency can be reconstructed perfectly using quadrature sampling with two sampling devices. Any bandpass signal can be generated by combining two amplitude modulated signals on carriers with a phase difference but with the same frequency. On the other side we can reconstruct the signals using sampling for modulation with a phase difference.

In digital transmission, data is modulated with an analog carrier frequency. Usually, two sinusoidal carrier frequencies are used to increase the efficiency of the channel. The modulating carriers called in-phase (I) and quadrature phase (Q) carriers, have a phase difference of 90 degrees. At the intermediate frequency, the information channel is filtered using a bandpass filter. After that, the signal is modulated separately into I- and Q-branches and lowpass filtered. The filtered signals are A/D-converted into digital data from which the symbols are detected at both branches. The discrete-time $\Sigma\Delta$-modulation based system includes a digital lowpass filter which is seen as a sharp bandpass filter at the intermediate frequency. The filter needed at the intermediate frequency is a low-order bandpass filter which attenuates the image frequencies of the previous modulation stages.

In portable battery-powered systems, the power consumption is one of the major issues. Most of the communication standards allow the sleeping mode for the terminal such that the time of the battery-powered operation can be increased. For this reason, the realization of sleeping functions are important. The analog mixers require a constant biasing and a considerable quiescent power is consumed. The realization of sleeping functions for mixers is difficult because of the relatively slow recovery from sleeping mode back to normal mixing operation. The $\Sigma\Delta$-modulator can be turned off and on within a short settling time. The digital implementation can be manufactured easily and no tuning is needed in the production. These features make $\Sigma\Delta$-modulation based systems very attractive for personal communication applications.

Fig.7 shows a test setup for emulating a digital transmission system [18]. The transmission is done using conventional mixers. The transmitted data signals are generated on a personal computer using an interface D/A-board for both I- and Q- branches. The receiver part consists of sigma-delta A/D-converters and a data collection arrangement in a personal computer. This system is capable of emulating a digital communication system which uses a quadrature based modulation at a 2 *MHz* carrier frequency.

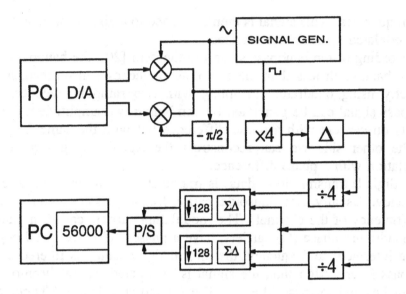

Figure 7: Emulation arrangement of digital communication system using ΣΔ-modulators for mixing in I- and Q- branch.

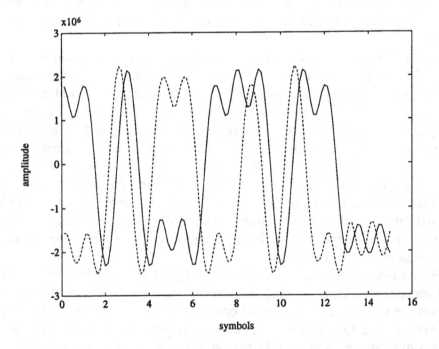

Figure 8: I- and Q- branch data received using the intermediate frequency sampling rate.

Sampling is done directly at a 2 *MHz* carrier frequency and filtering and detection are done digitally. The received data waveforms of the GMSK modulated (BT=0.3) data are shown in Fig.8. A spectrum of the received data is shown in Fig.9.

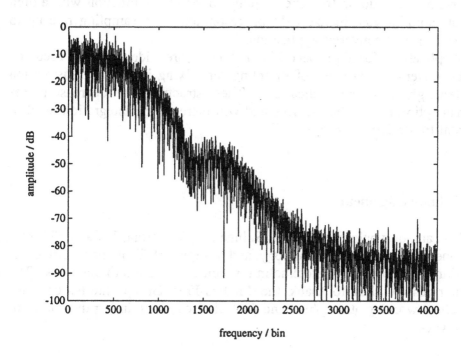

Figure 9: Spectrum of the received GMSK data, 8092-point FFT.

The measurements indicate that the sampled system operates as well as a conventional analog modulation based system. Compared to bandpass ΣΔ-modulation systems [19], the matching between the I- and Q- branch is not as good. Also the timing error between the branches is larger. The advantage of the proposed system compared to bandpass modulation is that a higher carrier frequency can be achieved. The bandpass ΣΔ-modulator requires a sampling frequency at least four times higher than the carrier frequency, and thus such high intermediate frequencies cannot be reached.

5. Conclusion

Current CMOS technology achieves 50 *MHz* sampling frequency for analog SC-structures. High sampling rates combined with high order $\Sigma\Delta$-modulators provide theoretically S/N_q of more than 90 *dB* at an oversampling ratio of 16. These architectures in combination with a high input sampling rate realize a 14-bit resolution for a sampling rate up to 5 *MHz* in A/D-converter applications.

Applications for high speed $\Sigma\Delta$-modulators are wideband high accuracy A/D-converters. The use of sampling for mixing in digital transmission systems gives us new applications. These structures will decrease power consumption in portable devices and will increase the integration level of digital transmission systems.

6. Acknowledgement

The author would like to thank S. Ingalsuo, V. Eerola, T. Saramäki from Tampere University of Technology, and professor H. Tenhunen from Royal Institute of Technology, Stockholm for their contribution to this work. The author also wishes to thank the Median-Free Group International for excellent working atmosphere and fruitful discussions during the course of this work.

References

[1] J.C. Candy, and G.C. Temes, et al., *Oversampling Delta-Sigma Data Converters,* IEEE Press, New York 1992.

[2] T. Ritoniemi, T. Karema and H. Tenhunen, "A Fifth Order Sigma-Delta Modulator for Audio A/D-Converter'', in *Proc. IEE 1991 International Conference on Analogue to Digital and Digital to Analogue Conversion*, Swansea, UK, September 1991, pp.153-158.

[3] B. Brandt, and B. Wooley, "A CMOS Oversampling A/D Converter with 12b Resolution at Conversion Rates Above 1 MHz'' in *Proc. IEEE International Solid-State Circuits Conference*, February 1991, pp.64-65.

[4] S. Ingalsuo, T. Ritoniemi, T. Karema, and H. Tenhunen, "A 50 MHz Cascaded Sigma-Delta A/D Modulator", to be included in *Proc. IEEE Custom Integrated Circuits Conference*, May 1992.

[5] F. Op 't Eynde, G. Ming, and W. Sansen, "A CMOS Fourth Order 14b 500 k-Sample/s Sigma-Delta ADC Converter" in *Proc. IEEE International Solid-State Circuits Conference*, February 1991, pp.62-63.

[6] W.L. Lee and C.G. Sodini, "A topology for higher order interpolative coders", in *Proc. IEEE International Symposium on Circuits and Systems*, May 1987, pp.459-462.

[7] D. Welland, B. Signore, E. Swanson, T. Tanaka, K. Hamashita, S. Hara, and K. Takasuka, "A Stereo 16-bit Delta Sigma A/D Converter for Digital Audio", *J. Audio Eng. Soc.*, Vol.37, June 1989, pp.467-486.

[8] P. Ferguson, A. Ganesan, R. Adams, S. Vincelette, R. Libert, A. Volbe, D. Andears, A. Charpertier, and J. Dattoro, "An 18b 20kHz Dual Sigma Delta A/D Converter", in *Proc. IEEE Solid-State Circuits Conference*, February 1991, pp.68-69.

[9] T. Ritoniemi, T. Karema and H. Tenhunen, "The Design of Stable High Order 1-Bit Sigma-Delta Modulators", in *Proc. IEEE International Symposium on Circuits and Systems,* May 1990, pp.3267-3270.

[10] Y. Matsya, K. Uchimura, A. Iwata, T. Kobayashi, M. Ishikawa, and T. Yoshitome, "A 16-bit Oversampling A-to-D Conversion Technology Using Triple-Integration Noise Shaping," *IEEE J. Solid-State Circuits*, Vol. SC-22, December 1987, pp.921-929.

[11] M. Rebeschini, N. Bavel, P. Rakers, R. Greene, J. Caldwell and J. Haug, "A High-Resolution CMOS Sigma-Delta A/D Converter with 320 kHz Output Rate", in *Proc. IEEE Int. Symp. on Circuits and Systems*, May 1989, pp.246-249.

[12] T. Karema, T. Ritoniemi and H. Tenhunen, "Fourth Order Sigma-Delta Modulator for Digital Audio and ISDN applications", in *Proc. IEE European Conference on Circuit Theory and Design*, September 1989, pp.223-227.

[13] S. Ingalsuo, T. Ritoniemi, and H. Tenhunen, "New Differential Amplifier for High Speed SC-Circuits", in *Proc. IEE European*

Conference on Circuit Theory and Design, September 1991, pp.1349-1358.

[14] Y. Tsividis, and P. Antogetti, et al., *Design of MOS VLSI for Telecommunications*, Prentice-Hall, New Jersey, 1985, pp.145-169.

[15] S. Masuda, Y. Kitamura, S. Ohya, and M. Kikuchi, "CMOS- Sampled Differential Push-Pull Cascode Operational Amplifier'' in *Proc. IEEE International Symposium on Circuits and Systems*, May 1984, pp.1211-1214.

[16] A. Baschirotto, R. Alini, and R. Castello, "Bicmos Operational Amplifier with Precise and Stable DC Gain for High-Frequency Switched Capacitor Circuits'', *IEE Electronics Letters*, Vol.27, July 1991, pp.1338-1340.

[17] D. Haigh, and B. Singh, "A Switching Scheme for Switch Capacitor Filters which Reduces the Effect of Parasitic Capacitances Associated with Switch Control Terminals,'' in *Proc. IEEE International Symposium on Circuits and Systems*, Los Angeles, USA, May 1983, pp.586-589.

[18] V. Eerola, H. Lampinen, T. Ritoniemi and H. Tenhunen, "Direct Conversion Using low pass Sigma-Delta modulation'', to be included in *Proc. IEEE International Symposium on Circuits and Systems* , May 1992. [19] R. Screier, and M. Snelgrove, "Bandbass Sigma-Delta Modulation'', *IEE Electronics Letters*, Vol.25, 1989, pp. 1560-1561.

[20] A. Jerry, "The Chignon Sampling Theorem - Its Various Extensions and Applications,'' *Proc. IEEE*, Vol.65, November 1977, pp.1565-1596.

Biography

Tapani Ritoniemi was born in Finland in 1964. He received the engineering degree from the Technical University of Tampere. He is currently completing doctoral studies. He has published more than 20 international articles, and he also holds some patents. He is starter of the VLSI Solution Ltd. company and he is a member of the Median-Free Group International. His research interests are sigma delta modulation, digital and analog median-free filtering and VLSI design.

Continuous Calibration, Noise Shaping D/A Conversion

Hans Schouwenaars, Wouter Groeneveld,
Corné Bastiaansen and Henk Termeer

Philips Research Laboratories
Eindhoven
The Netherlands

Abstract

The analog part of a current mode CMOS 5-bit bi-directional D/A converter for digital audio with 115 dB dynamic range and -90 dB distortion at 128 times oversampling is presented. A dynamic self-calibration technique based upon charge storage on the gate-source capacitance of CMOS transistors is used to obtain the required relative accuracy and absolute linearity of the current sources. A calibrated spare source is used to allow continuous operation. No laser or external trimming techniques are required.

1. Introduction

In digital audio applications there is a growing interest in highly oversampled noise shaping D/A converters. In these systems a high dynamic range can be obtained at reasonable oversampling ratios by using higher order noiseshapers which reduce the number of bits, and push the resulting quantization noise out of the frequency band of interest. The stability conditions of these noiseshapers is known and can be solved [1]. The main advantages of 1-bit implementations are obvious. First, the differential linearity is always guaranteed by the bitstream, allowing accurate reconstruction of low level signals. Second, there is no need for analog accuracy as no intermediate analog levels between the digital "0" and "1" are needed. Third, the 1-bit converter concept mates well with other digital circuitry. Most of the converter itself consists of digital

CMOS circuits, while the small analog part can also be designed in the same conventional CMOS process. Consequently, the integration of the converter together with complex functions such as digital volume and tone control becomes possible.

The main disadvantage of 1-bit conversion is the presence of many high frequency components with a relatively large amplitude in the 1-bit data signal. This often causes intermodulation distortion and interference in the 1-bit D/A converter, leading to whizzes and frizzles within the audio band and thus reducing the true dynamic range of the converter. However, the use of an 1-bit switched capacitor D/A converter [2] does result in a 16-bit performance. Others use multi-stage noise shaping and simple CMOS inverters [3]. To achieve a wider dynamic range, the digital and the analog converter part can be separated, eliminating on-chip interference and allowing the best process choice for both parts. This is reported in [4], in which true 18-bit performance has been achieved by an 1-bit D/A converter designed in a Bi-CMOS process.

Multi-bit noise shaping and D/A conversion enable the acquisition of an even wider dynamic range while keeping all converter parts on the same die. The use of more bits reduces the high-frequency noise and thus intermodulation problems, but increases the design effort on the multi-bit D/A converter.

In this paper a D/A converter architecture is presented which achieves a dynamic range of 115 dB in a conventional CMOS process, by using an oversampled 5-bit D/A converter [6].

2. General Design Considerations

In Fig.1 a conversion system which employs an oversampling D/A converter is shown. The input signal is a 20 bit digital sine wave at a sampling frequency of 44 kHz. The sampling frequency is increased to 5.6 MHz by using an oversampling factor of 128. The 20 bits at 5.6 MHz are then applied to a third-order noiseshaper which reduces the 20 bit input data to N-bit. Now, at the output of the noiseshaper the bit-stream is available to drive the oversampling D/A converter to produce V_{out}. Fig.2 shows the output spectrum of a 1-bit D/A converter with an almost full-scale output sine wave of about 1 kHz. The enormous quantization noise of the 1-bit converter is pushed to higher frequencies by the third-order noiseshaper at the rate of 60 dB per decade, as is shown in Fig.2. By integrating the output spectrum over the audio band of 20 Hz to 20 kHz, the RMS noise voltage is found and the dynamic range can be determined. A theoretical dynamic range of this conversion system of 120 *dB* can be obtained with an ideal D/A converter at these conditions as is shown by [1] and [2]. In practice, the dynamic range will be determined by intermodulation phenomena in the D/A converter.

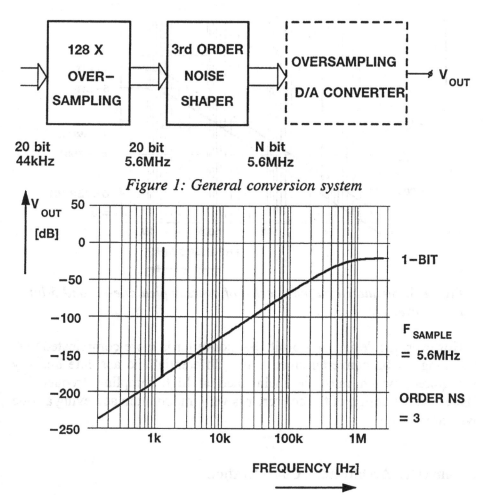

Figure 1: General conversion system

Figure 2: Simulated output spectrum of oversampled one-bit D/A converter

The quantization noise can be reduced using a multi-bit oversampling D/A converter ($N>1$). The noise reduction is determined by the available number of analog levels 2^N-1 and the noise transfer characteristic of the noise-shaping loop. This is discussed in [5]. For $N=5$, the quantization noise is reduced by 30 *dB*. In Fig.3 the output spectra of both a 1-bit and a 5-bit converter are shown. The latter is shifted 30 *dB* down compared to the 1-bit spectrum. This implies that theoretically a 30 *dB* wider dynamic range can be obtained using a 5-bit D/A converter, and at the same time possible intermodulation problems are reduced considerably.

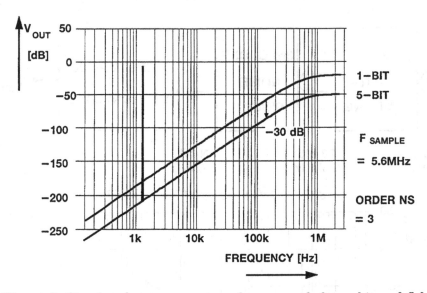

Figure 3: Simulated output spectra of oversampled one-bit and 5-bit D/A converters

However, the design of a 5-bit D/A converter is more complicated than designing a 1-bit converter since a high differential and absolute linearity is required, which must be maintained at high sampling frequencies. Therefore, the design of the converter is very important in achieving a wide dynamic range.

3. Converter Architecture Considerations

There are several possibilities for the converter architecture. In Fig.4 the well-known principle of binary-weighted current conversion is shown for the 3-bit case. The output current of such a converter I_{out} is drawn as a function of time t in Fig.4a. In Fig.4b the number of matched current sources which contribute to the output current is shown. To generate the top of the sine wave, called full-scale (*F.S.*), all the available current sources are needed. To generate the lowest part of the sine wave, no current sources are needed. Obviously, at the zero crossings of the sine wave, 0.5 *F.S.*, half of the available current sources are used.

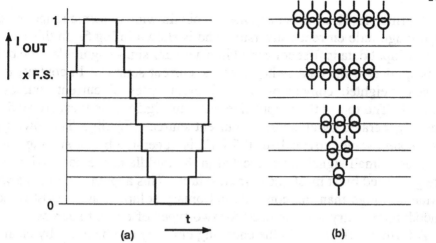

Figure 4: Binary-weighted current conversion

An advantage of this binary weighted current conversion is the easy decoding of the data. The MSB is simply connected to the data switches of half the total number of current sources. The main disadvantage occurs at low signal levels. These signals are often located at half full scale. Many current sources are needed to generate the required DC level and do not contribute to the audio information. These current sources generate noise and glitches which seriously affect the dynamic range.

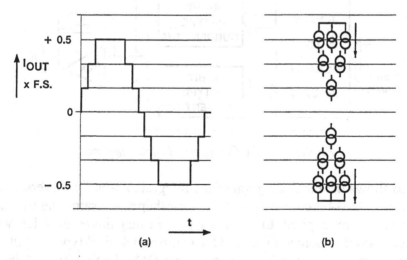

Figure 5: Sign-magnitude current conversion

A principle which is better suited to obtain wide dynamic range is called sign-magnitude current conversion, and is shown in Fig.5. In this case the binary input data is first converted into a binary sign-magnitude code before it is applied to the data switches of the current sources. In contrast to the binary weighted conversion, two different types of current sources are needed. The top of the output sine wave in Fig.5a is now called +0.5 *F.S.* and is generated by all sourcing current sources (Fig.5b). The lowest part of the sine wave, now called -0.5 *F.S.*, is generated by all sinking current sources. Small signal levels located in the middle of the conversion range are generated by at most one current source. This approach permits a wider dynamic range than the conventional binary solution, at the cost of some encoding circuitry and the need for two types of current sources.

A further refinement in the encoding circuitry can be made by changing the binary sign-magnitude code into a thermometer sign-magnitude code. This guarantees the differential linearity of the converter since the currents are switched on and off one by one. Furthermore, thermometer encoding minimizes the amplitude of the glitches that continuously occur since a noise-shaper generates the input code for the D/A converter.

4. D/A Converter Block Diagram

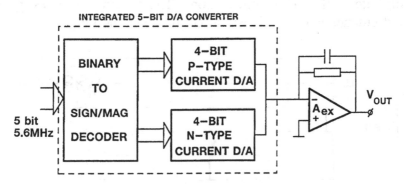

Figure 6: D/A Converter block diagram

Fig.6 shows the block diagram of the integrated 5-bit D/A oversampling converter. The 5-bit output data of the noiseshaper at a sampling frequency of 5.6 *MHz* are applied to a binary-to-sign-magnitude encoder which generates two thermometer codes. One controls a 4-bit P-type current D/A, and the other controls a 4-bit N-type current D/A. Each 4-bit D/A basicly consists of 15 current sources and current switches. The output currents of both converters are added and converted into voltage by means of an external operational amplifier A_{ex} and its feedback network, which also performs the first postfiltering action.

As was previously discussed, the absolute linearity of the converter at high signal levels will be limited by the internal linearity of both the P-type and the N-type D/A, and therefore by the matching of the 16 current cells. Straightforward component matching techniques are not adequate [7,8]. Therefore, our self-calibration technique is used to match the current sources. This continuous calibration technique does not require external components or laser trimming. Moreover, it is insensitive to element aging.

5. Basic calibration principle

First, the calibration principle for one single current source will be described, and later on this will be extended to a larger number of current sources.

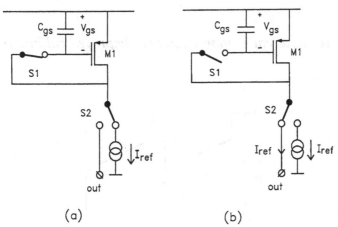

(a) (b)

Figure 7: Calibration principle. (a) Calibration (b) Operation

The basic circuit of a current cell is shown in Fig.7. When the switches S_1 and S_2 are in the depicted state, a reference current I_{ref} flows into transistor M_1, as it is connected as an MOS-diode. The voltage V_{gs} on the intrinsic gate-source capacitance C_{gs} of M_1 is then determined by the transistor characteristics. When S_1 is opened and S_2 is switched to the other position, the gate-to-source voltage V_{gs} of M_1 is not changed since the charge on C_{gs} is preserved. Provided that the drain voltage is not changed either, the drain current of M_1 will still be equal to I_{ref}. This current is now available at the I_{out} terminal, and the original reference current source is no longer needed.

6. Imperfections

In practice, the switches S_1 and S_2 are MOS-transistors as shown in Fig.8a. This gives rise to some disturbing effects, which cause changes in the gate voltage of M_1 during switching.

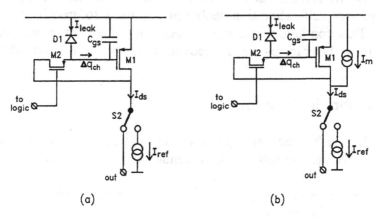

(a)   (b)

Figure 8: (a) Real calibration circuit (b) Improved calibration circuit

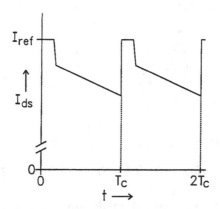

Figure 9: Drain current I_{ds} of M_1 versus time

When M_2 is switched off, its channel charge q_{ch} is partly dumped on the gate of M_1, and so the charge on C_{gs} of M_1 is decreased by an amount Δq_{ch}. The charge change implies a sudden decrease of V_{gs} of M_1:

$$\Delta V_{gs,q} = \frac{\Delta q_{ch}}{C_{gs}} \tag{1}$$

After switching, another effect influences V_{gs}. Although M_2 is switched off, the reverse-biased diode D_1 between its source and V_{dd} is still present. The leakage current I_{leak} of this diode decreases the charge on C_{gs} continuously. Assuming that calibration is done at $t=0$, the gate voltage equals:

$$V_{gs,leak}(t) = V_{gs}(0) - \frac{I_{leak}}{C_{gs}}t \tag{2}$$

The changes in the gate voltage of M_1 are transformed into changes in the drain current I_{ds} by its transconductance g_m. The output current can then be represented by the solid line in the graph of Fig.9. The voltage drop described by equation (1) causes a drop in the output current just after calibration:

$$I_{ds,q} = I_{ref} - g_m \Delta V_{gs,q} = I_{ref} - g_m \frac{\Delta q_{ch}}{C_{gs}} \tag{3}$$

The transconductance of M_1 equals:

$$g_m = \sqrt{2\mu C_{ox} \frac{W}{L} I_{ds}} \, , \tag{4}$$

in which μ represents the electron mobility and C_{ox} the oxide capacitance per square micron.

The gate-source capacitance can be rewritten as:

$$C_{gs} = \frac{2}{3} WL C_{ox} \, , \tag{5}$$

Substituting equations (4) and (5) into (3) yields:

$$I_{ds,q} = I_{ref} - \frac{3}{2} \sqrt{\frac{2\mu}{C_{ox}}} \cdot \frac{\Delta q_{ch}}{L} \sqrt{\frac{I_{ds}}{WL}} \tag{6}$$

The leakage effect in the drain current can be calculated in the same way. The time-dependent drain current due to the diode leakage (2) is:

$$I_{ds,leak}(t) = g_m V_{gs,leak}(t) = I_{ref} - g_m \frac{I_{leak}}{C_{gs}}t \tag{7}$$

Substitution of equations (4) and (5) into (7) yields:

$$I_{ds,leak}(t) = I_{ref} - \frac{3}{2} \sqrt{\frac{2\mu}{C_{ox}}} \cdot \frac{1}{L} \sqrt{\frac{I_{ds}}{WL}} I_{leak}t \tag{8}$$

This formula implies that after a certain time T_c the current cell has to be calibrated again to keep its output current within a specified range.

The results of equations (6) and (8) clearly indicate that the ratio μ/C_{ox} should be small to keep the changes in the drain current small, so an advanced CMOS process is preferable. In addition, both equations contain parameters that can be influenced by the actual design.

All the adjustable parameters have limits that depend on other considerations as well. The current I_{ds} is determined by the properties of the circuit in which the current cell has to be implemented. The transistor width and especially its length should be as large as possible for optimal calibration and $1/f$ noise behavior, but this is limited by layout size considerations that become more important when many cells are used. Furthermore, the W/L ratio has a minimum because of limits on V_{gs}.

A very simple addition to the circuit of Fig.8a is shown in Fig.8b. In parallel to the current source transistor M_1, a main current source I_m is added which has a value of about 90 % of the reference current. This decreases the value of M_1's current to about 10 %, so its transconductance is decreased by a factor of $\sqrt{10}$. Furthermore, the size of M_1 can now be optimized. More information on the feasibility of the self-calibration technique is given in [9].

7. Continuous Current Calibration

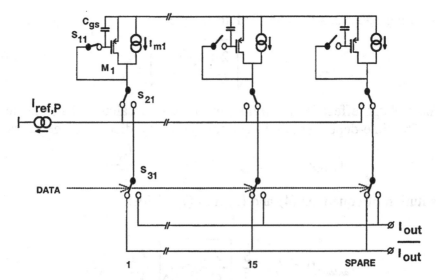

Figure 10: Basic diagram continuous current calibration

To make the calibration technique suitable for the design of a D/A converter, it must be extended to an array of current sources as is shown in Fig.10. For a 4-bit converter, 15 identical current cells are needed. Uncalibrated, the 15 main current sources I_m have an unacceptable difference in their absolute values. Calibration is performed as follows. Assume the switches are in the state shown in the figure. Only the first current cell is connected to the refence source $I_{ref,P}$. The difference between $I_{ref,P}$ and the first main current source I_{m1} is supplied by transistor M_1 as it is connected as a diode. The switch S_{11}, which is in series with the gate of M_1, is opened, and the two-way switch S_{21} changes its state. The intrinsic gate-source capacitance C_{gs} of transistor M_1 stores the voltage corresponding to the current difference, and the absolute value of the calibrated current remains unchanged. Thus a duplicate of the reference source is created which can be applied to the output line I_{out} or its complement through switch S_{31}. The second current cell is then connected to the reference source and calibrated, and so on, until all current cells of the D/A are calibrated. During calibration, a spare current cell is operating in place of the cell under calibration, so the D/A can operate continuously without the need for special calibration cycles which is the case in other calibration techniques [10]. After calibration of all 15 cells, the spare cell itself is calibrated as well. To avoid inaccuracies caused by leakage, the calibration process is cycled repeatedly which is controlled by a calibration pointer.

8. Reference Current Adjustment

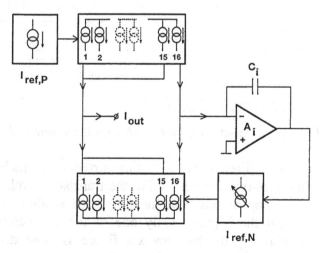

Figure 11: Reference current adjustment

Fig.11 shows the reference current adjustment of the two basic current D/A converters of Fig.6. The reference current $I_{ref,P}$ is duplicated 16 times in the upper P-type current source D/A converter. The outputs of 15 current source cells can be switched to the I_{out} terminal, and 15 outputs of the lower N-type current sink D/A converter can be switched to the same I_{out} terminal. The 16-th output current of both types of converters are compared and the current difference is integrated by means of an internal operational amplifier A_i and a capacitor C_i. The output voltage of the integrator is used to adjust the reference current source $I_{ref,N}$. In this way, sufficient matching between the P-type and the N-type converter is achieved.

9. Bi-directional Calibrating Current Cell

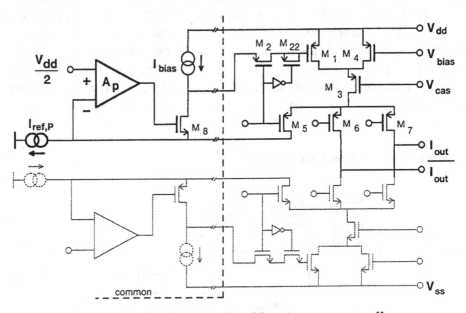

Figure 12: Bi-directional calibration current cell

Fig.12 shows the bi-directional calibrating current cell which consists of both a P-type and a N-type current cell at the transistor level. For clarity the P-type part is emphasized, and the N-type is shaded as it is fully complementary with the P-part. Only one of the 16 identical cells is drawn. The switch S_{In} of the previous figure is now drawn as the compensated transistor pair M_2/M_{22}. The main current source is transistor

M_4, which is biased by V_{bias} at about 90 % of the value of $I_{ref,P}$, which equals 70 μA. In parallel is transistor M_1 which will supply the 10% current difference between the reference source $I_{ref,P}$ and the main current source M_4. The drains of these transistors are connected to the source of a cascode transistor M_3, biased by V_{cas}. The combined current is routed through one of three transistors M_5-M_7, controlled by digital data. The two two-way switches of the previous figure are implemented as one three-way switch.

Left of the dashed lines in Fig.12 is the common circuitry that is needed only once for the 16 cells. It consists of the operational amplifier A_p driving the gate of a level-shift transistor M_8 which is biased by a small current source I_{bias}. The potential across the reference source $I_{ref,P}$ is set by A_p at $Vdd/2$. The potential of the two output lines are also set at $Vdd/2$ by the external operational amplifier A_{ex} of Fig.6. The operational amplifiers A_{ex} and A_p have unity gain bandwidths of 10 MHz and are Miller compensated.

To calibrate the current cell, transistors M_5 and M_2 are turned on for a small period of time, as previously described. The output current of the cell can then be used for the normal D/A function using the data switches M_6 and M_7.

The most important transistor dimensions of the current cell are given in Table 1

Transistor	M1	M2	M3	M4	M5	M6	M7	M8
W/L	100/35	2/1.6	30/5	400/16	10/1.6	10/1.6	10/1.6	20/1.6

Table 1: Current cell transistor dimensions

10. Measurement Results

All presented results are measured unweighted true RMS over a bandwidth of 20 kHz at a sampling frequency of 5.6 MHz.

Fig.13 shows a photograph of the typical output voltage of the oversampling multi-bit D/A converter.

A sine-wave of 20 kHz is generated at about -20 dB relative to full-scale. Only four current levels are needed; the signal in between two levels is not an oscillation but the oversampled and noise shaped input data. The Signal-to-Noise and Total-Harmonic-Distortion (S/(N+THD)) of this small signal is 90 dB.

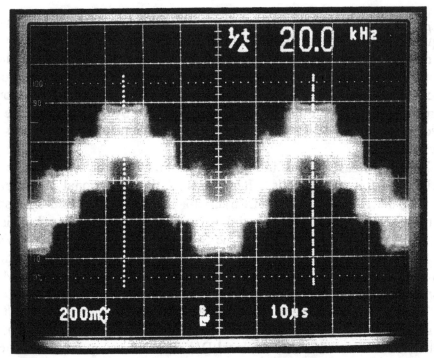

Figure 13: Typical output voltage oversampled multi-bit D/A converter

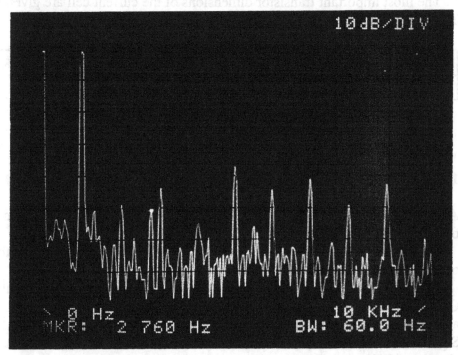

Figure 14: Output spectrum 5-bit D/A converter $f_{in}=1$ kHz -84 dB

Fig.14 shows the output spectrum of the oversampled 5-bit converter generating a 1 *kHz* signal at -84 *dB* relative to full-scale.

The converter is calibrated at 44.1 *kHz*, i.e. each current cell is refreshed at 44.1 *kHz* divided by 16, or 2760 *Hz*. That is where the marker is located. The amplitude of the frequency component which is caused by the self-calibrating technique is about 135 *dB* down from full-scale and therefore of no importance.

Only harmonic components of the 1 *kHz* signal contribute significantly to the measurements. The S/(N+THD) of this signal is 31 *dB*, corresponding to a dynamic range of 115 *dB*.

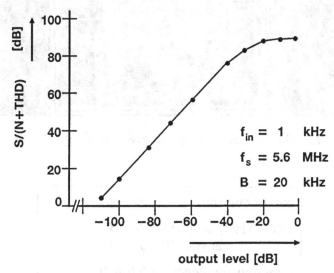

Figure 15: Measured S/(N+THD) as a function of output level

Fig.15 shows the measured S/(N+THD) as a function of output level of the converter. At small signal levels, up to -30 *dB*, 115 *dB* dynamic range is obtained. At full-scale a S/(N+THD) ratio of 90 *dB* is measured. This is due to distortion caused by the integral linearity of the P-type and the N-type converter, and by the matching between both converters.

A die photograph of the circuit with overlay is shown in Fig.16. The binary to sign-magnitude conversion is performed by means of a PLA which has been automatically generated. The most important specifications of the converter are listed in Table 2.

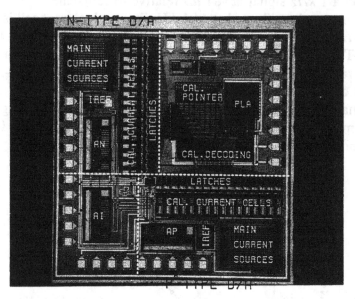

Figure 16: Die photograph of the oversampling D/A converter

output current	± 1 *mA*
S/(N+THD) at 0 *dB*	90 *dB*
at -60 *dB*	55 *dB*
sampling frequency	5.6 *MHz*
supply voltage	5 *V*
power dissipation	100 *mW*
temperature range	-10 to 70 °C
process	1.6 μ*m* CMOS
active chip area	7.5 *mm²*

Table 2: D/A converter specifications

11. Conclusion

The application of the multi-bit approach combined with a sign-magnitude decoding in an oversampled D/A converter not only increases the dynamic range of the converter but also reduces the intermodulation sensitivity. A dynamic self-calibration technique is used to construct the accurate bit currents without external components. Measurement results show that a dynamic range of 115 dB and a S/(N+THD) of 90 dB at full-scale has been achieved using a standard CMOS process.

Acknowledgment

The authors thank P.J.A. Naus and P.A.C. Nuyten for their help and advice concerning the third-order noise-shaper.

References

[1] E.F. Stikvoort, "Some remarks on the stability and performance of the noise-shaper or sigma-delta modulator", IEEE Transactions on Communications, Vol.36, No.10, Oct. 1988.

[2] P.J.A. Naus et al.,"A CMOS Stereo 16-bit D/A Converter for Digital Audio", IEEE J. Solid-State Circuits, Vol. SC-22, pp.390-395, June 1987.

[3] Y. Matsuya et al.,"A 17-bit Oversampling D-to-A Conversion Technology Using Multistage Noise Shaping", IEEE J. Solid-State Circuits, Vol. SC-24, pp.969-975, August 1989.

[4] B. Kup et al., "A Bitstream D/A Converter with 18-bit resolution", IEEE J. Solid-State Circuits, Vol.26, pp.1757-1763, Dec. 1991.

[5] P.J.A. Naus and E.C. Dijkmans, "Multibit Oversampled Sigma-Delta A/D Converters as Front End for CD Players", IEEE J. Solid-State Circuits, Vol. SC-26, pp.905-909, July 1991.

[6] H.J. Schouwenaars, D.W.J. Groeneveld C.A.A. Bastiaansen and H.A.H. Termeer, "An Oversampled Multibit CMOS D/A Converter for Digital Audio with 115 dB Dynamic Range", IEEE J. Solid-State Circuits, Vol.26, pp.1775-1780, Dec. 1991.

[7] H.J. Schouwenaars, D.W.J. Groeneveld and H.A.H. Termeer, "A low-power stereo 16-bit CMOS D/A converter for digital audio", IEEE J. Solid-State Circuits, Vol.SC-23, pp.1290-1297, Dec. 1988.

[8] M.J.M. Pelgrom, A.C.J. Duinmaijer and A.P.G. Welbers, "Matching properties of MOS transistors", IEEE J. Solid-State Circuits, Vol.24, pp.1433-1440, Oct. 1989.

[9] D.W.J. Groeneveld, H.J. Schouwenaars, H.A.H. Termeer and C.A.A. Bastiaansen, "A Self-Calibration Technique for Monolithic High-Resolution D/A Converters", IEEE J. Solid-State Circuits, Vol.SC-24, pp.1517-1522, Dec. 1989.

[10] H.S. Lee, D.A. Hodges and P.R. Gray, "A self-calibrating 15b CMOS A/D converter", IEEE J. Solid-State Circuits, Vol.SC-19, pp.813-819, Dec. 1984.

Biographies

Hans J. Schouwenaars was born in Rotterdam, The Netherlands, on October 9, 1954. He received the B.S degree in electrical engineering from the Hogere Technische School, Dordrecht, The Netherlands, in december 1978. and the degree of master of philosophy of the University of Wales, United Kingdom in 1989. In 1978 he joined the Analog Integrated Electronics Group of the Philips Research Laboratories, Eindhoven, The Netherlands where he is working as a member of the scientific staff on the design of integrated circuits in bipolar and CMOS processes such as operational-amplifiers, reference-sources, high-performance sample-and-hold amplifiers, high-resolution analog-to-digital and digital-to-analog converters for digital audio applications.

D. Wouter J. Groeneveld was born in Rotterdam, The Netherlands, on February 28, 1962. In 1986 he received the Ir. degree in electrical engineering from the Delft University of Technology, Delft, The Netherlands. In that same year he joined Philips Research Laboratories, Eindhoven, The Netherlands, where he is working in the field of high resolution D/A and A/D converters and associated circuits for digital audio.

Cornelis A.A. Bastiaansen was born in Breda, The Netherlands, on August 4, 1962. He received the B.S. degree in electrical engineering from the Technical College, Breda, The Netherlands, in 1985. In that same year he joined the Philips Research Laboratories, Eindhoven, The Netherland, where he is engaged in the design of AD/DA converters for video and digital audio applications.

Henk A.H. Termeer was born in Sint Oedenrode, The Netherlands, on February 29, 1948. He received the degree in electrical engineering from the Eindhoven Technical College, The Netherlands, in 1967. In 1970 he joined Philips Research Laboratories, Eindhoven, The Netherlands, where he is engaged in electrical engineering and IC design of MOS and bipolar circuits in the Analog Integrated Electronics Group.

Bandpass Sigma Delta A-D Conversion

A.M. Thurston, T.H. Pearce, M.D. Higman,
* M.J. Hawksford

GEC-Marconi Research Centre, Chelmsford
* University of Essex, Colchester
U.K.

Abstract

Modern high performance radio systems increasingly rely on digital signal processing techniques. In many cases performance is limited by the A-D conversion process, particularly where high linearity is required. In this paper three examples are presented of bandpass sigma-delta A-D converters which offer a cost effective means of encoding narrowband IF signals to a high linearity and with low spurious content.

1. Introduction

The technique of base-band sigma-delta Analogue to Digital (A-D) conversion is well established and is frequently used in speech and communications applications [1,2]. More recently, single-bit conversion schemes have found a wider application in digital audio systems for both A-D converters and Digital to Analogue (D-A) converters where enhanced linearity especially at low signal levels is a principal advantage [3,4].

For RF applications the base-band conversion scheme requires demodulation of the message using two matched mixers fed by inphase and quadrature carriers as shown in Fig.1. The output of each mixer is then low pass filtered and subsequently encoded with separate A-D converters. While this approach may be satisfactory for low resolution systems it has inherent performance limitations due to *I-Q* imbalance, and lack of sensitivity at the carrier due to masking by DC offsets.

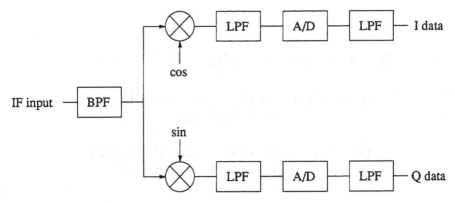

Figure 1: Conventional digital IF system architecture with analogue mix to baseband and separate A/D converters in the I and Q paths.

To ensure 40 *dB* I-Q image rejection, for example, an amplitude and phase match of better than 0.1 *dB* and 1 degree respectively is necessary. This would mean that a signal received at say 10 *kHz* above the IF centre frequency would produce an image component only 40 *dB* lower at a frequency 10 *kHz* below the IF centre frequency. For many applications this would present a serious limitation on resolution in the frequency domain, on single sideband reception for example. Similarly the presence of DC offsets and low frequency 1/*f* noise at the mixer outputs and A-D converter inputs will produce components which cannot be distinguished from signals at the IF centre frequency. This will place a restriction on reception and demodulation of signals requiring the extraction of a carrier component. The DC component may be removed by high pass filtering the digital *I,Q* signals, but this will produce a null in the response around the IF centre frequency.

A new design of sigma-delta A-D converter is presented which can encode the signal at the intermediate frequency and then by digital post processing convert to baseband *I* and *Q* as shown in Fig.2. This enables the quadrature mixing to be achieved with a high degree of accuracy. Orthogonality is guaranteed by sampling at four times the IF centre frequency (f_{IF}). The two local oscillator signals are then of the form 1, 0, -1, 0 and 0, 1, 0, -1 representing samples of cos $2\pi f_{IF}$ and sin $2\pi f_{IF}$. The resultant *I,Q* image rejection and DC offset is then set by the number of bits in the digital processing, which can be made arbitrarily larger than that of the A-D converter.

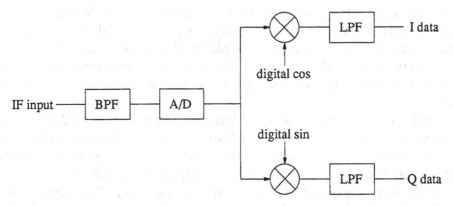

Figure 2: Alternative digital IF system architecture with single A/D converter encoding the IF signal and subsequent digital mix to baseband.

The technique of band-pass sigma-delta conversion is described and a method for designing band-pass A-D converters from existing baseband modulator designs is given. Practical results are described with the performance specified in conventional RF circuit terms.

The novel techniques described here are the subject of worldwide granted and pending patents.

2. Baseband Sigma-Delta Modulation

It is desirable in the transformation of a baseband sigma-delta modulator to its bandpass equivalent that certain of the basic properties of the coder remain unchanged. With the existence of a direct mapping technique from baseband to bandpass this condition is easily met and consequently much of the wealth of information regarding baseband coders which has been acquired over the last few decades may be applied almost directly to the bandpass equivalents. As a result parameters such as overload levels, noise power densities, signal shaping properties and statistical information about the signal levels within the coders already exist and the process of designing bandpass coders is greatly simplified.

In this section the basic properties of baseband sigma-delta modulators will be explained and several simple equations will be derived to characterize key features of an example baseband coder. Later, when the conversion from baseband to bandpass is applied, these equations will be modified accordingly such that the effect of the transformation may be seen.

Fig.3 shows an example of a baseband sigma-delta modulator. A delayed version of the output signal $Y(Z)$ is subtracted from the input signal $X(Z)$ to produce the error signal $E(Z)$. The error signal is filtered by the loop filter $A(Z)$ to give a filtered error signal $F(Z)$, where $F(Z)$ is the weighted sum of the first and second integrals of $E(Z)$. The filtered error signal is quantized, often quite coarsely, to produce the output signal. The stability of the closed loop depends on the design of the loop filter and it is here that the subtleties of different designs exist. This is particularly true in the case of single bit modulators in which the open loop gain is not defined and consequently stability is much harder to predict.

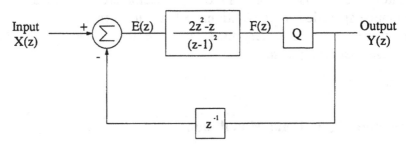

Figure 3: Example of Baseband Sigma Delta Modulator comprising loop filter A(z) and quantizer Q in feed forward path and single delay element in feed back path.

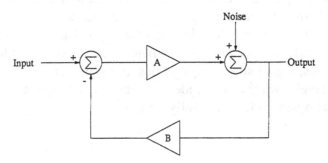

Figure 4: Standard Feedback Loop Model of Sigma Delta Modulator.

The basic properties of the coder can be seen by considering the standard feedback loop model shown in Fig.4. In the figure, 'A' represents the filters in the feed forward path which would correspond to the loop filter and the gain of the quantizer, and 'B' represents the filters in the feedback path, which in the example modulator is just a simple delay. The additive noise source is a crude model of the noise added by the quantizer which is assumed to be white, though in practice this assumption is not accurate. The two properties of interest are the ways in which the modulator filters the signal and shapes the noise.

Using the feedback loop model it is simple to establish the transfer functions between each of the signal and noise inputs and the output. These transfer functions may be called the Signal Transfer Function (STF) and the Noise Transfer Function (NTF) respectively and are given by:

$$STF = \frac{A}{1 + AB} \tag{1}$$

$$NTF = \frac{1}{1 + AB} \tag{2}$$

It is desirable that the signal be encoded without phase or amplitude variation across the passband, and that the noise be heavily attenuated. At first glance both of these seem possible over a wide frequency range by making 'A' very large, but this cannot be achieved in practice because the loop would be unstable. However, 'A' can be made large over a narrow band of frequencies by making its gain frequency dependent. In baseband coders the loop filters have high gain at low frequencies and reduced gain at high frequencies to maintain stability. As a result, at low frequencies, the STF is virtually flat and the NTF heavily attenuates the quantizing noise.

Returning to the example modulator given in Fig.3, the STF and NTF may now be derived. Both derivations assume that the quantizer has unity gain, which for the single bit quantizer is not accurate. However, since the properties of the quantizer are unaffected by the baseband to bandpass transformation this will not affect the final results. With the unity gain assumption, the transfer functions are given below:

$$STF(Z) = \frac{2(Z-0.5)}{Z} \tag{3}$$

$$NTF(Z) = \frac{(Z-1)^2}{Z^2} \tag{4}$$

At low frequencies the zero in the STF introduces very little phase and amplitude ripple to the signal, whilst the double zero in the NTF heavily attenuates low frequency noise at the output of the coder. Fig.5 shows the output spectrum of the example modulator with a sinusoidal input at $f_s/100$, where f_s is the sampling rate. The noise shaping properties of the coder are clearly seen.

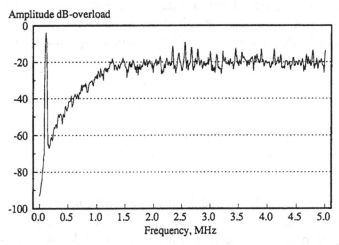

Figure 5: Output Spectrum of Baseband Sigma Delta Modulator.

Fundamental to the noise shaping properties of the coder is the open loop transfer function between the input and the output of the quantizer. This fact is readily confirmed in (2), the generalized formula for the NTF. When considering purely digital modulators it is best described as a simple Open Loop Transfer Function (OLTF), but when mixed digital and analogue architectures are considered, as in the case of sigma-delta A-D converters, this is more usefully described in terms of the Open Loop Impulse Response (OLIR). For practical reasons delay is implemented in the digital feedback path since it is cumbersome to achieve in an analogue filter, and consequently the digital loop filter in the feedforward path will have numerator and denominator of equal order. The OLIR then neglects the digital delay terms and describes only the true filtering properties of the loop filter.

For the example modulator the open loop transfer function is given by:

$$OLTF(Z) = \frac{2Z-1}{(Z-1)^2} \tag{5}$$

Separating the delay from the filter elements gives:

$$OLTF(Z) = \frac{2Z^2 - Z}{(Z-1)^2} \cdot Z^{-1} \tag{6}$$

for which the open loop impulse response can be shown to be:

$$OLIR(n) = (2 + n) \qquad n = 0, 1, 2, 3, \ldots. \tag{7}$$

In virtually all baseband coders the OLIR takes the form of:

$$OLIR(n) = (P + Qn) \tag{8}$$

where P and Q are the levels of the first and second order components of the impulse response respectively. Such separation of the different order components becomes highly relevant when analogue loop filters are designed, as will be seen in section 5.

This then concludes the overview of the basic features of baseband modulators and the conversion to bandpass will now be considered.

3. Conversion to Bandpass

It is possible to design sigma-delta modulators centred on any frequency from DC to $f_s/2$, and indeed any frequency above this, since the sampled data has many aliased images reflected about multiples of the sampling rate. Coders in which the IF lies between DC and $f_s/2$ are referred to as sampling coders, whilst those operating at an image frequency are referred to as subsampling coders.

This paper is concerned only with coders in which the IF is positioned at an odd multiple of the quarter sampling frequency since they have two distinct advantages over their counterparts. Firstly, the analogue loop filters required in the A-D converters are rather simpler in design, and secondly the digital mix to baseband I and Q signals is trivial to implement since the digital mixing signals are easily generated as $\cos = \{1,0,-1,0\}$ and $\sin = \{0,1,0,-1\}$.

Consider the conventional design of IF encoder shown in Fig.1. Two separate A-D converters are used, one each to encode the I and Q components of the IF signal. The aim is to design one A-D converter to encode the IF signal directly.

A first step towards this can be made by repositioning the mix to baseband from the pre-conversion analogue domain to the post-conversion digital domain. As a result the I and Q signals to be encoded will no longer be at baseband but will be modulated by cosinusoidal and sinusoidal carriers respectively, each at half the sampling frequency. The effect of this is that the original baseband signals will now alternate in polarity from sample to sample, and the modified A-D converters must be able to accommodate this. The modification is easily made to the example baseband modulator of Fig.3. To achieve a polarity inversion with every clock cycle, an invertor is placed immediately following every delay element, or quite simply $-Z$ is substituted for Z in the original architecture.

It is then noted that the I-Q separation in the system architecture is achieved by sampling each A-D converter alternately, with a 180° phase difference between them. The two A-D converters are otherwise identical. It is therefore possible to arrange for one converter to encode both I and Q signals in a time multiplexed fashion by simply doubling the sample rate and replacing each delay element with a two stage shift register to keep I and Q data separate. Thus, for example, on even numbered sampling instants the in-phase component of the IF signal is presented to the modulator, and the corresponding in-phase data is encoded by the arithmetic operators of the modulator. Quadrature data is stored halfway down the shift registers within the modulator and is not affected. On odd numbered sampling instants the quadrature data is encoded whilst the in-phase data is stored in the shift registers. The conversion is made by substituting Z^2 for Z in the recently modified coder. Combining the two substitutions, i.e. $-Z^2$ for Z, gives the necessary substitution to convert the baseband coder into its quarter sampling rate bandpass equivalent.

Fig.6 shows the resulting modulator after $-Z^2$ is substituted for Z in the example coder of Fig.3. It is noted that the signals at the input summing node are now added, though negative feedback is still achieved since the two sample period loop delay corresponds to a 180° phase shift at the IF, and hence the signals are added out of phase to achieve cancellation.

The signal and noise transfer functions of the new modulator are given below.

$$STF(Z) = \frac{2(Z^2 + 0.5)}{Z^2} \qquad (9)$$

$$NTF(Z) = \frac{(Z^2 + 1)^2}{Z^4} \tag{10}$$

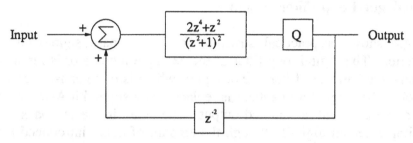

Figure 6: Bandpass Sigma Delta Modulator obtained by substituting -Z^2 for Z in the baseband modulator of Fig. 3.

Noise is now heavily attenuated around the quarter and three quarter sampling frequencies allowing IF signals to be encoded, whilst the signal transfer function is identical to that of the baseband coder except shifted to the IF position.

The open loop impulse response is still of importance in designing actual A-D converters, and is now given by:

$$OLIR(n) = \left(2 + \frac{n}{2}\right) \cdot \cos\frac{\pi n}{2} \qquad n = 0,1,2,3, \dots \tag{11}$$

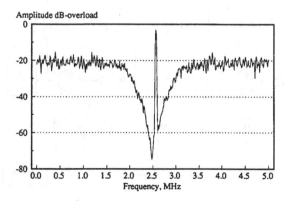

Figure 7: Output Spectrum of Bandpass Sigma Delta Modulator.

Fig.7 shows the output spectrum of the bandpass modulator for an input sinusoid at $f_s/100$ above the quarter sampling frequency. All the properties of the baseband coder are preserved but are translated to the IF position. The necessary loop filters are all that is now required before an actual A-D converter can be designed.

4. Analogue Loop Filter Design

Fig.8 shows a practical structure for a bandpass sigma-delta A-D converter. The digital loop filter of the example modulator is replaced by an analogue bandpass filter, the design of which is to be considered in this section. The single bit quantizer is replaced by a single bit A-D converter, i.e. comparator with a sampled output. The delay in the feedback path is still implemented digitally though the amount of delay introduced here is not the total required since delay already exists in the analogue elements of the circuit and the DACs will have some inherent delay between the arrival of data at their inputs and the temporal centre of the energy of their output pulses. Typically one half to one whole sample period of loop delay will be accumulated in these areas, leaving the remainder for the digital delay line. The single bit DACs must be included in the feedback path as the output of the digital delay line will not be sufficiently linear in most applications. For optimum linearity in the DACs a return to zero output pulse is required to minimize non-linear memory effects and consequently the output pulses from the DACs will be shorter than the sampling interval.

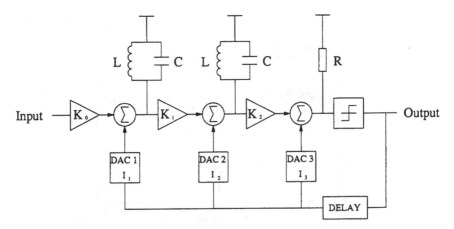

Figure 8: Practical Implementation of Bandpass Sigma Delta A-D Converter.

The key to the design of the converter lies in the design of the loop filter, and for this two options exist; switched capacitor filters and true analogue filters. For lower performance applications switched capacitor implementations will suffice and allow complete integration of the converter, however, since the analogue input is effectively sampled at the input to the converter the normal I-Q mismatch and non-linearity problems associated with sample and hold circuits will be replicated in the output of the A-D converter. For more demanding applications true analogue filters offer the best solution since the need for a sample and hold is removed and any errors introduced by the sampling process, which now takes place after the loop filter, are subjected to the same noise shaping processes which govern the quantizing noise. This paper is concerned with the latter option.

For the coder to operate correctly the response of the loop filter to both the analogue input signal and the pulsed waveform from the DACs must match that of the digital loop filter. For the analogue signal the match need only occur in and around the passband since out-of-band components are not of interest and will have been removed before the converter by filtering. This match may be achieved simply by considering the required amplitude and phase response of the filter within the passband. For the digitized signal from the DACs the match must be broadband and is very complicated if considered in the frequency domain because the cyclic repetition of the digital filters frequency response is not matched by that of the analogue filter. For this reason a time domain analysis must be used to match the pulse response of the analogue filter at the sampling instants to the impulse response of the digital filter, usually referred to as 'impulse invariant design'.

Consider again the architecture shown in Fig.8. The twin buffered parallel LC filter has virtually identical phase and amplitude responses as the digital loop filter, both within the passband and considerably beyond. The main difference is that the original digital filter offered infinite gain at the resonant frequency whereas that of the LC filters will be determined by the Q-factors. Equations (1) and (2) show that this will have little effect on the STF but will limit the extent of noise shaping at the frequencies where the mismatch occurs. In practice this implies that in the centre of the passband, approximately 1% to 2% of Nyquist in width, the noise power spectrum flattens out rather than continuing to fall as the centre frequency is approached.

The desired impulse response is achieved with three DACs. The output of DAC1 is filtered by both LC filters and is used to generate the second order component of the impulse response, whilst the output of DAC2 is filtered only once and generates the first order component. DAC3 is a correction DAC and is required to compensate for effects caused by the finite width of the pulses of the two main DACs. This can be made clearer by considering the mathematics.

Firstly the impulse response of the digital loop filter (11) must be converted to a continuous time waveform by substituting t/T for n, where T is the sampling interval of the coder. The coefficient m comes from the sampling rate to IF ratio and is given by $m = 4 f_{IF}/f_s$.

$$OLIR(t) = \left(2 + \frac{t}{2T}\right) \cdot \cos \frac{\pi tm}{2T} \qquad m = 1,3,5,... \qquad (12)$$

This then is the waveform which must be generated. Laplace analysis of the pulse waveform from the DACs and the response of the loop filters allows two expressions to be written which describe the filters pulse response both during and after the application of the pulses. If the pulse width is τ and is centred on $t = 0$ then the two equations for the open loop pulse response are

$$OLPR(t) = \left[\frac{I_1 K_1 K_2}{2\omega_n C^2}\left\{t + \frac{\tau}{2}\right\}\sin\omega_n\left\{t + \frac{\tau}{2}\right\} + \frac{I_2 K_2}{\omega_n C}\sin\omega_n\left\{t + \frac{\tau}{2}\right\} + I_3\right]R \qquad -\frac{\tau}{2} \le t < \frac{\tau}{2} \qquad (13)$$

and

$$OLPR(t) = \left[\frac{I_1 K_1 K_2}{2\omega_n C^2}\left\{\tau\cos\frac{\omega_n\tau}{2}\sin\omega_n t + 2t\sin\frac{\omega_n\tau}{2}\cos\omega_n t\right\} + \frac{2I_2 K_2}{\omega_n C}\left\{\sin\frac{\omega_n\tau}{2}\cos\omega_n t\right\}\right]R \qquad t \ge \frac{\tau}{2} \qquad (14)$$

where K_1 and K_2 are the gains of the buffer amplifiers in amps per Volt, ω_n is the IF frequency in radians per second and C and R are the values of the capacitors and the resistor shown in Fig.8. Values for I_1, I_2, K_1, K_2, C, R, and τ are selected to equate terms in $\cos\omega_n t$ and $t \cdot \cos\omega_n t$ in (14) to those in (12). Finally I_3 the current for DAC3, is selected to equate (13) and (12) at $t = 0$. The error term in $\sin\omega_n t$ found in (14) is cumbersome to remove, but by pulsing the DACs very slightly earlier than originally intended a phase shift can be introduced to compensate for this and no loss of performance is found. A more detailed analysis may be found in [5].

A loop filter has now been designed which closely matches the phase and amplitude response of the example digital filter, and an arrangement of DACs is provided to generate the desired open loop pulse response. This general approach may be used to design several different types of bandpass sigma-delta A-D converters and in the following section examples will be given of sampling, subsampling and interpolative coders.

5. Bandpass Sigma-Delta A-D Conversion

The techniques demonstrated in the previous section may be applied generally to the design of several different types of converter and three examples will now be given. The first is a standard sampling converter in which the IF is positioned at one quarter of the sampling frequency and will allow direct use of the equations given earlier. The second example is a subsampling A-D converter in which the IF is positioned at three quarters of the sampling frequency. To implement this it is necessary to adjust the ratio of first and second order components of the open loop impulse response to accommodate the reduction in oversampling ratio. The third example is of an interpolating A-D converter, similar to the subsampling converter but in which a two bit quantizer is used in conjunction with a single bit DAC to extend the dynamic range whilst maintaining the linearity advantages of single bit D-A conversion.

6. Sampling Bandpass Sigma-Delta A-D Converter

The design of this type of A-D converter, the most basic in the bandpass sigma-delta family, has already been extensively described in the preceding sections and will not be covered further. The measured performance of a prototype converter will instead be described in some detail.

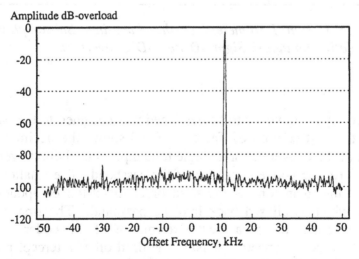

Figure 9: Decimated Output Spectrum of Sampling Coder for Single Tone Input 3 dB below overload.

A-D converters are usually characterized in terms of resolution, differential non-linearity etc., but this convention will not be adopted here since conventional analogue terms such as signal to noise ratio and third order intercept points are more relevant both to the applications of the converters and to the performance of the converters themselves. In all cases the reference power level used for measurements will be the overload point of the converter in question, and figures will be quoted in *dBO*.

Design Parameter	Sampling	Subsampling	Interpolative
IF	2.5 *MHz*	7.5 *MHz*	7.5 *MHz*
O/P word size	1 bit	1 bit	2 bits
O/P sampling rate	10 *MHz*	10 *MHz*	10 *MHz*
DAC size	1 bit	1 bit	1 bit
DAC rate	10 *MHz*	10 *MHz*	30 *MHz*
Bandwidth	100 *kHz*	100 *kHz*	100 *kHz*
Q-factor	100	300	300

Table 1: Design Parameters of the Sampling, Subsampling and Interpolative Bandpass Sigma-Delta A-D Converters.

The prototype converter to be described was designed to the parameters given in the first column of Table 1. Fig.9 shows the decimated output spectrum of the converter for a single tone input 3 *dB* below overload and positioned 10 *kHz* above the centre of the passband. The smaller signal to the left of the main tone is the third order distortion component of the converter which is aliased back into the passband. The first column of Table 2 describes the measured performance of the coder, giving the in-band noise power spectral density, the third order intercept point for a twin tone input and the spurious free dynamic range of the converter in a 1 *Hz* noise bandwidth for a single tone input.

Performance	Sampling	Subsampling	Interpolative
Noise Power Density/Hz	-117 *dBO*	-114 *dBO*	-130 *dBO*
Third Order Intercept	26 *dBO*	24 *dBO*	32 *dBO*
Spurious Free Dynamic Range in 1 *Hz* Single tone input	95 *dB*	92 *dB*	108 *dB*

Table 2: Performance of the Sampling, Subsampling and Interpolative Bandpass Sigma-Delta A-D Converters.

This performance may be compared to that of conventional 'flash' A-D converters in which conversion is achieved with a large number of comparators with uniformly spaced switching thresholds. Within the passband the noise floor is equivalent to that of an 8 bit flash A-D converter sampling at 10 *MHz*; at the decimated output it is equivalent to an 11 bit flash A-D sampling at 200 *kHz*. However the linearity and spurious free dynamic range of the sigma-delta A-D converter are very much better than equivalent flash A-D converters. With flash converters distortion products tend not to fall with decreasing signal level and typical spurious free dynamic ranges are around 55 *dB*.

7. Subsampling Bandpass Sigma Delta A-D Converter

This is an example of a converter in which the IF is positioned in the second quantizing null, i.e. at three quarters of the sampling frequency. The design procedure is very similar to that of a sampling converter except for two main differences. Firstly, in order to encode a narrow passband at three quarters of the sampling frequency analogue loop filters with enhanced Q-factors will be required compared to those of a sampling A-D converter since the IF is three times higher in this example. Consequently the Q-factors will need to be exactly three times as large in order to achieve the same in-band performance. Secondly the impulse invariant design of the loop filter must now accommodate the higher IF of the loop filters and accordingly 'm' is set to 3 in equation (12).

274

The complete set of design parameters and measured results for this converter are given in the second columns of Tables 1 and 2 respectively. Fig.10 shows the output spectrum of the prototype converter with a twin tone input 3 *dB* below the twin tone overload level. In theory the performance of the sampling and subsampling converters should be identical, and a very close match is seen in the measured results. The main cause for the fall in performance is that the required Q-factor of 300 was difficult to achieve in practice and consequently a lower value had to be accepted. This resulted in in-band noise and distortion components rising slightly and degrading the performance, however, the resulting linearity and spurious free dynamic range was still far better than that obtainable from equivalent flash converters.

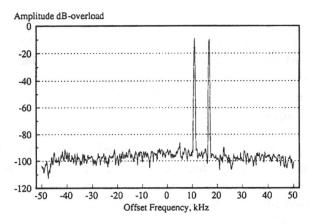

Figure 10: Decimated Output Spectrum of Subsampling Coder for Twin Tone Input 3 dB below overload

8. Interpolating Bandpass Sigma-Delta A-D Converter

The aim of the interpolating converter is to increase the dynamic range of subsampled converters by operating their DACs at four times the IF and to compliment this with extra levels in the quantizer rather than increasing the sampling rate. An interpolating look up table may be used to interface the low-speed multi-level data generated by the quantizer to the higher-speed single-bit DAC in a pulse density format.

In this example the data is interpolated by a factor of 3 to convert it from a two-bit to a one-bit data stream. With careful consideration of the phasing requirements the necessary interpolating function is found to be as shown in Fig.11. The combined response for all three pulses must then be used in the impulse invariant design of the loop filter. This will now be considered in more detail.

Interpolation Look-Up Table				
Input	Output Sequence			
3	1	-1	1	
1	1	1	1	
-1	-1	-1	-1	
-3	-1	1	-1	

2-bit data from quantizer at f_s

1-bit data to DACs at $3f_s$

Figure 11: Definition of Interpolating Look-up Table to convert 4 level data to a 2 level bit stream.

Generalizing the design of the loop filter somewhat, its pulse response to the three pulses from the DAC must take the form of:

$$K\left(P + \frac{Qn}{2}\right) \cdot \cos\frac{3\pi n}{2} \tag{15}$$

where K is a scaling factor equal to the output levels of the quantizer and must therefore take on one of the values ± 1 or ± 3. If the response of the filter to a single pulse from the DAC is given by:

$$\left(R + \frac{Sn}{2}\right) \cdot \cos\frac{3\pi n}{2} \tag{16}$$

then the composite response to the code representing $+3$ is given by:

$$\left(R + \frac{Sn}{2}\right) \cdot \cos\frac{3\pi n}{2} - \left(R + \frac{S}{2}\left(n + \frac{2}{3}\right)\right) \cdot \cos\frac{3\pi}{2}\left(n + \frac{2}{3}\right) + \left(R + \frac{S}{2}\left(n + \frac{4}{3}\right)\right) \cdot \cos\frac{3\pi}{2}\left(n + \frac{4}{3}\right) \tag{17}$$

$$= 3\left(\left(R + \frac{1}{3}S\right) + \frac{Sn}{2}\right) \cdot \cos\frac{3\pi n}{2} \tag{18}$$

whilst the composite response to the code representing $+1$ is given by:

$$\left(R+\frac{Sn}{2}\right)\cdot\cos\frac{3\pi n}{2}\ +\ \left(R+\frac{S}{2}\left(n+\frac{2}{3}\right)\right)\cdot\cos\frac{3\pi}{2}\left(n+\frac{2}{3}\right)\ +\ \left(R+\frac{S}{2}\left(n+\frac{4}{3}\right)\right)\cdot\cos\frac{3\pi}{2}\left(n+\frac{4}{3}\right) \quad (19)$$

$$=\left(\left(R+\frac{1}{3}S\right)+\frac{Sn}{2}\right)\cdot\cos\frac{3\pi n}{2} \quad (20)$$

These results are clearly of the correct form to achieve the desired response given in (15). Two aspects must be considered here. Firstly, in the interpolation lookup table I and Q separation must still be maintained and so the pulses relating to the inphase component of the signal must be time multiplexed with those relating to the quadrature component. Secondly a more detailed analysis of the circuit shows that the desired response of DAC3, the correction DAC, is now non-linear across the four coding levels, but this is simple to implement in practice.

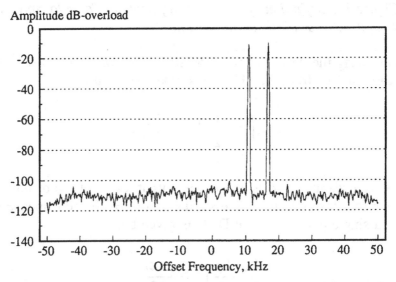

Figure 12: Decimated Output Spectrum of Simulated Interpolative coder for Twin Tone input 3 dB below overload.

The design specifications and the measured performance of the interpolative converter are given in the third columns of Tables 1 and 2 respectively whilst Fig.12 shows the output spectrum of the simulated converter for a twin tone input 3 *dB* below overload. Comparing the results with those of the subsampling coder the improved performance is soon observed. The overload point of the interpolative converter is 10 *dB* higher than that of its counterpart and this increase accounts for most of the improvement in the in-band noise power density, which is referred to the overload point of the converter. The linearity of the converter has also improved and consequently the third order intercept point has risen. Another improvement which is not shown in Table 2 is that as the overload condition is approached the in-band noise power density of the interpolative converter rises much less than in the subsampling and sampling coders. Comparing these results with those of flash converters the in-band noise performance is equivalent to 10 bits at 10 *MHz*, or at the decimated output equivalent to a 13 bit converter sampling at 200 *kHz*. Once again the linearity and spurious free dynamic range far exceed those of flash converters.

9. Implementation

All the designs presented can be implemented using discrete components. This allows easy experimentation with new ideas, though obviously an integrated version is more suitable for a final production model. Unlike switched capacitor implementations which allow the possibility of complete integration, the analogue filters discussed in this paper are not suitable for integration but do offer superior performance.

As with all analogue to digital conversion schemes using feedback, the performance of the converter is limited to the performance of the component producing the cancellation signal. The use of a single bit DAC eases the design, reducing it to a switch, thus optimising linearity. The noise floor of the DAC and hence the dynamic range of the converter is also limited by the quality of the clock signal driving it. The DAC output must possess minimum jitter at transition, needing close attention to clock phase noise and pattern-dependent charge-storage behaviour in the DAC itself.

One other factor that affects the noise floor of the converter is the Q factor of the loop filters. By separating the poles of the loop filters wider bandwidths may be encoded to a restricted dynamic range, and in this case low Q-factors will suffice for the loop filters. However, to encode high dynamic range, narrow bandwidth signals, the poles of the loop filters would typically be coincident and their Q-factors would be maximized to obtain the full dynamic range. Practically, Q-factors in excess of a few hundred are difficult to achieve and this then presents a limitation on the maximum attainable dynamic range.

For correct operation of the converter, the total open loop propagation delay, including any clock delay, must be approximately two sample periods. The effect this has on the implementation depends on the required sample clock frequency. This delay can be achieved by a combination of analogue and digital delay.

Once the input signal has been converted it is usually necessary to decimate the output data. This is not only to produce a multi-bit output, but also to reduce the data rate to one commensurate with the converters bandwidth. This allows further processing to be undertaken using general purpose devices.

Due to the speed and complexity of the decimation process an Application Specific Integrated Circuit (ASIC) is usually required. Cost and power consumption make CMOS ASICs attractive although they may not be fast enough for high sample frequencies. Operating at higher frequencies allows wider bandwidths to be encoded, but the signal processing requirements for decimation are more severe. For wide bandwidths the output from the decimator will be at a correspondingly high rate and if any further processing is required, this rate may be too high for a general purpose digital signal processor to handle. The solution would then be either complex discrete circuits or more ASICs adding considerable cost.

10. Conclusions

The sigma-delta A-D conversion technique has traditionally been applied at baseband where, by exploiting the linearity advantages of oversampled single bit D-A converters, high linearity decimated code may be produced. The technique may easily be adapted to operate at an intermediate frequency allowing high linearity conversion of IF signals without the need for pre-conversion mix to baseband I and Q components. With post-conversion mix to baseband and decimation the normal system limitations of I-Q mismatch and masking by DC offsets are virtually eliminated.

Three examples of bandpass sigma-delta A-D converter have been presented. The first is a sampling converter in which the IF was positioned at one quarter of the sampling frequency whilst the second is the subsampling equivalent with the IF positioned at three quarters of the sampling frequency. In both cases the decimated output is roughly equivalent to 11 bit PCM coding at 200 kHz but with substantially improved linearity and spurious free dynamic range compared to multi-comparator flash conversion. The third example is an interpolative converter in which extended dynamic range is achieved with two bit quantization whilst a one bit DAC operating on an interpolated pulse density code is used to achieve maximum linearity. In this case the decimated output is roughly equivalent to 13 bit pcm at 200 *kHz* but again with significantly improved linearity and with lower spurious content than can be achieved with flash converters.

The converters may be employed for any application where high linearity conversion of narrowband IF signals is required and are particularly suited to high performance radio systems.

The novel techniques described in this paper are the subject of worldwide granted and pending patents.

References

[1] T.H. Pearce and A.C. Baker, "Analogue to Digital HF Radio Receivers", IEE Colloquium on Systems Aspects and Applications of ADC's for Radar, Sonar and Communications, November 1987, Digest no.1987/92.

[2] T.H. Pearce and M.J. Jay, "Performance Assessment for Digitally Implemented Radio Receivers", IEE Conference on Radio Receivers and Associated Systems, Cambridge, July 1990, Conference Publication no.325, pp.113-117.

[3] P.J.A. Naus, E.C. Dijkmans, E.F. Stikvoort, A.J. McKnight, D.J. Holland, W. and Bradinal, "A CMOS Stereo 16-bit D-A Converter for Digital Audio", *IEEE J. Solid-State Circuits*, vol. SC-22, no.3, June 1987, pp.390-395.

[4] M.J. Hawksford, "Multi-Level to 1-bit Transformations for Applications in Digital-to-Analogue Converters Using Oversampling and Noise Shaping", *Proc. Institute of Acoustics*, vol.10, part 7, 1988.

[5] A.M. Thurston, T.H. Pearce, and M.J. Hawksford, "Bandpass Implementation of the Sigma-Delta A-D Conversion Technique", IEE International Conference on Analogue to Digital and Digital to Analogue Conversion, Swansea, September 1991, Conference Publication no.343, pp.81-86.

280

Biographies

Andrew M. Thurston received the B.A. Degree in Engineering from Trinity College, Cambridge University, in 1987, and the M.Sc. Degree in telecommunications and information systems from the University of Essex in 1989. He is currently working towards the Doctor of Philosophy Degree on the subject of bandpass sigma-delta A-D conversion at the University of Essex.

In 1987 he joined GEC-Marconi Research Centre as a research engineer in the Signal Processing Division, and is currently leader of the A-D Conversion Section. His areas of interest include oversampled conversion techniques and the design and implementation of analogue filters.

Timothy H. Pearce received the Masters Degree in semiconductor electronics from the University of Southampton in 1968. Since then he has been with GEC-Marconi Research Centre, and gained a wide range of experience in signal processing techniques, with a special interest in high performance analogue to digital conversion methods. He is currently manager of the Signal Processing Division within the Communications Research Laboratory. He is a member of the Institution of Electrical Engineers and a Chartered Engineer.

Malcolm D. Higman received the B.Eng. in Electronic Engineering from Reading University, England, in 1990. In December 1990 he joined the Signal Processing Division at the GEC-Marconi Research Centre where he is currently working on the development of high performance sigma-delta analogue to digital converters.

Malcolm Omar Hawksford is a Reader in the Department of Electronic Systems Engineering at the University of Essex, where his principal interests are in the fields of electronic circuit design and audio engineering. Dr. Hawksford studied at the University of Aston in Birmingham and gained both a First Class Honours B.Sc. and Ph.D. The Ph.D. programme was supported by a BBC Research Scholarship, where the field of study was the application of delta-modulation to colour television and the development of a time compression/time multiplexed system for combining luminance and chrominance signals.

Since his employment at Essex, he has established the Audio Research Group, where research on amplifier studies, digital signal processing and loudspeaker systems has been undertaken. Since 1982 research into digital crossover systems has begun within the group and more recently, oversampling and noise shaping investigated as a means of analogue-to-digital/digital-to-analog conversion.

Dr. Hawksford has several AES publications that include topics on error correction in amplifiers and oversampling techniques. His supplementary activities include writing articles for *Hi-Fi News and Record Review* and designing commercial audio equipment. He is a member of the IEE, a chartered engineer, a fellow of the AES and of the Institute of Acoustics and a member of the review board of the *AES Journal*. He is also technical adviser for *HFN* and *RR*.

Analog Computer Aided Design

Introduction

Analog design has become a difficulty because it takes too much time and it contains too many errors. On the other hand, system integration requires mixed-signal Analog-Digital integrated circuits. Digital tools already exist to some extent. This is not the case however for analog tools: only very few of them are around.

Can analog designers be convinced to use such tools or will they continue to rely on their own magic? For what applications would they use such tools? Are high-speed, low-noise, low-distortion circuitry ever going to be designed by means of such tools or are they going to be used only for so-called lower level tasks such as layout?

In this text, answers are tried on all these questions. In the first text, which was presented by E. Malavasi of U.C. Berkeley, a top-down design methodology is described, starting with behavioral simulation. In the second text, which was presented by R. Carley of Carnegie Mellon University, the synthesis is presented of analog cells. Circuit behavior is described by equations. The third text, which was presented by G. Beenker of Philips Eindhoven, describes several packages which allow knowledge acquisition and synthesis. Constraint-driven routing is addressed as well.

The fourth text was presented by M. Degrauwe of CSEM, Neuchatel. It describes a wide variety of commercially available tools, to assist the designer in all design tasks. In the fifth paper, which was presented by E. Nordholt of Catena, Delft, design strategies are outlined towards more systematic design. Finally in the sixth and last paper, presented by G. Gielen of the K.U. Leuven, an open design system is presented based on declarative models.

All texts provide examples to make their point. Let us hope that the efforts described will lead to design tools that benefit all types of analog designers.

Willy Sansen

A Top-Down, Constraint-Driven Design Methodology for Analog Integrated Circuits

Enrico Malavasi, Henry Chang, Alberto Sangiovanni-Vincentelli,
Edoardo Charbon, Umakanta Choudhury, Eric Felt, Gani Jusuf,
Edward Liu & Robert Neff

Department of Electrical Engineering and Computer Sciences
University of California, Berkeley
U.S.A.

Abstract

This paper describes a top-down, constraint-driven design methodology for analog integrated circuits. Some of the tools that support this methodology are described. These include behavioral simulation tools, tools for physical assembly, and module generators. Finally, examples of behavioral simulation with optimization and physical assembly are provided to better illustrate the methodology and its integration with the tool set.

1. Introduction

The analog section of a complex mixed-signal system is often small, but it is its design that often consumes the most amount of time. Unlike its digital counterpart fully automatic synthesis tools are not available. Several attempts have been made to speed up the process by automating parts of the design such as the synthesis of modules and the automatic layout of often used components. Methods for low-level physical layout generation and schematic synthesis using architectural selection paradigms have been proposed [1,2]. Methods for very specific mid-to-low level circuits have also been proposed [3]. More recently automatic support tools for the experienced designer have been developed [4]; however, none of these provide integration of all of these features and missing from them are some of what we believe to be very fundamental ones. In this paper, we describe a new design methodology and the necessary set of tools that supports it for effective analog design.

The key points of our methodology are:

- top-down hierarchical process starting from the behavioral level based on early verification and constraint propagation;

- bottom-up accurate extraction and verification;

- automatic and interactive synthesis of components with specification constraint-driven layout design tools;

- maximum support for automatic synthesis tools to accommodate users of different levels of expertise but not the enforcement of these tools upon the user; this is not an automatic synthesis process;

- and consideration for testability at all stages of the design.

The top-down process implies a well-defined behavioral description of the analog function. The behavioral characterization of analog circuits is quite different from the digital one; the analog characterization is composed of not only the function that the circuit is to perform, but also the second order non-idealities intrinsic to analog operation. In fact, errors in the design often stem from the non-ideal behavior of the analog section, not from the selection of the "wrong" functionality. To shorten the design cycle, it is essential that design problems be discovered as early as possible. For this reason, behavioral simulation is an essential component of any methodology. This simulation can help in selecting the correct architecture to implement the analog function with bounds (constraints) on the amount of non-idealities that is allowable given a set of specifications at the system level.

Constraints on performances of the selected architecture are propagated down to the next level of the hierarchy onto the components that can be designed following the same paradigm until all the leaves of the design space are reached. These leaves can either be transistors, other atomic components, or library objects. Since models have to be estimated at high levels in the hierarchy, a bottom-up verification is also essential to fully characterize components, interconnects, and parasitics.

The physical assembly of basic blocks at all levels of the hierarchy is time-consuming and rarely very creative. This step can be effectively accomplished with automatic synthesis tools. We recognize that the layout parasitics do affect the behavior of analog circuits and as such have to be controlled carefully. The amount of parasitics allowed on the interconnects is often estimated by the designer who usually is not able to guarantee the accuracy of the estimation when fitted with the final circuit. Often to guarantee proper functionality, designers will overconstrain the allowed parasitics. Proposed here is an approach where layout tools are directly driven by constraints on performances of the design components.

The testing of analog circuits requires a great deal of time as well as expensive equipment. This problem increases with the complexity of the circuits. It can be solved in part by taking into account the testing problem during all stages of the design, unlike the common practice of considering testing only after the design is finished.

Finally, we do not believe that full automation is achievable for all analog circuits. The amount of creativity and complexity needed to master the design of analog circuits is high. We do believe, however, that the creative task of the designer can and should be fully supported by a set of automatic and interactive tools that allow him/her to explore the design space with ease and full understanding of the trade-offs involved. Analytical tools play an important role in our methodology. We also maintain, though, that some components of an analog design could indeed come from module generators and some from libraries both of which embody the experiences of other designers. Thus our methodology does accommodate tools that favor design-reusability in its general framework.

This paper is organized as follows: In Section 2, the proposed methodology is presented. In Section 3, the tool set is described. In Section 4, examples are shown. And finally, these sections are followed by our conclusions in which future work is discussed.

2. The Methodology

Fig.1 shows the standard design hierarchy. At the top of the design hierarchy is a top node which could represent an entire chip or just part of a chip. From this top node there is a set of direct descendent lower nodes, representing a first level decomposition of the upper node. From these lower nodes the graph continues to be expanded. The decomposition may cease at any given node at any given level. The graph is completed with the decomposition of all of the nodes.

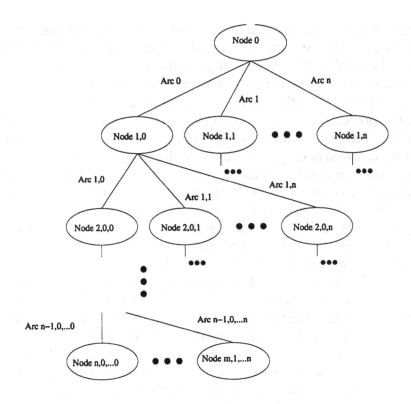

Figure 1: Design Hierarchy

An example decomposition is shown in Fig.2. Here, a mixed-mode chip is our top node. This diagram shows a way in which this top node can be decomposed. It also shows how decomposition may cease at any level. For example, a voltage reference circuit is found in our cell library. Thus, the decomposition of this node is not necessary. A solid line under the terminating node indicates this. In some cases the decomposition does not cease until the transistor or passive element level, as indicated by the capacitors and the MOSFETs in this figure.

Our methodology, illustrated in Fig.3, assists the hierarchical generation of the design by providing a rigorous procedure based on interactive and automatic tools. The large bubble encompassing most of the diagram represents the mapping. Given a set of circuit specifications (circuit characteristics, the design rules, the technology, and user options), a mapping is made to schematics or to layout.

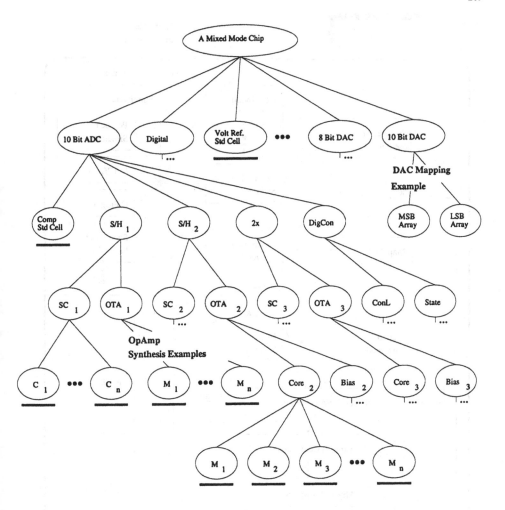

Figure 2: Example Design Hierarchy for a chip

Given a library of *n* architectures, the first operation that must be performed is architecture selection. Simulators and optimizers are used to aid the decision-making process. For very high-level blocks, a behavioral simulator may be employed. For low-level circuits a circuit simulator such as SPICE may be run. For an architecture where no simulators exist (e.g. a pre-made cell) the "simulator" could just be a list of performance specifications. If a suitable architecture cannot be found, then this selector must return to the upper node the fact that the selection has failed. This would mean that the specifications cannot be met; they must be changed in order to continue. If a standard cell (pre-designed cell) was chosen then a successful return to the upper node is made.

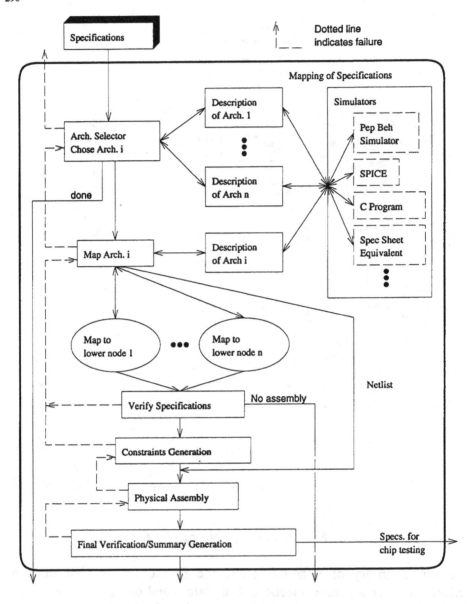

Figure 3: Design Methodology

Given specifications for a particular architecture we proceed to map the chosen architecture to the detailed specifications of the component (lower) blocks. This task can be very difficult. Architectures can be mapped automatically using non-linear optimizers; however, when this procedure fails the user must do the partitioning. If a solution to the mapping problem cannot be found, another architecture must be chosen.

The lower blocks are expanded in the same manner as this diagram depicts, thus recursively expanding the design hierarchy. From these lower blocks a set of the component specifications is returned. If the returned specifications fail to meet the criteria set by the mapping function, then the flow control is returned to the mapping function.

On a successful return from all of the components, the component specifications are compiled to form a list of specifications corresponding to the original list. If this compiled list fails to meet the expected specifications, then a new mapping is attempted. If it is successful, we have two options; we can either proceed with the layout or stop here returning only the schematics.

In creating the layout, the first step is the generation of the layout constraints for the assembly.

This can be done either manually or automatically. As before, if this step fails, the flow control is returned to the mapping function. If it is successful, then a constraint-driven physical assembly is performed. If this fails to meet the requirements then an alternate set of assembly constraints is derived. If the physical assembly is successful, then the final verification step/summary generation step is required. This step includes the extraction of the circuit and simulation with the same simulator used in architectural selection. The "extraction" process spans the entire range from a simple net-list extraction to a complete extraction with parasitics. Finally, a summary of the expected performances is generated. If all of the specifications are met, the flow control is returned to the upper node with the generated specifications.

This final block also formulates the test set, based on the extracted specifications, technology considerations, and the final layout. If the desired test coverage cannot be obtained then possible hardware modifications or alternative architecture suggestions are fed back to the mapping function or the architectural selection function, as appropriate. Since the majority of analog circuit failures is due to parametric rather than catastrophic device malfunctions, determining the test set is an essential part of the design methodology which cannot be split into a separate post-processing step. Testing is based on the same functional model as the high-level behavioral simulation. This model allows the circuit to be parametrically characterized by a minimum number of well-chosen tests. The test set is derived and ordered to minimize test expense (time and equipment) while meeting the test coverage constraints. The results of the high-level behavioral simulation are used to determine the relative value and cost of testing each behavioral component of the circuit. The test coverage constraints are formulated to include both catastrophic and parametric failures, although the latter account for the vast majority of chip failures. Test specifications and constraints are thus incorporated into the design of the circuit in a manner very similar to the manner in which performance constraints are accommodated. An appropriately optimized test set is part of the final chip specifications.

We believe that our design methodology is significantly different from the ones currently employed by circuit designers. Today, a typical design cycle starts with a set of specifications for an integrated circuit drawn in conjunction with the customer. Then the designer takes these specifications and performs a first level decomposition with little simulation. The partitioning is accomplished based mostly on the experience of the designers. Typically, today, designers resort to a bottom-up approach. Low level circuits are built, tested, verified, and assembled hierarchically from the bottom-up until the first level decomposition blocks are reached. The main problem with this design technique is that if the final blocks do not meet the specifications, the entire circuit has to be redesigned, possibly all the way from the bottom again. This can be very time consuming and costly. A typical solution to this is designing with overconstraints on the lower blocks. This, however, is also costly, because a sub-optimal solution is usually reached. Our methodology attempts to prevent these problems. We use behavioral simulations for early verification and design space exploration to guide the experience and the expertise of the designer. Each of the lower blocks is constrained as the tree is descended to match the performance specifications. Thus, at any time, we are reasonably sure that the original specifications are being met. We need not go to the bottom of our design and back up before realizing that a costly mistake has been made.

3. The Tool Set

The tool set that we present here can be divided into three categories: Behavioral Simulation Tools, Physical Assembly Tools, and Module Generation Tools. As a point of policy, we state here that all present and future tools we design will integrate into our methodology, thus providing continuity in our tool set. We hope for the adoption of our methodology by other tool writers to accelerate the process of expanding the tool set.

3.1. Analog Behavioral Simulation

The objective of behavioral simulation and modeling is to represent circuit functions with abstract mathematical models that are independent of circuit architectures or schematics. Behavioral simulation and modeling are useful because they help designers reduce design time. For example, in top-down design verification, designers can verify system design early with behavioral models of system components before investing time in detailed circuit implementation. Therefore, designers can explore the system design space rapidly. In bottom up design verifications, designers can often verify complex system behavior efficiently because evaluations of behavioral models are quick, resulting in fast system simulations. For digital circuits, behavioral modeling and simulation can be performed using hardware description languages such as VHDL or VERILOG. However, these languages do not provide any features for an adequate simulation of analog blocks in the system. Nonetheless, there is an increasing need for behavioral simulation of the analog blocks as well. One reason for this is the tendency in ASICs and VLSICs to integrate total systems, including both analog and digital blocks, onto a single chip. Another reason is that even "simple" analog blocks such as A/D and D/A converters are too complicated for a complete transistor-level simulation within reasonable amounts of CPU time. Therefore, both applications require a simulation environment in which the analog blocks can be represented by some behavioral model in order to reduce the CPU time needed for the simulation.

For the behavioral simulation of analog circuits, the following features are essential:

- The simulator and the behavioral models have to be *general*. The behavioral model of a given analog block must describe the behavior of that block considered as a black box, i.e. only describing its input-output behavior in terms of a set of model parameters to be supplied by the designer. As the behavioral model describes the analog block as a black box, it also has to hide all internal architectural details of the block as much as possible. This explains the term, *generic models*. Also, the simulation engine must operate independently of any particular models. In this way, it must be possible to simulate for instance any type of A/D or D/A converter in the same environment instead of having a different dedicated simulator for each specific architecture as is the present practice.

- The behavioral models for the analog circuits must not only include the first-order behavior of the circuit, but also the *second-order effects* in order to get a realistic idea of the performance of the overall system, including characteristics such as noise and distortion. Also, the statistical variation of the circuit performance has to be taken into account. In addition, analog circuits are sensitive to driving and loading impedances and may have "electrical" terminals where Kirchoff's voltage and current laws have to be satisfied.

- The simulation has to be performed in whatever *domain* is of interest to the designer or whatever domain is necessary to obtain the simulation results. For analog circuits, these certainly include the time and frequency domains, and ways to switch between both domains and to deal with noise, distortion, and statistical parameters. This also indicates that the degree of abstraction is different for the behavioral simulation of digital and analog circuits, i.e. the concept of "time", and the required "waveform accuracy" are different in both cases.

In order to realize the above design environment, the following research actions have been distinguished:

1. The set-up of generic behavioral models for common analog functional blocks (including the second-order effects), implementation, and testing of these models.

2. Automatic extraction of the actual parameters for the behavioral models from a given design (bottom-up path). This also allows us to investigate the performance of different architectures, in order to be sure that the generic models cover all (or as much as possible) architectures and as much as possible second-order effects in a realistic way.

3. The implementation of an efficient behavioral simulator and an analog behavioral description language, possibly based on an extension of existing digital languages such as VHDL or VERILOG, which includes all of the features and requirements discovered during the previous two actions.

Analog circuit functions are usually simple. The designers rarely choose the wrong function; instead, circuits tend to fail due to second order effects of the functions chosen. As a result, analog behavioral models must be independent of circuit architectures, yet capture all second order effects. To meet this goal, our strategy is to:

- Find the best mathematical representation for specific types of analog circuits, and develop realistic models by using appropriate differential equations, difference equations, transfer functions, statistical distributions, tables, or arithmetic expressions.

- Develop techniques to extract model parameters from lower level models that comprise the circuit, thereby building a model hierarchy shown in Fig.4.

According to the proposed strategies, we have developed a behavioral model for the class of Nyquist rate A/Ds [5]. The behavioral representation captures all the relevant A/D behavior without information concerning the actual implementation. The behavior of an A/D is affected by two basic statistical effects: noise and process variations. Noise can cause the same die to behave differently even when the same inputs are applied. Process variations can cause different dice to have different behavior. Circuit designers usually model noise effects by adding an input-referred noise to the input of an ideal, noiseless A/D. This approach is problematic because the A/D is nonlinear. So, noise cannot be referred to the input in general. They usually model process variations effects by several parameters such as offset error, gain error, integral nonlinearity (INL), and differential nonlinearity (DNL). This approach is inefficient because these parameters are not totally independent. The traditional CAD approach to A/D behavioral modeling is to model the A/D transfer function which maps a continuous input value to an output code. However, this approach is problematic because it does not represent noise effects. Furthermore, evaluating the transfer function is time consuming because it usually depends on a large number of circuit elements each with process variations. In our approach, all the statistical variations are captured in the model [5]. The variations are classified into noise and process variations according to how these non-idealities affect the A/D behavior. To describe noise effects we use a joint probability density function. To describe behavioral effects due to process variations we use a covariance matrix, Σ_t, whose rank characterizes the testability of an A/D; its decomposition yields efficient strategies for A/D testing.

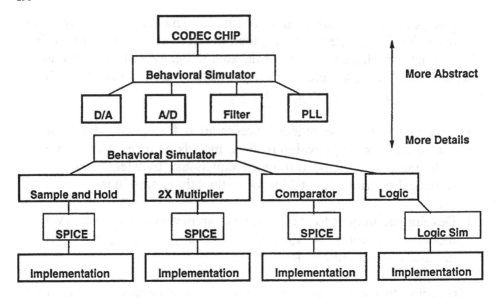

Figure 4: Model Hierarchy of Mixed Signal System

To have a useful behavioral simulation environment, we need to develop models for most of the high-level analog functions. To this end, we developed behavioral representations for voltage-controlled oscillators (VCO) and detectors that are essential circuit components in any phase-lock system. VCOs can be classified into two classes: sinusoidal VCOs and square wave VCOs. Sinusoidal VCOs have a near sinusoidal output signal and include crystal and LC tuned oscillators. In contrast to traditional macromodels [6] that do not model distortion or phase jitter, the proposed sinusoidal VCO model captures distortion effects by a nonlinear dynamic block represented by Volterra series, which is a generalization of the power series expansion for distortion analysis. Phase jitter effects are represented by a random phase parameter. On the other hand, square wave VCOs have a square wave output and include relaxation and ring oscillators. In general, the VCO architecture consists of m delay elements connected in a ring structure with m outputs. As the output is a square wave, we model the output by its zero-crossings (low-to-high and high-to-low transition times). The zero-crossings have both a deterministic component due to the input voltage control and a random component due to the jitter caused by noise in the delay elements. With this model, both first order and second order effects such as phase jitter are captured.

Another essential element in any phase-lock system is the detector which compares two signals (input waveform and VCO waveform). Traditionally, circuit designers represent a detector by its phase characteristic which plots the average product of the input waveform and the VCO waveform as a function of the phase difference between the two waveforms. One problem with this modeling approach is that the phase characteristic is defined only when the input waveform and the VCO waveform have the same frequency. As a result, this approach cannot be used during phase-locked loop acquisition. Another problem is that the phase characteristic depends on the shape of the waveforms, so the characteristic is not defined until input signals are applied. Therefore, the phase characteristic is not appropriate as a behavioral representation. Behaviorally, detectors are classified into two broad categories: multipliers (zero memory circuits) and sequential circuits (with memory). In our approach, we represent sequential circuits by state machines with analog inputs, a and b, analog outputs $g(a,b)$, and clock triggered by zero crossings of a and b. Moreover, the same representation can be used for multipliers if they are considered as state machines with a single state and no clock. In general, all detectors can be conveniently represented by such a state transition table, so the representation is general and independent of the circuit architecture.

In addition to developing analog behavioral models, we investigated parameter extraction techniques to identify the models from a circuit description or from measured data. For Nyquist rate A/D converters we developed a novel noise extraction technique. Noise effects depend on parameters such as capacitor sizes and transistor thermal and flicker noise, and pose as fundamental limits on converter resolutions. For example, due to noise the probability of getting a wrong A/D output is non-zero. Currently, circuit simulators such as SPICE cannot evaluate noise effects in A/D converters because the transfer function of A/D converter is discontinuous. One solution is the Monte Carlo technique that injects random numbers into the signal path of an A/D during a time-domain simulation. Due to the injected random numbers in the signal path, the output code may become incorrect by chance. The drawback of such statistical technique is that the error rate of A/D converters is very low. As a result, a large number of samples is needed. We have developed a technique based on probability theory to compute the error probability directly from power spectral density descriptions of electronic noises. The advantage is that the algorithm is efficient for A/D with low error rate and can handle high correlated noise such as flicker noise.

3.2. Physical Assembly

Digitally targeted tools are often inadequate to handle the critical and specific requirements of analog layout. The performances of analog circuits are much more sensitive to the details of physical implementation than the digital ones. Custom design requires great flexibility, and layout synthesis is often a multiple-objective optimization problem, where, along with area, wiring length and delay, two groups of relevant issues must be taken into account:

1. topological constraints, i.e. symmetries and matching;

2. parasitics associated with devices and interconnections.

If the synthesis is performed with no explicit reference to the performance specifications of the circuit, a large number of time-consuming layout-extraction-simulation iterations may be necessary to meet the original specifications.

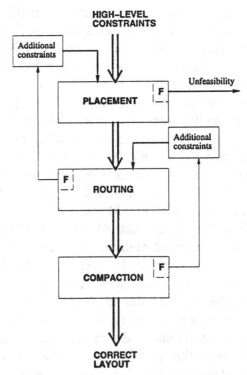

Figure 5: Performance-driven synthesis of analog layouts

Our constraint-driven approach for the layout of analog ICs consists of a direct mapping of performance specifications to a set of bounds on parasitics, which can then be controlled during the synthesis. High-level performance constraints are expressed in terms of the maximum degradation allowed on each performance with respect to its nominal value. Degradations are due to parasitics, i.e. stray resistances, capacitances toward substrate, and cross-coupling capacitances. Therefore, during the mapping process the sensitivity of performances with respect to the parasitic elements have to be considered.

The layout synthesis is a complex process consisting of several phases as depicted in Fig.5. The main phases, i.e. placement, routing, and compaction, have been designed to meet the constraints on parasitics and to account for topological constraints. Feedback is provided at all levels of the synthesis, thus providing a mechanism to transmit a failed specification mapping to the previous level so that a resynthesis attempt can be made.

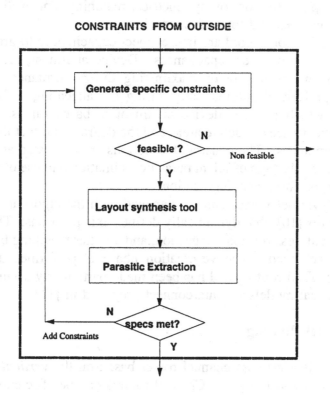

Figure 6: Internal structure of analog layout tools.

A general flow diagram valid for any of our layout tools is described in Fig.6. A set of specific parasitic constraints is generated by PARCAR [7] and used during the synthesis phase to drive the layout towards a configuration, which is likely to meet the performance specifications. The synthesis phase is followed by an extraction/verification phase that checks if the performance specifications have been met. Feedback paths are provided to allow partial or total resynthesis if necessary.

In what follows the layout tools and their applications are described. In Section 4.2 an example is shown to prove the suitability of the approach.

3.2.1. Placement

The optimization algorithm used in this tool is Simulated Annealing [8] and has been implemented in a tool called PUPPY [9]. The objective is not only area and wiring minimization, but also the elimination of parasitic constraint violations. The cost function to be minimized by the annealing accounts for parasitic control, symmetries, matching, and well separation as well as chip area and total wire length.

The parasitics considered are interconnect capacitances to ground, stray resistances, and crossover capacitances. Device abutment, often used by expert designers as a way of maximizing the performances of analog circuits, is performed automatically during the annealing. The decision mechanism which governs device abutment is based on estimations of diffusion capacitances of active areas and the degradation that these induce in the performances. The constraint violations are evaluated at every step of the annealing through efficient parasitic estimation and use of models for the resulting performance degradation.

Multiple symmetry axes can be managed with the algorithm of *Virtual Symmetry Axes* [10], that dynamically defines axis positions. Thus a more compact layout, less routing congestion, and a better matching between the modules is achieved. The verification phase is performed after either placement or final routing and has been made particularly accurate by use of the analytical models for interconnect proposed in [11].

3.2.2. Channel Routing

ART [12], is a gridless channel router based on the *vertical-constraint graph* (VCG) algorithm [13]. Channel routers can be efficiently used for detailed routing in the case where layout placement is generated to form a slicing structure and routing channels are isolated. The VCG is a graph whose nodes represent the horizontal wiring segments and whose edges represent constraint relations between the nodes. In the classical problem formulation the edges can be of two types. An undirected edge links two nodes if the associated segments have a common horizontal span. A directed edge links two nodes if one segment has to be placed above the other because of pin constraints. The weight of an edge is the minimum distance between the center lines of two adjacent segments. The routing problem consists of the minimization of the longest directed path in the VCG.

In the new constraint-driven formulation additional parasitic line coupling, i.e. crossover between orthogonal and adjacent parallel wiring lines, is controlled by means of an insertion in the VCG of additional directed edges and modification of the weights of existing edges. Unavoidable crossovers can be determined directly by the terminal placement on the top and bottom edges. The contribution of such crossovers cannot be reduced, and therefore, must be considered for the computation of lower bounds for the parasitics in the constraint generation phase. Coupling between adjacent lines that cannot be sufficiently spaced is reduced by the insertion of lateral or vertical shielding conductors.

Symmetric mirroring of interconnect nets is achieved by constraining their vertical position to the same value. Special connector configurations are used to guarantee good parasitic matching between nets crossing each other over the symmetry axis. The same number of corners and vias and the same interconnect length are maintained for both wires.

3.2.3. Area routing

ROAD [14] is a maze router based on the A* algorithm [15], on a relative grid with dynamic allocation. The A* algorithm is a heuristic improvement of the Lee-Moore algorithm [16]; in the wave propagation step only one element of the wave front is propagated at a time. The propagated element is the one minimizing an estimate of the length of a wire constrained to pass through it.

A cost function is defined for each net on the edges of the grid. The path found by the maze router is the one minimizing the sum of the values of the cost function. In ROAD, the cost function is a weighted sum of several items, thus expressing a multiple target optimization problem. The items considered are the local wire crowding, stray resistances, and capacitances, cross-coupling capacitances introduced by wire segments.

Weights define the relative criticality of each item in the cost function. Hence, the quality of the layout depends dramatically on the way weights are defined. In our approach, performance sensitivities to parasitics are

302

used to generate the weights for the cost function. After the weight-driven routing phase, parasitics are extracted and performance degradations are estimated and compared with specifications. If for some performance the specifications are not met, the weights of the most sensitive parasitics are increased and routing is repeated. If the weights of all the sensitive parasitics cannot be increased further, the tool returns an unfeasibility message for the layout with the given set of constraints.

3.2.4. Compaction

SPARCS [17] is a mono-dimensional compaction tool which combines two methods, one based on the *constraint graph* (CG) and a second on *linear programming* (LP). The compaction process is partitioned into two phases. In a preliminary step the parasitic constraints are mapped onto a graph which is solved using the longest path algorithm [18]. The configuration obtained from the solution of the CG is used as starting point for the linear program. New constraints accounting for symmetries are introduced in the linear program which is solved using the simplex method.

The longest path algorithm, originally implemented for digital compaction [19,20], has been modified to account for parasitics. Control over cross-coupling capacitances is enforced by adding directed edges to the original graph, so as to maintain minimum distances between pairs of critically coupled wiring segments. In geometrically complex wiring structures spacing constraints are distributed among all parallel segments forming the wires according to the relative length of the segments and whether they belong to the critical path. In case of overconstraints introduced by the spacing operation, the set of minimum distance constraints are adjusted by reducing the minimum distance between overconstrained pairs and by increasing the minimum distance between non-critical pairs. The resulting set of new constraints is more likely to pose a feasible problem to the compactor. If no legal and/or feasible set of minimum distance constraints are found, additional shielding lines are introduced.

The LP approach of the second step uses the graph solution of the previous step as a starting point, but introduces a set of additional constraints to account for device and wiring symmetries. The resulting linear program is solved through the simplex algorithm and the locations of the elements not lying in the critical path are rounded off, thus eliminating the need for use of a much slower integer programming algorithm.

3.3. Module Generation

Unlike digital circuits, analog circuits in general require more design freedom in order to be applied effectively. They often exploit the full spectrum of capabilities exhibited by individual devices. In an analog circuit the individual devices often have substantially different sizes and electrical characteristics. These circuits require optimization of various performance measures. As an example, among the performance measures for operational amplifiers are gain, bandwidth, noise, power supply rejection, dynamic range, offset voltage, etc. The importance of each performance measure depends on *circuit applications*.

For this reason, a performance-driven module generator is a better approach. Performance based module generators can be implemented using an expert system approach or a parameterized schematic of known architecture approaches. Expert system implementation for simple analog blocks such as opamps and comparators have been reported in earlier years [21]. Unfortunately, switched-capacitors and A/D converters are much more complex functions than opamps or comparators. There is no clear way as how to set the rules for this implementation so that the resulting ADCs are of any use at all.

Unlike the expert system implementation, a parameterized schematic of known architecture implementation is a better approach, because it can generate highly-effective circuits based on known architectures for complex analog functions in a flexible environment by optimizing the device sizes with respect to various performance measures [22,23,24]. This approach fits nicely into our methodology with minor modification.

In the remainder of this section, we will briefly discuss the existing module generators, OPASYN (opamp synthesis) [22], ADORE (switched-capacitor synthesis) [25], and CADICS [24] (ADC synthesis), all of which were developed at Berkeley.

3.3.1. Opasyn

To generate the lowest level of an analog block such as an Operational Amplifiers, OPASYN [22], an automatic synthesis tool for the generation of opamps based on analytic circuit models can be used. Given a set of specifications, this program performs the necessary optimizations to properly perform transistor sizing. It outputs this information to a layout generator, and a performance summary is also presented. Originally, this program only considered the generation of operational amplifiers.

However, it has been expanded so that it is now capable of generating comparators as well as output buffers.

OPASYN synthesizes layouts from specifications in two steps. In the first step, a circuit topology is selected based on the specifications, and the circuit is optimized to meet the required specifications. For the selected topology, design parameters are selected to reduce the complexity of the circuit synthesis problem, without reducing the range of possible results. For example, parameters may define the sizing of several devices or a bias current used throughout the circuit. The circuit specifications can be computed as a function of the design parameters, through a set of equations which have been derived beforehand. A typical computation may be the frequency of the dominant pole in an opamp. In this case the design parameters are used to compute small signal models of the devices in the circuit, and these small signal parameters may be used to express the location of the dominant pole. OPASYN can use these equations to optimize the parameters to solve a constrained optimization problem, meeting all required specifications, and minimizing a cost function. The cost function may be a combination of area and power.

For layout generation OPASYN uses leaf cell generators, to make the building blocks required for the layout. A floorplanning algorithm using slicing trees places the devices, and then custom routing and layout spacing strategies finish the opamp layout [26].

3.3.2. Adore

ADORE is a switched-capacitor filter module generator that places a special emphasis on the physical design aspect of module generation. ADORE separates the switched-capacitor filter design problem into a filter synthesis problem and a layout problem. The FILSYN [27] program is used to automate the design of a switched capacitor filter, using methods and making decisions as a filter designer might. In the layout phase, ADORE uses a sophisticated algorithm to generate the capacitor arrays required, and then uses standard opamp and switches to complete the layout. Opamps, switches, and capacitors are ordered to optimize the wiring required.

This program is currently being modified to better fit into our design methodology and to expand its capabilities such as implementing fully differential switched-capacitor filter architectures.

3.3.3. Cadics

Unlike switched capacitor circuit synthesis [25,28], A/D converter synthesis does not yet have a well-defined procedure. The module generation of ADC functions is a particularly difficult problem because of the wide spectrum of resolutions and sampling rates encountered in real applications, the tradeoff of area versus power dissipation, and component matching requirements for the different ADC architectures. The approach we plan to pursue for the module generation of the ADC function, is to focus on *techniques* that allow design of circuits with efficiency comparable to the ones manually designed by skilled designers.

Because of the wide range of parameter spreads in analog integrated circuits, analog designers over the years have developed circuits which cancel out the first order variations in key parameters. This means, however, that the analog circuits then become sensitive functions of second order variations of such parameters, for example the matching of input devices in differential pairs, or capacitor matching in a gain of two block. We try to use as much analog circuit design techniques as possible to further reduce or overcome this second order effects, for example auto-zeroing circuit [29], trim capacitor array [30], and fully differential architecture so that the module generation procedures can be simplified.

In the ADC function, the realm of required performance maps into four distinct classes of realizations spanning the conversion rate spectrum (integrating, oversampling, successive approximation, and flash/pipeline architectures), and two distinct regions of accuracy/linearity (less than 10 bits and greater than 10 bits). Each of these requires a different implementation in order to be reasonably competitive with hand-crafted designs in terms of power dissipation and silicon area. Therefore, if *not carefully planned*, ADC synthesis can require the development of *an extremely large number* of such module generators.

The number of required generators can be reduced by careful selection of the architectural approach. For example, the use of binary-weighted capacitor approaches [31] requires that a complex self-calibration controller be added to the module for resolutions above 10 bits. The technique also does not scale gracefully to video speed or to very high resolution. In contrast, the use of oversampling techniques [32,33] at the low speed end, digitally-corrected algorithmic techniques in the mid-range [30,34], and pipelined techniques at the high speed end [35] potentially allow spanning of the entire speed spectrum with basic elements based on the same differential switched capacitor integrator/amplifier block.

A performance-driven CMOS ADC module generator named CADICS [24] has been developed. Given a set of specifications and all the necessary files such as technology, design-rule, device-matching, and floorplan, CADICS will generate complete netlist as well as layout of CMOS ADCs. The implementation of this ADC module generator is divided into two parts: circuit synthesis and layout synthesis.

The circuit synthesis takes as its inputs a set of specifications for an ADC such as resolution, conversion-speed, maximum input signal frequency, maximum differential non-linearity (DNL), maximum integral non-linearity (INL), total power dissipation, supply voltage, reference voltage, area, layout aspect-ratio range, and technology. In addition to these specifications, a user can also provide statistical information for capacitor and transistor matching. The *core* circuit synthesis which is the device sizing generation can be implemented using *a global* or *hierarchical* optimization approach.

A global optimization approach attempts to optimize an ADC circuit all at once. An example of this approach is to use a circuit optimization program such as DELIGHT.SPICE [36]. Unfortunately, this is a time consuming approach and often there will be convergence problems since it requires repetition of SPICE runs. Another method is to formulate the ADC performances in closed-form equations as a function of electrical characteristics of individual devices. This approach is known as an optimization based on a standard schematic approach [22,23]. This approach is suitable for small and simple analog circuit blocks such as opamps, comparators, or output buffers since the closed-form equations for this blocks can be formulated. ADCs, on the other hand, consists of many functional blocks and some of its performances such as DNL and INL have no closed-form equation so that a straight approach like opamp synthesis cannot be applied.

Recognizing that ADC circuits are realized using different functional blocks such sample/hold blocks, gain-of-two circuits, comparators, digital circuitry, and others, we chose to implement the core of circuit synthesis of CADICS by using a hierarchical optimization approach. This approach is desirable because it optimizes the ADC performances by doing local optimization for each subblock individually. As a result, this circuit synthesis step will consume very little time. Another advantage for this approach is the expansion capability of using other available low-level synthesis tools to optimize low-level subblocks and of allowing incorporation of standard cell blocks if desired. The flow chart of circuit synthesis with the hierarchical optimization procedure is shown in Fig.7.

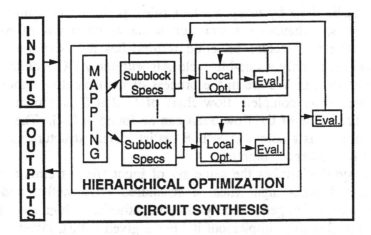

Figure 7: Hierarchical Optimization Procedure

Referring to Fig.7, for a given set of specifications, circuit synthesis undergoes a sequence of operations. First, the ADC specifications and its respective relative priorities will be mapped into the subblock specifications and their respective priorities. This mapping function can be realized by a set of rules given by experienced analog designers or by running a behavioral simulation of the ADC circuit or a combination of both. Then the specifications and priorities of each subblock are fed into the local optimizer to generate the device sizes. This local optimizer can be a low-level synthesis block such as OPASYN to generate opamps or LAGER [37] to generate digital circuitry. One can also use library cells.

When all the subblocks have been optimized locally, the ADC is evaluated by calling a behavioral simulator to obtain the ADC performance. A behavioral simulator is included, because device level simulation like SPICE requires too much time when simulating the entire ADC circuit. Since CADICS predated the development of the general behavioral model and simulator described previously, it still uses a special purpose simulator consisting of discrete time, macro-models of each functional block. If the performance is not good enough, the entire operation will be repeated by relaxing the given specifications until the desired performances are obtained or the iteration limit of optimization cycles is exceeded.

The operation described above is referred to as ADC Netlist Module Generation (NMG). It is architecture-specific with the only exception that the optimization routine is shared by all of the NMGs. Therefore, if there is more than one ADC architecture NMG stored in the data base, the circuit synthesis needs a routine to select which ADC architecture to be generated depending on the given set of specifications. The architectural selection process can be done by using a set of rules, performances summaries of previous runs, exhaustive search, or some combination of all three. Currently, the architecture selection routine has not been implemented completely since only a one-bit-cycle Algorithmic ADC architecture has been implemented. We are in the process of implementing another ADC architecture. The complete flow-chart of CADICS is shown in Fig.8. Detailed discussion of the flow-chart can be found in [24]. The output of circuit synthesis is a complete SPICE netlist, device structure information files, and performance summary.

Layout synthesis takes the same set of input files as circuit synthesis with two additional input files: a structured ADC netlist and device structure information generated by circuit synthesis. A hierarchical layout procedure is chosen to implement it. For a given netlist, layout synthesis carries out a sequence of operations such as netlist rearrangement, circuit partitioning, floorplan optimization, leaf-cells generation, subblock place and route, and global place and route to generate the complete layout. It is a mixed top-down and bottom-up approach [38].

In CADICS, we do customized circuit partitioning with a fixed floorplan, because of its suitability for this architecture. The circuit partitioning of the ADC circuit is reflected in the netlist by using subcircuit definition available in SPICE. A slicing structure floorplan design is chosen here so that Stockmeyer's algorithm [39] can be used to determine the optimal transistor slicing structure for a given area and aspect ratio.

Analog circuit performance depends on *fine details of layout*. As a result, many analog layout techniques to improve the performance are implemented in layout generation. To isolate capacitors from noisy signals, capacitors are placed on the top of wells connected to analog ground line. Depending on the floorplan and netlist, well and substrate contacts will be created during placement. Fully differential architecture and symmetry in layout are maintained. Subblock routing is done with ROAD [14], a constraint-driven analog maze-router, and currently we are incorporating ART [12], a constraint-driven analog channel router. The global routing and pads placement are carried out by MOSAICO [40]. The complete flow chart of the ADC synthesis is shown in Fig.8. Notice that the area optimizer is being shared by the circuit and layout synthesis. The reason is that we allow the user to modify the resulting netlist generated by circuit synthesis. If the netlist is modified, then the area optimization needs to be carried out again in order to determine the shape of the leaf-cells, subblocks, and the entire ADC layouts.

309

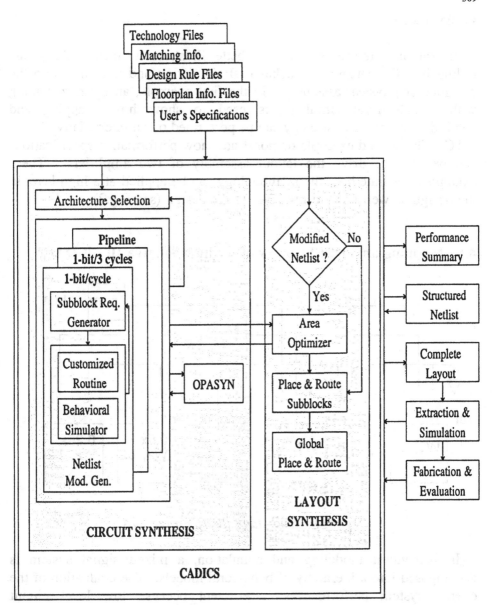

Figure 8: Flow-Chart of CADICS

4. Examples

Two examples are presented here to better illustrate our methodology, one falling into the category of behavioral simulation and the other into the category of physical assembly. In the first example, an optimizer along with a behavioral simulator is used to show how mapping and determination of a test strategy can be performed on an interpolative 10 bit DAC. The second example demonstrates how performance specifications can be used to drive the layout assembly of two amplifiers. These examples illustrate how the methodology can be applied at a high level in the design as well as at a low level in the design (see Fig.2).

4.1. Mapping and Test Strategy of an Interpolative 10 Bit DAC

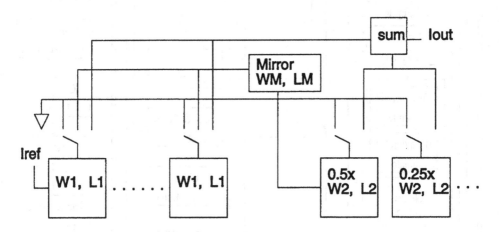

Figure 9: Two stage interpolative D/A

In behavioral modeling and simulation, a mixed signal system is decomposed into a hierarchy of behavioral models. The evaluation of the overall system performances is efficient because low level circuit simulations are not necessary. Rather, the system performances are predicted directly from the performances of the behavioral models of the components. To illustrate the efficiency of this approach to computing overall system specifications such as INL and DNL of a high resolution D/A, a simple example of performance-driven hierarchical synthesis [41] is briefly described. In the following example, numerical optimization with behavioral simulation is used for high level architecture selection, as opposed to worst case analytical analysis in [31] or ruled based systems

[42]. We used an existing optimizer to find the optimal resolutions of the stages of a two stage, interpolative 10 bit D/A (Fig.9) for a performance constraint of $\pm 3\sigma$ INL and DNL less than 2 and 0.5 lsb, respectively, where lsb refers to the smallest output step. The D/A behavioral model consists of two non-ideal stages and a non-ideal current mirror connecting the first to the second. The first stage, with N_1 bits of resolution, consists of 2^{N_1} non-ideal, equally weighted transistors, and the second stage consists of N_2 non-ideal, binary weighted transistors. The transistors are abstract representations of the real current sources, and they contain all referred mismatches. The transistor dimension, W/L, is constrained to be larger than four for bias requirements, and the referred mismatches, $\sigma_L = 0.02\mu m$ and $\sigma_W = 0.02\mu m$, are assumed to be independent for all transistors. From optimization runs using a DEC 5000, the total active area is minimal for a first stage with resolution of 5 bits. Furthermore, it can be shown by Monte Carlo integration that nearly all of the realizations of its INL and DNL will fall within the $\pm 3\sigma$ INL, DNL bounds shown on Fig.10.

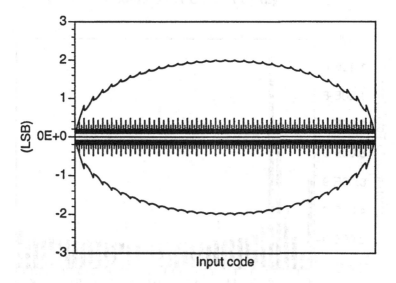

Figure 10: $\pm 3\sigma$ INL, DNL bounds

Fig.11 shows a magnified section of the DNL bounds and clearly shows the typical DNL magnitude characteristics of the second stage's binary weighted architecture.

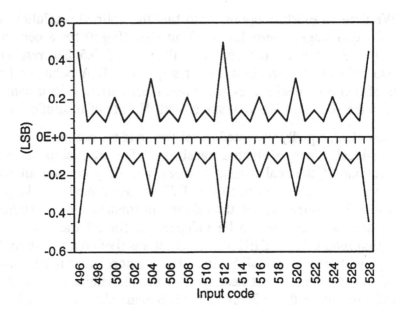

Figure 11: ± 3σ DNL bounds

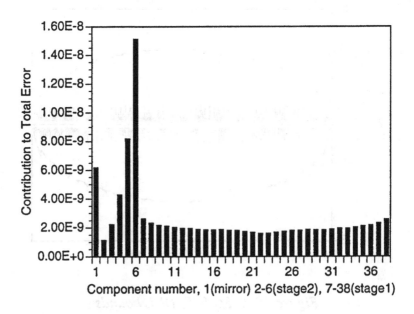

Figure 12: Relative importance of component errors

The behavioral simulation results are also used for testing and fault diagnosis. To minimize test time, it is desirable to minimize the number of tests required to obtain the desired test coverage and to order the test set such that the components that are expected to cause the most errors are tested first. This goal is accomplished for the example circuit by computing the cross-correlation between each of the components and the A/D INL error vectors. The result of such computation is shown in Fig.12, where the relative importance of each of the current sources is plotted. It is observed that the system performance is most sensitive to the largest binary weighted current source. The testing information conveyed through Fig.12 is used throughout the design process to ensure that the components to which the output is most sensitive can be both exercised and observed and that the testing time is minimized within the test coverage constraints.

4.2. Physical Assembly

4.2.1. An Operational Amplifier

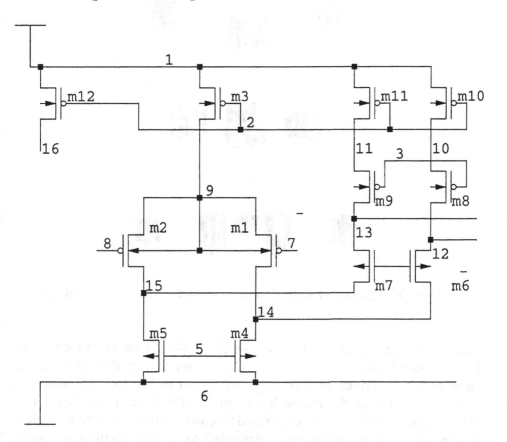

Figure 13: Schematics of the op amp used as example

314

Fig.13 shows the schematic of a folded cascode opamp. The amplifier was placed and routed imposing constraints on three performances, *Low Frequency Gain*, *Phase Margin*, and *Unity Gain Bandwidth*. The performance constraints, listed in Table 1, were translated onto bounds of parasitics by PARCAR. Topological constraints were also considered during the process. The resulting layout is shown in Fig.14.

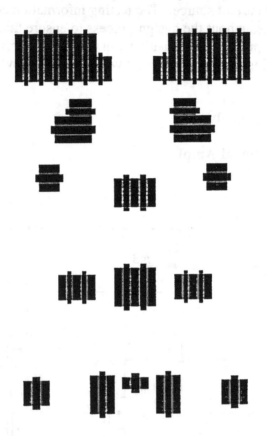

Figure 14: Placement with Topological and Parasitic Constraints

After performing the placement, PUPPY predicted no violations in the imposed constraints. The circuit was then routed with ROAD using the same constraint-driven methodology, as shown in Fig.15. All parasitics were extracted from the routed layout and a SPICE based verification on the performance degradation confirmed that no violations occurred. Table 1 shows the resulting performance degradations of the circuit synthesized with and without enforcement of the specifications.

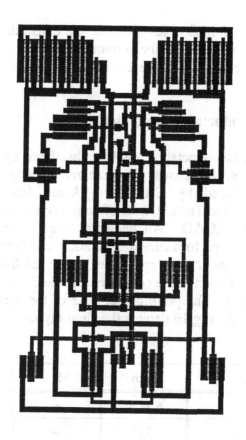

Figure 15: Complete Layout

Performances	Specs.	Not Enforcing Specs	Enforcing Specs.
Bandwidth	50 *kHz*	200 *kHz*	35.0 *kHz*
Phase Margin	0.5°	2.0°	0.46°
Gain	0.5 *dB*	0.3 *dB*	0.3 *dB*

Table 1: Specifications and the Resulting Performance Degradations of an Operational Amplifier synthesized using different techniques

The place and route procedure required 1600 seconds of CPU time on a DECsystem 3100. The results are consistent with an analysis of the layout. The most critical nets, 10, 11, 12, 13, 14, 15, (see Fig.13) have been carefully minimized in length when the constraints were introduced. Also no crossings were necessary to route these nets. In addition abutment was used in four cases to remove constraint violations.

4.2.2. A Transconductance Amplifier

Similarly a transconductance amplifier has been synthesized using the same methodology. Transistor elements and biasing were automatically generated and placed by CADICS. Again from the performance constraints, reported in Table 2, a set of bounds on parasitics was generated and used to drive ROAD. All parasitics were extracted and a similar verification on the performance degradation confirmed the absence of violations. The layout was compacted using SPARCS and again the performance degradations were computed. Table 3 shows the results of the compaction process. The performance data of the synthesized layout are reported in Table 2. Fig.16 depicts the final layout.

Performance	Min	Max	Result
Bandwidth	6 MHz	∞	6.06 MHz
Phase Margin	75°	∞	76°
Gain	60 dB	66 dB	63.8 dB

Table 2: Transconductance Amplifier Synthesis Results

	original	compacted
number of instances	39	39
area	102 480	79 800
area compression	100 %	77.9 %
wiring length	7333	5883
CPU time	-	56.3 sec

Table3: Transconductance Amplifier Compaction

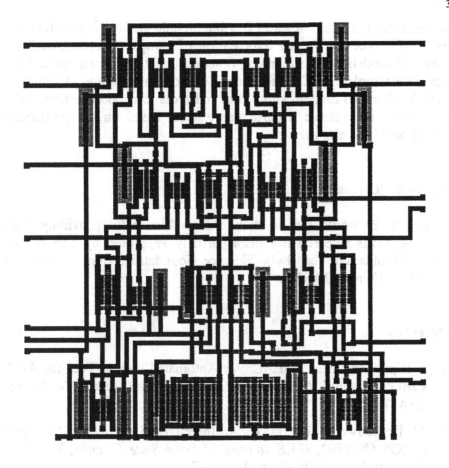

Figure 16: Final Layout of the Transconductance Amplifier

5. Conclusion

A novel design methodology has been presented for analog circuit design. The methodology is aimed at reducing substantially the design cycle of complex analog and mixed analog-digital systems while maintaining or even improving the performance of the final design. This goal is achieved through the use of a hierarchical approach where early verification and constraint propagation is favored. High-level behavioral simulation is used for immediate high-level verification of circuit performances and used for supporting the constraint-driven layout tools. The tools presented are a first cut at the implementation of the design methodology advocated here. Future work includes the use of more optimization tools such as ECSTASY [43] to improve the performance of a given architecture; a new set of flexible module generators to offer a wide variety of choices in the implementation of complex systems; additional behavioral models for basic analog functions, such as filters, phase-locked loops, oversampled ADCs, and phase detectors; a complete mixed-signal, mixed-level mixed-domain simulator; and tools to support testing such as one which will evaluate a particular test set based on the probabilistic description of the circuit.

Acknowledgements

The authors would like to thank Micro Linear for providing design examples. This research has been partially supported by The National Science Foundation; the Swiss Science Foundation, Bern, Switzerland; SRC; MICRO; and DARPA.

References

[1] J. Rijmenants et al., "ILAC: An Automated Layout Tool for Analog CMOS Circuits", *IEEE Journal of Solid State Circuits*, Vol.24, no.2, pp.417-425, April 1989.

[2] M. Degrauwe et al., "IDAC: An Interactive Design Tool for Analog CMOS Circuits", *IEEE Journal of Solid State Circuits*, Vol.SC-22, No.6, pp.1106-1116, December 1987.

[3] J.M. Cohn, D.J. Garrod, R.A. Rutenbar and L.R. Carley, "KOAN/ANAGRAM II: New Tools for Device-Level Analog Placement and Routing", *IEEE Journal of Solid State Circuits*, Vol.26, no.3, pp.330-342, March 1991.

[4] S.W. Mehranfar, "A Technology-Independent Approach to Custom Analog Cell Generation", *IEEE Journal of Solid State Circuits*, Vol.26, n.3, pp.386-393, March 1991.

[5] E. Liu, A. Sangiovanni-Vincentelli, G. Gielen and P. Gray, "A Behavioral Representation for Nyquist Rate A/D Converters", in *Proc. IEEE ICCAD*, pp.386-389, November 1991.

[6] Mark Sitkowski, "The Macro Modeling of Phase Locked Loops for the Spice Simulator", *IEEE Circuits and Devices Magazine*, Vol.7, No.2, pp.11-15, March 1991.

[7] U. Choudhury and A. Sangiovanni-Vincentelli, "Constraint Generation for Routing Analog Circuits", in *Proc. Design Automation Conference*, pp.561-566, June 1990.

[8] S. Kirkpatrick, C. Gelatt and M. Vecchi, "Optimization by simulated annealing", *Science*, Vol.220, n.4598, pp.671-680, May 1983.

[9] E. Charbon, E. Malavasi, U. Choudhury, A. Casotto and A. Sangiovanni-Vincentelli, "A Constraint-Driven Placement Methodology for Analog Integrated Circuits", in *Proc. IEEE Custom Integrated Circuits Conference*, May 1992.

[10] E. Malavasi, E. Charbon, G. Jusuf, R. Totaro and A. Sangiovanni-Vincentelli, "Virtual Symmetry Axes for the Layout of Analog IC's", in *Proc. 2nd ICVC, Seoul, Korea*, pp.195-198, October 1991.

[11] U. Choudhury and A. Sangiovanni-Vincentelli, "An Analytical-Model Generator for Interconnect Capacitances", in *Proc. IEEE Custom Integrated Circuits Conference*, pp.861-864, May 1991.

[12] U. Choudhury and A. Sangiovanni-Vincentelli, "Constraint-Based Channel Routing for Analog and Mixed-Analog Digital Circuits", in *Proc. IEEE ICCAD*, pp.198-201, November 1990.

[13] H.H. Chen and E.S. Kuh, "A Gridless Variable-Width Channel Router", *IEEE Trans. on CAD*, Vol.CAD-5, n.4, October 1986.

[14] E. Malavasi, U. Choudhury and A. Sangiovanni-Vincentelli, "A Routing Methodology for Analog Integrated Circuits", in *Proc. IEEE ICCAD*, pp.202-205, November 1990.

[15] G.W. Clow, "A Global Routing Algorithm for General Cells", in *Proc. Design Automation Conference*, pp.45-51, 1984.

[16] C. Lee, "An algorithm for path connections and applications", *IRE Trans. Electron. Computer*, Vol.EC-10, pp.346-365, Sep 1961.

[17] E. Felt, E. Charbon, E. Malavasi and A. Sangiovanni-Vincentelli, "An Efficient Methodology for Symbolic Compaction of Analog IC's with Multiple Symmetry Constraints", in *subm. to the European Design Automation Conference*, 1992.

[18] A.R. Newton, "Symbolic Layout and Procedural Design", in *Design Systems for VLSI Circuits*, pp.65-112. De Micheli et al. Eds., Martinus Nijhoff, 1987.

320

[19] J.R. Burns and A.R. Newton, "SPARCS: A New Constraint-Based IC Symbolic Layout Spacer", in *Proc. IEEE Custom Integrated Circuits Conference*, pp.534-539, 1986.

[20] J.L. Burns, "Techniques for IC Symbolic Layout and Compaction", Memorandum UCB/ERL M90/103, UCB, November 1990.

[21] R. Harjani et al., "A Prototype Framework for Knowledge-Based Analog Circuit Synthesis", in *Proc. Design Automation Conference*, pp.42-49, 1987.

[22] H.Y. Koh, C.H. Séquin and P.R. Gray, "Automatic synthesis of operational amplifiers based on analytic circuit models", in *Proc. IEEE ICCAD*, pp.502-505, 1987.

[23] M.G. DeGrauwe et al., "An Analog Expert Design of Analog Integrated Circuits", in *Proc. IEEE International Solid-State Circuits Conference*, pp.212-213, 1987.

[24] G. Jusuf, P.R. Gray and A. Sangiovanni-Vincentelli, "CADICS - Cyclic Analog-To-Digital Converter Synthesis", in *Proc. IEEE ICCAD*, pp.286-289, November 1990.

[25] H. Yaghutiel, A. Sangiovanni-Vincentelli and P.R. Gray, "A Methodology for Automated Layout of Switched-Capacitor Filters", in *Proc. IEEE ICCAD*, pp.444-447, 1986.

[26] H.Y. Koh, C.H. Séquin and P.R. Gray, "Automatic Layout Generation for CMOS Operational Amplifiers", in *Proc. IEEE ICCAD*, pp.548-551, November 1988.

[27] G. Szentirmai, "FILSYN - A General Purpose Filter Synthesis Program", in *Proc. of the IEEE*, pp.65:1443-1458, October 1977.

[28] D. Lucas, "Analog Silicon Compiler For Switched Capacitor Filters", in *Proc. IEEE ICCAD*, pp.506-513, November 1987.

[29] M.G. DeGrauwe et al., "A Micropower CMOS Instrumentation Amplifier", in *IEEE Journal of Solid State Circuits*, June 1985.

[30] H. Ohara et al., "A CMOS Programmable Self-Calibrating 13-bit 8-channel Data Acquisition Peripheral", *IEEE Journal of Solid State Circuits*, pp.362-369, December 1987.

[31] P.E. Allen and P.R. Barton, "A Silicon Compiler for Successive Approximation A/D and D/A Converters", in *Proc. IEEE Custom Integrated Circuits Conference*, pp.552-555, 1986.

[32] M.W. Hauser et al., "MOS ADC-Filter Combination That Does Not Require Precision Analog Components", in *Proc. IEEE International Solid-State Circuits Conference*, February 1985.

[33] B.H. Boser and B.A. Wooley, "The Design of Sigma-Delta Modulation Analog-To-Digital Converters", *IEEE Journal of Solid State Circuits*, pp.1298-1308, December 1988.

[34] G. Jusuf and P.R. Gray, "An Improved 1-Bit/Cycle Algorithmic A/D Converter", Memorandum UCB/ERL M90/69, University of California at Berkeley, August 1990.

[35] S.H. Lewis, "Video-Rate Analog-To-Digital Conversion Using Pipelined Architectures", Ph.d. thesis, University of California at Berkeley, November 1987.

[36] W. Nye, D.C. Riley, A. Sangiovanni-Vincentelli and A.L. Tits, "DELIGHT-SPICE: An Optimization-Based System for the Design of Integrated Circuits", *IEEE Trans. on CAD*, Vol.7, n.4, pp.501-519, April 1988.

[37] R.W. Brodersen et al., *Users' Manual*, Dept. of EECS, University of California at Berkeley, 1990.

[38] G. Jusuf, P.R. Gray and A. Sangiovanni-Vincentelli, "A Compiler for CMOS Analog-To-Digital Converters", in *TECHCON*, pp.347-350, October 1990.

[39] L. Stockmeyer, "Optimal Orientation of Cells in Slicing Floorplan Designs", *Information and Control*, Vol.57, No.2/3, pp.91-101, May 1983.

[40] J. Burns, A. Casotto, M. Igusa, F. Marron, F. Romeo, A. Sangiovanni-Vincentelli, C. Sechen, H. Shin, G. Srinath and H. Yaghutiel, "Mosaico: An integrated Macro-cell Layout System", in *VLSI '87, Vancouver, Canada*, Aug 1987.

[41] H. Chang, A. Sangiovanni-Vincentelli, F. Balarin, E. Charbon, U. Choudhury, G. Jusuf, E. Liu, E. Malavasi, R. Neff and P. Gray, "A Top-down, Constraint-Driven Design Methodology for Analog Integrated Circuits", in *Proc. IEEE Custom Integrated Circuits Conference*, May 1992.

[42] L.R. Carley et al., "ACACIA: The CMU Analog Design System", in *Proc. IEEE Custom Integrated Circuits Conference*, 1989.

[43] J.M. Shyu and A. Sangiovanni-Vincentelli, "ECSTASY: A new Environment for IC Design Optimization", in *Proc. IEEE ICCAD*, pp.484-487, November 1988.

Biographies

Enrico Malavasi received the "Laurea" degree Summa cum Laude in Electrical Engineering from the University of Bologna, Italy, on Dec. 12, 1984. In 1986 he started working at the Dept. of Electrical Engineering and Comp. Science (DEIS) of the University of Bologna on research topics related to CAD for analog circuits. In 1989 he joined the Dept. of Electrical Eng. and Comp. Science of the University of Padova, Italy, as assistant professor.

In 1990 and 1991 he was at the University of California, Berkeley as a visiting scholar and graduate student. In collaboration with the research group of Prof. Sangiovanni-Vincentelli, he worked on performance-driven CAD methodologies for analog layout. His research interests include several areas of analog design automation: layout, design methodologies, optimization and circuit analysis.

Henry Chang received his Sc.B. degree, Magna cum Laude, with honors, in Electrical Engineering from Brown University in 1989. Since Fall 1989, he has been working towards his MS degree and Ph.D. in Electrical Engineering at U.C. Berkeley. In Summer 1991, he worked at Micro Linear in San Jose, California pursuing his research.

His research interests include analog design methodologies, optimization of analog ICs, and module generators for analog ICs.

As an undergraduate student, during Summer 1988, he was a recipient of a Westinghouse Research Grant. He is currently being supported by a National Science Foundation Graduate Fellowship.

Alberto Sangiovanni Vincentelli is a Professor of Electrical Engineering and Computer Sciences at the University of California at Berkeley where he has been on the Faculty since 1976. He holds a Dr. of Eng. degree from the Politecnico di Milano, Italy in 1971. He was Visiting Scientist at the IBM T.J.Watson Research Center in 1980 and Visiting Professor at MIT in 1987. He has held a number of visiting professor positions at the University of Torino, University of Bologna, University of Pavia,

University of Pisa and University of Rome. He is a corporate Fellow at Harris and Thinking Machines, has been a director of ViewLogic and Pie Design System, and on the Technical Advisory Board of Cadence Analog Division and Synopsys, companies that he helped funding. He has consulted for several major US, (including ATT, IBM, DEC, GTE, NYNEX, General Electric), European (Alcatel, Bull, Olivetti, SGS-Thomson, Fiat), and Japanese companies (Kawasaki Steel). In 1981 he received the Distinguished Teaching Award of the University of California. He has received three Best Paper Awards (1982, 1983, and 1990) and a Best Presentation Award(1982) at the Design Automation Conference and is the co-recipient of the Guillemin-Cauer Award (1982-1983), the Darlington Award (1987-1988), and the Best Paper Award of the Circuits and Systems Society of the IEEE (1989-1990). He has published over 250 papers and three books in the area of CAD for VLSI. Dr. Sangiovanni-Vincentelli is a fellow of the Institute of Electrical and Electronics Engineers and has been Technical Program Chair of the International Conference on CAD for 1989 and the General Chair for 1990.

Edoardo Charbon received the Diploma in Electrical Engineering from the Swiss Federal Institute of Technology (ETH) at Zurich with a thesis on Statistical Estimation Theory of Noisy Signals. In 1989 he worked as a research assistant at the Micro-Sensor and Actuator Research Lab at ETH where he designed A/D converters for instrument applications. In 1989, he joined the University of California at San Diego, where he earned a M.S. degree in Communication Theory and Systems. He also worked at the University of Waterloo, Canada, where he designed nano-Tesla devices. Since 1991 he has been with the University of California at Berkeley, where he is pursuing his PhD. His present research activities include Analog IC Layout Synthesis, Parasitic Extraction and Modeling and Ultra-Low Noise Circuit Design for Sensor Applications.

Umakanta Choudhury received his B.Tech degree in Electrical Engineering from Indian Institute of Technology, Kharagpur in 1986. He received his M.S. degree in Electrical Engineering from the University of California at Berkeley in 1988, and is currently working towards his Ph.D. degree. He worked during the Summer of 1988 at HP, Santa Rosa on interconnect modeling for high-frequency circuits. His research interests include CAD for mixed analog/digital and high frequency circuits. He is the recipient of Tong Leong Lim Predoctoral Prize from University of California at Berkeley in 1988.

Eric Felt graduated from Duke University in 1991 with a B.S.E. in electrical engineering, Summa Cum Laude, with Distinction. He is currently working towards his M.S. and Ph.D. degrees in electrical engineering at the University of California, Berkeley. His research interests include analog layout compaction and analog testing. As a first year graduate student he received a California MICRO fellowship.

Gani Jusuf received the BSEE degree with high honors and the MSEE degree from U.C. Berkeley, in 1986 and 1990 respectively. Currently, he is a doctoral candidate at U.C. Berkeley. His research interests include data communications, telecommunications, and signal processing IC design particularly in analog/digital interface in VLSI technology, CAD for mixed analog/digital IC design, and device and technology modeling for analog/digital circuit design. Since 1987, he has been with Electronics Research Laboratory as a graduate student researcher. In 1991, he was a visiting lecturer teaching a Microelectronics Devices and Circuits course at U.C. Berkeley.

Edward Liu received the BSEE degree with highest honors from U. C. Berkeley in 1987, and the MSEE degree from Stanford University in 1988. In 1989, he was with VLSI Technology, Inc as an IC Designer. In Summer 1990, he was with Cadence Design Systems, Inc as a software engineer. Since Fall 1989, he has been a doctoral candidate at U.C. Berkeley. His research interests include analog behavioral simulation and modeling, data converter design, and noise analysis for mixed mode sampled-data systems. As an undergraduate student, he received the Fong Award and the EECS Departmental Citation for academic excellence in his junior and senior years, respectively. As a designer at VLSI Technology, Inc, he received an innovation award for a CMOS comparator design. As a graduate student, he received fellowships from Stanford University, California MICRO, and the Semiconductor Research Corporation.

Robert Neff received the B.S. in Engineering, with distinction, from Swarthmore College in 1985. He received the M.S. in Electrical Engineering from the University of California, Berkeley, in 1987. From 1987 to 1989 he worked for IBM Corp, in San Jose, California, designing mixed analog digital circuits for a disk drive servo system. Since 1989 he has been studying toward his Ph.D. degree at the University of California at Berkeley. His research interests include analog/digital conversion calibration techniques, and synthesis of analog/digital converters.

Analog Cell-Level Synthesis using a Novel Problem Formulation[1]

L. Richard Carley, P.C. Maulik,
Emil S. Ochotta, and Rob A. Rutenbar

Department of Electrical and Computer Engineering
Carnegie Mellon University
Pittsburgh
U.S.A.

Abstract

A novel formulation of the cell-level analog circuit synthesis problem is presented in this chapter. This novel formulation is characterized by the use of *encapsulated device evaluators*, the use of *dc-free biasing*, and the use of *Asymptotic Waveform Estimation* to compute low order pole/zero models of small signal transfer functions. This synergistic combination of elements makes it possible to design high performance analog circuits using high accuracy device models while requiring greatly reduced expert designer input in the form of equations describing circuit behavior.

1. Introduction

The ever increasing complexity of integrated circuits (ICs) leads inevitably in the direction of having complete systems on a single application specific IC (ASIC). However, because the world is analog in nature, an ever increasing number of ASIC designs will incorporate some amount of analog circuitry. Unfortunately, as many companies in the ASIC design business have found, it is often the analog portion of the design which consumes a majority of the design resources and which limits the

[1]This work was funded in part by the National Science Foundation under grant MIP-8657369, the CMU-NSF Engineering Design Research Center, Semiconductor Research Corporation under Contract 91-DC-068, and the International Business Machines Corporation.

total design time. In order to cope with the ever increasing demand for new analog integrated circuit designs, in the face of a limited supply of analog design engineers, it is desirable to develop analog CAD tools that can increase the efficiency of analog circuit and system designers. This chapter focuses on the development of a new problem formulation for cell-level analog circuit synthesis - the task of going from performance specifications for analog circuit cells (e.g., opamps and comparators) to sized transistor schematics. Analog synthesis tools can leverage the designer's time both by automating tedious numerical calculations and by acting as a repository of circuit-specific design knowledge.

Development of our new formulation of the cell-level analog circuit synthesis problem was driven by the desire to synthesize high performance analog circuits, while minimizing the amount of expert designer input required. This formulation strikes a compromise between simulation-based optimization approaches and equation-based optimization approaches, drawing elements from both. The three primary ideas that distinguish this problem formulation are: (1) it uses simulation quality device models, (2) it does not required full dc solution of every proposed circuit, and (3) it uses an efficient numerical technique to evaluate linearized circuit transfer functions.

Section 2 provides a brief review of several approaches to solving the analog circuit synthesis problem. In section 3, the new formulation of the analog circuit synthesis problem used by OBLX is presented in more detail. In section 4, we briefly describe OBLX, which uses simulated annealing to solve the new problem formulation. In section 5, we demonstrate that high performance analog CMOS circuits can be automatically synthesized with very little expert designer input.

2. Background

Most methods of analog circuit synthesis rely on having some underlying method to predict the performance of a circuit as a function of the design variables; e.g., device sizes and bias currents. Thus, the two distinguishing characteristics of most analog synthesis systems are how performance is predicted and how the values of the design variables are determined. One approach to performance prediction is to use a circuit simulator. [8,4,17] Simulation-based optimization has a long history. [8] Brayton et al. [3] contains a good survey. The accuracy of circuit performance predicted by simulation is excellent, and is presumably limited only by the accuracy of the device models and the inherent limitations of the simulation algorithm.

Unfortunately, due to the large computational cost of circuit simulation, only a limited number of points can typically be explored in the design space, the space whose dimensions are the design variables. Therefore, if not provided with a good initial guess for the design variables (e.g., device sizes), the optimization may fail to converge to a good solution, even when one exists. For analog circuits, simulation-based optimization methods are typically used to improve an existing design, rather than to determine the design variables from scratch.

Recent approaches to solving the cell-level analog circuit synthesis problem have generally employed some form of analytical equations to predict the approximate behavior of both analog circuits and devices - much like the approach of human circuit designers. In many cases these analytical equations are partially solved symbolically, which decreases the number of independent variables that must be solved for by optimization techniques. The computational cost of evaluating these approximate analytical equations is extremely low, making it possible to explore a large number of points in the design space. Methods based on using analytical equations to evaluate performance have been able to successfully synthesize analog circuits without requiring any starting point information (e.g., [2, 16, 9, 1, 6, 7, ,13]).

There are two barriers to using equation-based approaches for cell-level analog circuit synthesis: acquiring tractable equations that accurately describe the device characteristics and acquiring tractable equations that accurately describe the circuit behavior. All of the above equation-based synthesis tools employ device model equations that are simpler than the device models used in circuit simulation. In addition, as device dimensions continue to scale downward, analytically tractable device model equations for small devices become even more difficult to develop and verify, increasing the likelihood that synthesized designs will not meet the required performance specifications. For example, a majority of published results to date have given synthesis examples in 2μ or 3μ CMOS technologies. [16, 9, 2, 7] Although simulation-based optimization techniques can be used to refine a design generated using an equation-based approach (e.g., [2]), the initial design generated using equation-based methods may or may not serve as a reliable starting point for simulation-based optimization because of model inaccuracies, and more importantly, the expert designer must still provide the equations.

The difficulty in acquiring tractable equations that accurately describe the circuit behavior is exacerbated because they must be acquired separately for each desired circuit topology, unlike device model equations which need only be acquired for each fabrication process. Frequently, tractable equations will only accurately describe circuit behavior over a limited portion of the possible design space. Therefore, substantial time is required of an expert analog circuit designer in order to create a set of equations that describes a new circuit topology. This is a major barrier to the practical engineering applications of equation-based synthesis. One approach to partially automating the acquisition of these equations is symbolic analysis or symbolic simulation. [7] However, to date, symbolic simulation techniques have been limited to linearized circuits or circuits with only mild nonlinearities. [7,18]

3. OBLX Formulation

Our new formulation of the cell-level analog circuit synthesis problem strikes a compromise between simulation-based optimization approaches and equation-based optimization approaches, drawing elements from both. This formulation, like other optimization-based approaches, encompasses a numerical optimization tool wrapped around a method for performance prediction. What distinguishes the new formulation is a novel method for predicting performance and the fact that the node voltages of the circuit being designed are included in the set of design variables. There are three cornerstones to the new analog synthesis formulation: the use of *encapsulated device evaluators*, the use of *dc-free biasing*, and the use of *Asymptotic Waveform Estimation* (AWE) [15]. In the next three subsections, we describe each of these three components of the new formulation in more detail and in the final subsection we will put them all together to describe how performance is predicted.

3.1. Why Encapsulated Device Evaluators

To design high performance circuits, we must be able to accurately evaluate the performance characteristics of these circuits, such as phase margin, which can depend on small signal device parameters in a very complex manner. In order to achieve this goal we concluded that we needed very accurate device models. In order to handle the complexity of very accurate device models, we treat them just as circuit simulators do, as black boxes which we call *encapsulated device evaluators*. All knowledge about device behavior is embedded within the *encapsulated device evaluator* which makes the synthesis approach independent of the specific device type and specific technology, and enables accurate design of high performance analog circuits in advanced technologies such as short-channel CMOS. [10]

A benefit of the above approach is that no equations relating to the devices must be provided by an expert designer. In addition, with device equations encapsulated, migration of an existing design to a new technology requires only a new encapsulated device evaluator. Finally, the encapsulated device evaluator can be derived directly from a circuit simulation (e.g. SPICE) model - which means that developing device models for synthesis in a new technology is trivial as long as accurate circuit simulation device models exist for that technology. For example, the results presented in Section 5 make use of the BSIM MOS device model from SPICE3. [5]

In general, circuit simulator device models determine the device currents and a complete small signal model from information about the device geometry and node voltages. For example, a performance prediction for the $Gain(x)$ of an MOS common-source amplifier would require that we provide the encapsulated device evaluator with the widths (W's) and lengths (L's) of the devices, and all of the circuit node voltages (V's). If we let the vector x be the set of (W's, L's, and V's), then the gain might have the following form:

$$Gain(x) = g_{m1}(x) \left(r_{o1}(x) \parallel r_{o2}(x) \right)$$

where $g_{m1}(x)$, $r_{o1}(x)$, and $r_{o2}(x)$ are the transconductance and output resistances respectively of the transistors in a circuit. These small signal parameters of the transistors are in turn dependent on the dimensions of the transistor and its bias voltages; i.e., they are dependent only on the vector x. Therefore, they can be determined by calling an encapsulated device evaluator with x as the argument.

3.2. Why the DC-Free Biasing?

Unfortunately, the idea of encapsulated device evaluation is in direct conflict with the approach for finding dc node voltages adopted in most equation-based synthesis formulations. In order to write equations that predict a circuit's performance, it is necessary to first predict circuit voltages and currents, from which the device small signal models can be determined. Prediction of approximate circuit voltages and currents can be done symbolically when using tractable equations to represent the device I-V characteristics. However, a symbolic solution for determining the node voltages is impossible when using encapsulated device evaluators because the designer is not expected to know the device I-V characteristics.

The simulation-based approach to determining node voltages is to solve for a dc equilibrium of the nonlinear circuit using a nonlinear equation solving technique (e.g., Newton-Raphson). Unfortunately, each dc solve can require a large amount of computational time. In addition, the nonlinear equation solver may not find an answer. Solving for the dc equilibrium is one of the more difficult aspects of circuit simulation. If we were to adopt the simulation-based approach we might as well just use a circuit simulator to predict performance.

A key strategy of the new formulation is to avoid doing a full dc solution for each performance evaluation, without requiring any knowledge about the device characteristics. Insight as to how this might be done comes from noting that precise solution of the dc problem may not be necessary to predict performance. In fact, as long as node voltages are not too much in error (less than a few 10's of mV), the device currents and small signal model parameters will be near their correct values. This observation led us to consider having the numerical solver not only solve for the device sizes, but also solve for the circuit voltages. This avoids the need for the expert designer to formulate dc bias equations.[10] In addition, distributing one dc solve over the entire optimization process, instead of performing a full dc solve on each circuit whose performance is being predicted, makes it practical to explore a large portion of the design space using an acceptable amount of computer time.

Note, it is easy to switch from using only node voltages to using branch currents, node voltages, or any mixture of the two which forms an independent basis set. This may be desirable for bipolar transistors in which a small change in base-emitter voltage causes a large change in collector current. However, for compatibility with circuit simulator device models, we have employed only node voltages to date.

It is very important to remember that the actual node voltages are unknown; they are determined by the optimizer. Therefore, early in the design process, the sum of the currents from all devices attached to a given node may not equal zero, a violation of KCL, because the optimizer does not determine the precise dc equilibrium voltages for each set of device sizes. Envision a circuit in which every node has an ideal current source attached; and the current flowing in each current source is set to make the sum of all currents into that node equal zero. In this vision, one of the goals of the optimization process is to drive the values of all of these imaginary current sources to zero, as well as meeting all other performance constraints and maximizing (or minimizing) the objective function.

If information is to be gleaned from predicting the performance of a small signal model derived at these approximate node voltages, the KCL violating currents must be limited in size. In this case performance predictions will be accurate enough to guide device sizing. Maintaining a limit on the size of the KCL violating currents is one of the tasks that the optimizer must perform. In addition, as a final answer is approached, the KCL violating currents must be driven to approximately zero and the performance predictions will then become more precise. One method which we have adopted that facilitates keeping the node voltages near their correct dc solution is to have a desired operating region (e.g., saturation or forward active) for each device, and to limit voltage excursions using these operating region constraints.

3.3. Why AWE for Computing Small Signal Performance?

As mentioned above, in order to accurately predict performance characteristics of high performance analog circuits, such as phase margin which can depend on many poles and zeros, we must have an accurate and efficient method for solving small signal transfer functions. A major innovation of this approach to analog circuit synthesis is its use of direct numerical solution for analysis of linearized circuits. In our experience, deriving highly complex equations to predict linear circuit performance (e.g., an equation to predict phase margin) by hand can be both tedious and error prone. Using direct numerical solution therefore, both decreases the amount of expert designer knowledge required and improves the accuracy of the performance predictions.

Symbolic simulation [7] is an alternative to direct numerical solution, but the current state of the art may not be sufficient for many high performance analog circuit synthesis tasks. If realistic device models are used, symbolic expressions can become very complex; e.g., over 10,000 terms for a circuit with 10 devices. If the device models themselves are pruned before symbolic analysis, the resulting expressions may lack the accuracy required for high-performance designs. If the original expressions are pruned symbolically (as can be done using ISAAC [7] around a given design point x, then the result is usually only accurate in some limited range about x. Therefore, we chose to use a direct numerical solution method.

Analysis of linearized circuits is based on an efficient numerical solution technique, AWE [14]. AWE employs a moment matching strategy to fit a low order rational transfer function to the response of an arbitrary linear system. Although this is in its own way an approximation, the approximation is made at each design point, and so the errors can be controlled quite accurately. The use of AWE to evaluate small signal transfer functions greatly reduces the required computer time over direct solution of the small signal matrix at each desired frequency making it practical to explore a large portion of the design space.

The CPU-intensive step in AWE is the LU factorization of the small signal matrix. However, we can exploit the LU factored matrix a second time in order to facilitate the convergence of the KCL violating currents to zero. Using the linearized circuit and the LU factors it is straightforward to compute changes in node voltages that would be used in the first step of a gradient-descent dc solution method. Various optimizers can generally make use of this information in one way or another. This is an example of the synergy between the various components of the new formulation.

The use of AWE to handle linearized transfer functions of all types greatly frees the expert analog designer, when creating a synthesis tool for a new circuit topology, from the need to derive, manipulate, invert, or guide the symbolic pruning of expressions for small signal transfer functions. This is particularly advantageous for high performance analog designs in which a number of poles and zeros may play a complicated role in determining performance.

3.4. Predicting Performance

Fig.1 illustrates how, in general, analog circuit performance is predicted. We start with the current design variable vector, $\mathbf{x}$, which contains all of the device sizes (e.g., W's and L's) and the current estimate of each node voltage (V's). Using the circuit topology, the node voltages can be converted trivially into relative terminal voltages for each device and the *encapsulated device evaluator* can then be called to compute terminal currents and small signal model element values (the top box in Fig.1).

The KCL violating currents are determined directly from the topology of the circuit by simply summing all of the currents into each node. While this is not itself a performance measure, they must be computed during performance prediction because the optimizer must drive the values of the KCL violating currents to zero.

Any desired transfer function can be computed by placing the small signal model for the circuit into a linear test circuit and calling AWE, which will create a rational approximation to that transfer function with p poles and $(p-1)$ zeros. p can be selected by the user or determined automatically by AWE based on the "goodness" of the fit. Performance measures such as gain, phase margin, unity gain frequency, -3 dB frequency, input referred noise, power supply rejection ratio, common mode rejection ratio, etc. can all be determined by simple manipulations of the results of AWE analyses.

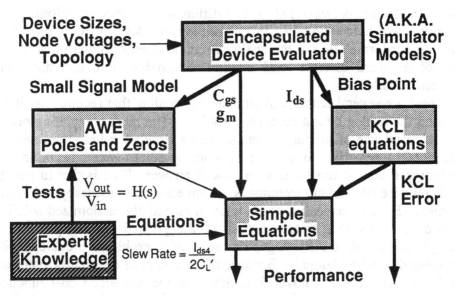

Figure 1: Block diagram showing how analog circuit performance is predicted.

Some performance specifications are difficult to predict by any of the above methods, notably those involving transient responses (e.g. slew rate) or large signal behavior (e.g. output swing). Therefore, the expert designer must still provide equations. These equations can be expressed in terms of any of the above information; i.e., small signal parameters, the dc bias information, or the results of AWE analyses.

4. Solving the Optimization Problem

We have explored two different optimization approaches to solving the new formulation of the analog circuit synthesis problem. Our first approach used a nonlinear constrained optimizer that employed sequential quadratic programming [10] (though in this case small signal transfer functions were predicted using expert designer provided equations). Making the KCL violating currents equal to zero was simply included as a constraint along with performance constraints. In order to aid in the search for the feasible region, we added each constraint sequentially, starting with the KCL constraint. We have created automatic synthesis tools using this method for over 40 different opamp topologies, 4 different comparator topologies, and a sample-and-hold topology. In this section, we describe an approach to solving our formulation of the analog circuit synthesis problem that makes use of a mixed discrete- and continuous-variable simulated annealing optimizer. [14]

There are several features of simulated annealing that recommend it for solving the analog circuit synthesis problem. The most compelling one is its independence of starting point. Existing simulation-based approaches are notably sensitive to the initial starting point, generally failing to converge if started too far from a "good" answer. This is due in part to their reliance on gradient techniques, which easily become trapped in local minima. Because annealing starts with an essentially randomized solution (during the high-temperature regime of the gradual cooling process) it completely avoids the starting point sensitivity problems prevalent with gradient-descent techniques. Because of the nature of the analog synthesis problem formulation, a novel formulation of the annealer's cost function and move generation mechanism was required. Specifically, we have a mixture of continuous (e.g., node voltages) and discrete variables (e.g., device sizes) which must be optimized. Further, the weights for the various terms in the cost function must be modulated dynamically with temperature in order to control the KCL violation terms and satisfy the performance constraints while optimizing the objective function.

The heart of the annealer is the cost function which is used to give a scalar value to the "cost" or "value" of every design point visited. In the case of OBLX the cost function has the following form:

$$cost(x) = w_o\, E_{obj}(x) \;+\; w_k(T)\, E_{kcl}(x) \;+\; w_s(T)\, E_{spec}(x)$$

where the optimizer must find x to minimize E_{obj} such that E_{kcl} and E_{spec} equal zero. E_{obj} is the object function of the minimization. It can be any combination of desired goals; e.g., a weighted sum of device area and settling time. E_{spec} is the cost of not meeting a performance specifications. Note, once a performance specification is met, its contribution to this term goes to zero. E_{kcl} is the cost associated with not satisfying KCL in the circuit. For accurate performance prediction, this error must be driven to zero. Note that the weights w_k and w_s are explicitly shown as functions of temperature, T. In order to satisfy the requirements that E_{kcl}, and E_{spec} be driven to zero it is necessary to modulate these weights as a function of the temperature during annealing. Fig.2 presents an example of the dynamic weight modulation scheme. The weights are adjusted to keep all three of the terms in the cost function comparable in size. However, the weights w_k and w_s have a minimum value, and once they reach that minimum, the associated cost terms are then driven to zero by the annealing action. For a more detailed description of this simulated annealer and more results [13].

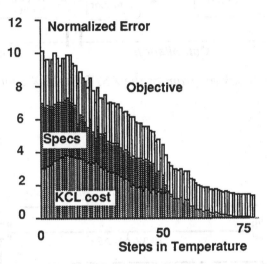

Figure 2: Relative size of terms in the cost function during the design of a fully differential opamp. The vertical axis represents the normalized cost of each of E_{obj} E_{kcl} E_{spec} and it is unitless. The x axis corresponds to the number of steps (i.e., temperature changes) since the start of the annealing run.

5. Synthesis Examples

In addition to the simulated annealer, OBLX, that solves the new formulation of the analog synthesis problem, we have implemented a front end interface/compiler called ASTRX that takes an unsized circuit topology in the form of a SPICE netlist and device model files and generates the code required to compute all of the various cost terms used by OBLX (see Fig.3).

In this section we present examples of the application of ASTRX and OBLX to the synthesis of the high slew rate fully-differential opamp topology shown in Fig.4. Note, this is a reasonably complex circuit, and a hand design using this topology has been published. [11,12]

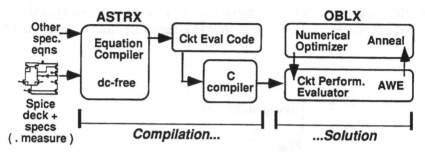

Figure 3: Block diagram of the ASTRX/OBLX synthesis tool.

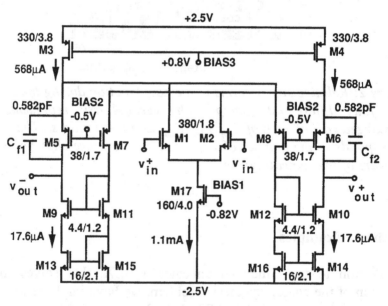

Figure 4: Schematic diagram of fully differential opamp design example.

5.1. Example Design

The device sizes and bias voltages shown on Fig.4 correspond to an example design whose performance characteristics are summarized in Table 1. The excellent agreement between the performance predicted by

OBLX and the actual performance measured by simulation is a tribute to the use of simulation quality device models and AWE. In order to accurately estimate the phase margin, AWE employed a 6 pole and 5 zero model of the opamp's transfer function. It is interesting to note that during the design process the performance of 80,000 possible circuits was predicted. The total design time required was 37 minutes on a 60 MIP workstation. A majority of the computational time was used to perform AWE evaluations (approximately 92 %).

Performance Measure		Objective	Spec	OBLX	HSPICE
V_{DD} / V_{SS}	(V)	Fixed	+2.5/-2.5	+2.5/-2.5	+2.5/-2.5
Total C_{Load}	(pF)	Fixed	1.25	1.25	1.25
dc Gain	(dB)	$\geq$	84	83.5	83
UGF	(MHz)	Maximize	≥ 90	138	140
Phase Margin	(deg)	$\geq$	45	72.4	72
Slew Rate	$(V/\mu s)$	$\geq$	150	150	145
Output Swing	(V)	$\geq$	2.5	2.6	2.5
0.1% Settling Time	(ns)	Minimize	≤ 100	22	55
Input CM Range	(V)	$\geq$	1.0	1.74	1.75
Area	(μm^2)	Minimize		16.54	
Power	(mW)			5.68	5.72

Table 1: Specifications, OBLX performance prediction, and simulation performance prediction for the fully differential opamp design example.

ASTRX was able to automatically generate performance predictions for all of the specifications listed in Table 1 except for the slew rate, the output voltage swing and the common mode input range. For these performance measures, we added the following simple equations.

$$Slew\ Rate = (I_{ds3} + I_{ds4})\ /\ 4\ (C_L + C_{db6} + C_{db10} + C_f)$$

$$V_{OL} = V_{DS13/14} + V_{DSAT9/10}$$

$$V_{OH} = V_{DD} - V_{DS3/4} - V_{DSAT5/6}$$

$$V_{IL} = V_{DD} - V_{DS3/4} - V_{DSAT1/2} + V_{GS1/2}$$

$$V_{IH} = V_{GS1/2} + V_{DSAT17}$$

Note that all of the right hand side variables are known: they either come directly from the optimizers choices of node voltages, or from the device evaluator which generates currents, small signal model parameters, and other device information such as V_{DSAT} for each device.

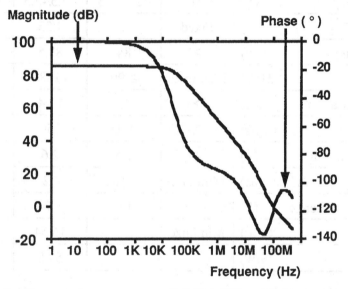

Figure 5: Frequency response of fully differential opamp design example.

As can be seen in Fig.5, the frequency response of this circuit is quite complex, particularly the phase response. However, the phase margin as predicted by AWE is within 1 degree of that predicted by simulation. The one noteworthy difference between the predicted and simulated performance is the settling time to 0.1%. The settling time is predicted by using the AWE poles and zeros and assuming a linear system response, with the additional provision that the predicted slew rate is used as an upper bound on the actual rate of change of the output. A careful look at the transient response shown in Fig.6 clearly indicates that while the upward going output of the opamp initially rises at a high slew rate, when it reaches +2.7 V it changes to a much lower slew rate. Examination of the simulation reveals that the upper current source transistor, *M3*, drops into the linear region during the slewing transient. This is a result of the ΔV_{GS} of *M5* and *M8* increasing dramatically as their drain currents try to increase from a quiescent value of 17.6 µA to a peak of 284 µA during the slewing transient.

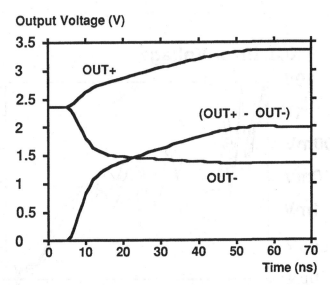

Figure 6: Transient response of fully differential opamp design example undergoing a full scale output swing.

One solution to this problem is to add one more performance constraint; one that guarantees that *M5-M8* are large enough to handle the slewing current without allowing *M3/M4* to enter the linear region. That is,

$$V_{DD} - V_{BIAS2} \geq V_{DSAT3/4} + V_{GS5/6}(@I_{DS5/6} = 0.5 \times I_{DS3/4})$$

Note that to determine $V_{GS5/6}(@I_{DS5/6} = 0.5 \times I_{DS3/4})$ we again make use of the device evaluator, iteratively searching for the V_{GS} that will cause the desired I_{DS} to flow. This and several other common routines for determining device voltages and other parameters are provided as primitive operators for use in expressing equations by the designer. Another constraint on the sizes of *M11/M15* and *M12/M16* to make sure that *M7* and *M6* do not drop into the linear region should also be added. A general observation is that the process of entering a new topology frequently proceeds in this kind of an iterative fashion. Often, when an anomaly is discovered between the performance as predicted by OBLX and that predicted by simulation, a new equation must be added or an old equation must be modified in order to improve the accuracy of the performance prediction.

5.2. DC-Free Validity

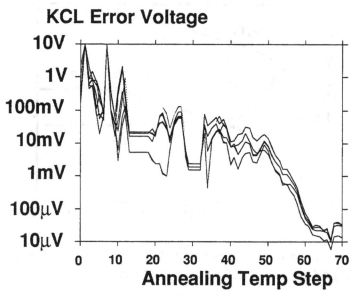

Figure 7: The "first-order" voltage error as predicted during the design of a fully differential opamp.

An important question is "How valid was the dc-free biasing approach?". Fig.7 shows average value of the first order predicted voltage error at each node for all circuit trials at a given temperature. Note, the "first order" predicted voltage error at a node is just the vector of KCL violating currents at each node multiplied by inverse of the small signal conductance matrix. The x axis corresponds to the number of steps (i.e., temperature changes) since the start of the annealing run. One should not be surprised that the node voltages are greatly in error during the first 10-15 temperature steps, since at this point the annealer is accepting a large fraction of proposed moves, even ones that make KCL further from correct. However, from step 15 on the average voltage error stays below 100 *mV* at all nodes, hovering around 10 *mV* for most of the annealing run. During this middle part of the annealing run the performance predictions are reasonably accurate, and the approximate devices sizes are being determined. During the last 20 steps of the run, fine adjustments to device sizes are being made, and the KCL voltage error is being driven down to less than 100 μV.

5.3. Stochastic Nature of Annealing

		Spec	1	2	3	4
			Pred /Sim	Pred /Sim	Pred /Sim	Pred /Sim
V_{DD} / V_{SS}	(V)	+2.5 / -2.5				
C_{Load}	(pF)	1.0				
Gain	(dB)	83	86/87	84/86	84/84	84/80
UGF	(MHz)	90	99/92	120/121	145/146	237/239
Ph.Marg.	(deg)	45	91/91	90/90	88/88	86/86
Area	(μm²)	-	5500	7000	6600	9100
Power	(mW)	-	1.27	1.87	2.30	5.66
Moves	(10^3)	-	60	60	63	72
CPU	(hours)	-	1.0	1.0	1.0	1.2

Table 2: Four folded-cascode designs with varying UGF. The CPU hours are on a 20 MIP workstation. These designs are in a 1.2 μ CMOS process.

Table 2 shows several fully-differential opamps designed by ASTRX/ OBLX that have varying unity-gain frequencies and areas. These designs are part of an experiment to explore the tradeoff between device area and unity gain frequency (UGF). The cost function being minimized in this case was:

$$E_{obj} = area + 1/UGF$$

where area was normalized to 1000 μm^2 and UGF was normalized to 1 *MHz*. The table presents four typical designs resulting from different random starting points, ordered by UGF. Note that there is a general trend of increasing area and CPU time for higher unity gain frequencies. The table also emphasizes one of the important points about simulated annealing. Although it is able to escape local minima, it is not guaranteed to find the global optimum. The second design is slightly inferior to the other three in terms of the area required for the performance obtained. It is standard practice to synthesize a number of circuits to the same specifications and then to pick the best ones. The excellent correspondence between the predicted and simulated values of UGF and phase margin are a testimony to the accuracy of the device models and the ability of AWE to characterize the small signal transfer function.

6. Conclusions

A new approach for the synthesis of cell-level analog circuits using SPICE-quality device models and simulated annealing optimization techniques was presented. The synergistic combination of using encapsulated device evaluators, using dc-free biasing, and using AWE to compute low order pole/zero models of small signal transfer functions leads to many benefits. First, this combination allows accurate prediction of the performance of high performance analog circuits - thus making it possible to apply synthesis with confidence to some classes of high performance analog circuits (those in which critical performance parameters can be measured using small signal analyses). Second, dc-free biasing, by eliminating the need for a full dc solution at every design point, makes it practical to explore a large portion of the design space using an acceptable amount of computer time. The use of AWE to evaluate small signal transfer functions also greatly reduces the required computer time over direct solution of the small signal matrix at each desired frequency. The final advantage of this combination is a dramatic decrease in the level of expert designer input required to create a tool for the synthesis of a new circuit topology. Use of dc-free biasing removes the need for the designer to provide equations relating to the dc operating point of the circuit. Use of encapsulated device evaluators removes the need for any simplified equations relating to the device model. And, use of AWE removes the need to create equations that predict small signal transfer functions.

References

[1]   E. Berkcan and M. d'Abreu and W. Laughton, Analog Compilation Based on Successive Decompositions, in *Proceedings of 25th Design Automation Conference*, pp.369-375. ACM/IEEE, 1988

[2]   M.G.R. Degrauwe et al., IDAC: An Interactive Design Tool for Analog CMOS Circuits, IEEE Journal of Solid-State Circuits, Vol. SC-22, No.6, pp.1106-1116, Dec. 1987

[3]   R.K. Brayton et al., A Survey of Optimization Techniques for Integrated-Circuit Design, *Proceedings of the IEEE*, Vol.69, No.10, pp.1334-1362, Oct. 1981

[4]   W.T. Nye et al., DELIGHT.SPICE: An Optmization-Based System for the Design of Integrated Circuits, *IEEE Transactions on Computer-Aided Design*, Vol.7, No.4, pp.501-519, April 1988

[5]   B.J. Sheu et al., BSIM: Berkeley Short-Channel IGFET Model for MOS Transistors, *IEEE Journal of Solid-State Circuits*, Vol. SC-22, No.4, pp.558-566, Aug. 1987

[6]   G.G.E. Gielen and H.C.C. Walscharts and W.M.C. Sansen, ISAAC: A Symbolic Simulator for Analog Integrated Circuits, *IEEE Journal of Solid-State Circuits*, Vol.24, No.6, pp.1587-1597, Dec. 1989

[7]   G.G.E. Gielen and H.C.C. Walscharts and W.M.C. Sansen, Analog circuit design optimization based on symbolic simulation and simulated annealing, *IEEE Journal of Solid-State Circuits*, Vol.25, No.3, pp.707-714, June 1990

[8]   G.D. Hachtel and R.A. Rohrer, Techniques for the Optimal Design and Synthesis of Switching Circuits, *Proceedings IEEE*, Vol.55, No.11, Nov. 1967

[9]   H.Y. Koh and C.H. Sequin and P.R. Gray, OPASYN: A Compiler for CMOS Operational Amplifiers, *IEEE Transactions on Computer-Aided Design*, Vol.90, No.2, pp.113-125, Feb. 1990

[10] P.C. Maulik and L.R. Carley, Automating analog circuit design using constrained optimization techniques, *Proceedings of ICCAD*, Nov. 1991

[11] K. Nakamura and L. R. Carley, A current-based positive-feedback technique for efficient cascode bootstrapping, *Proc. VLSI Circuits*

Symposium, June 1991

[12] K. Nakamura and L.R. Carley, An Enhanced Fully-Differential Folded-Cascode Op Amp, *IEEE Journal of Solid-State Circuits*, Vol.27, No.4, April 1992

[13] Z-Q. Ning and T. Mouthaan and H. Wallinga, SEAS: A Simulated Evolution Approach to Analog Circuit Synthesis, *Proceedings of Custom Integrated Circuits Conference*, IEEE, 1991

[14] E.S. Ochotta and R.A. Rutenbar and L. R. Carley, Equation-Free Synthesis of High-Performance Linear Analog Circuits, *Proc. Brown/MIT Conference on Advanced Research in VLSI and Parallel Systems*, March 1992

[15] L.T. Pillage and R.A. Rohrer, Asymptotic waveform evaluation for timing analysis, *IEEE Transactions on Circuits and Systems*, Vol. CAD-9, No.4, April 1990

[16] R. Harjani and R.A. Rutenbar and L.R. Carley, OASYS: A Framework for Analog Circuit Synthesis, *IEEE Transactions on Computer-Aided Design*, Vol.8, No.12, pp.1247-1266, Dec. 1989

[17] J-M. Shyu and A. Sangiovanni-Vincentelli, ECSTASY: A New Environment for IC Design Optimization, *Proceedings of ICCAD*, pp.484-487, 1988

[18] K. Swings and S. Donnay and W.Sansen, HECTOR: A Hierarchical topology-Construction Program for Analog Circuits Based on A Declarative Approach to Circuit Modeling, *Proceedings of Custom Integrated Circuits Conference*, IEEE, 1991

Biographies

L. Richard Carley received the S.B. degree from the Massachusetts Institute of Technology in 1976 and was awarded the Guillemin Prize for the best EE Undergraduate Thesis. He remained at MIT where he received the M.S. degree in 1978 and the Ph.D. in 1984.

He has worked for MIT's Lincoln Laboratories and has acted as a consultant in the area of analog circuit design and design automation for Analog Devices and Hughes Aircraft among others. In 1984, he joined Carnegie Mellon University, Pittsburgh PA, where he is currently an Associate Professor of Electrical and Computer Engineering. His research is in the area of analysis, design, and automatic synthesis of mixed analog/digital systems.

He received a National Science Foundation Presidential Young Investigator Award in 1985, and a Best Paper Award at the 1987 Design Automation Conference. At the 1991 ICCAD conference he received a Distinguished Paper Mention for work on analog circuit layout. He is a senior member of the IEEE.

P.C. Maulik received the B.Tech. degree in Electronics and Electrical Communication Engineering from the Indian Institute of Technology, Kharagpur, India and the M.E. degree in Electrical Engineering from the University of Florida, Gainesville, Florida, U.S.A. in 1983 and 1986 respectively. He is currently a Ph.D. candidate in the Department of Electrical and Computer Engineering at Carnegie Mellon University, Pittsburgh, Pennsylvania, U.S.A. His research interests are in VLSI CAD, analog circuit design, optimization algorithms, and parallel processing.

Emil Ochotta received the B.Sc. degree in Electrical Engineering (Honors with Distinction) from the University of Alberta, Edmonton, Canada, in 1987, and the M.S. degree in Electrical and Computer Engineering from Carnegie Mellon University, Pittsburgh, U.S.A, in 1989. He is currently a Ph.D. candidate at Carnegie Mellon, where his work focuses on automatic synthesis of analog integrated circuits. He is the principal architect of the OASYS-VM and ASTRX/OBLX synthesis systems. His research interests include knowledge-, language-, and optimization-based approaches to high-performance circuit synthesis, physical CAD, and software environments for VLSI design. He is a student member of IEEE.

Rob A. Rutenbar received the B.S. degree in Electrical and Computer Engineering from Wayne State University, Detroit, U.S.A., in 1978, and the M.S. and Ph.D. degrees in Computer Engineering (CICE) from the University of Michigan, Ann Arbor, U.S.A., in 1979 and 1984, respectively.

In 1984, he joined the faculty of Carnegie Mellon University, Pittsburgh, U.S.A., where he is currently an Associate Professor of Electrical and Computer Engineering, and of Computer Science. His research interests include VLSI layout algorithms, parallel CAD algorithms, and applications of automatic synthesis techniques to VLSI design, in particular, synthesis of analog integrated circuits, and synthesis of CAD software.

In 1987, Dr. Rutenbar received a Presidential Young Investigator Award from the National Science Foundation. At the 1987 IEEE-ACM Design

Automation Conference, he received a Best Paper Award for work on analog circuit synthesis. In 1989, he was guest editor of a special issue of IEEE Design and Test. Since 1990 he has been on the Executive Committee of the IEEE International Conference on CAD; over the last several years he has also been on the program committee of numerous conferences and workshops in the area of VLSI CAD. At the 1991 ICCAD conference he received a Distinguished Paper Mention for work on analog circuit layout. He is currently on the editorial board of IEEE Spectrum.

Dr. Rutenbar is a Senior member of IEEE, and a member of ACM, Eta Kappa Nu, Sigma Xi and AAAS.

Analog CAD for Consumer ICs

Gerard F.M. Beenker, John D. Conway,
Guido G. Schrooten, André G.J. Slenter

Philips Research Laboratories
Eindhoven
The Netherlands

Abstract

Although CAD tools for analog IC design have a relatively long history, the automation of the analog IC design process is still in its infancy. This is mainly due to the fact that during the last ten years most emphasis has been placed on digital CAD. Currently analog CAD is receiving a lot of attention from universities, commercial CAD suppliers and CAD R&D centres in leading IC companies. In this paper we will show that automating parts of the analog IC design process is feasible, but far from being simple. We will mainly focus on how the problems associated with automated analog design are tackled by Philips' CAD teams by describing our activities on automated analog circuit and layout synthesis. It will become obvious that IC design expertise is indispensable in successfully solving these problems. We will conclude by summarising a number of possible directions for future analog CAD research.

1. Introduction

With the advent of electronic integration, the euphoria arose that all electronics could be implemented using digital circuitry. Therefore, the CAD world mainly focused on digital VLSI design tools for placement and routing, logic synthesis, high level synthesis, mixed level simulation, etc. Combinatorial optimisation techniques like (non)linear programming and simulating annealing found their way into many CAD applications.

Today, digital CAD tools still receive most of the attention. Scanning the proceedings of the main CAD conferences reveals that only a few papers deal with analog CAD. In audio and video consumer applications, signal processing ICs have to communicate with the analog world, so analog to digital (A/D) and digital to analog (D/A) convertors remain indispensable. In current consumer applications, very high demands are put on those convertors. For example, in Compact Disc and Digital Compact Cassette applications, convertors with high resolution combined with excellent linearity are required. In contrast, for HDTV applications, convertors must be extremely fast combined with excellent signal-to-noise (S/N) ratios. In addition, due to miniaturisation, analog and digital blocks are often combined on one IC, reducing the PCB area while dramatically increasing the IC design complexity. Although the area occupied by an analog convertor on a consumer IC is typically only 10% of the total area, the design time of these high-end convertors is about 75% of the total design time. However, besides the time-consuming design of these convertors, one is also faced with problems due to using both analog and digital circuitry on one IC. For example, one must take special precautions to minimize the interference (like substrate coupling) between the analog and digital circuitry. This problem is specifically difficult since it is often only discovered when one has actually fabricated the relevant IC.

The tools which are currently used by analog designers include schematic editors, circuit simulators and layout editing tools. New tools are being developed. Mixed level and mixed signal simulation tools [1] are successfully being introduced nowadays. These tools allow early feedback in the design process in order to efficiently generate the appropriate architecture. They allow designers to analyze the implications of specific circuit level choices and provide the means to simulate the behaviour of both analog and digital building blocks. Currently, these tools only simulate ideal circuit behaviour. They do not take a number of important effects into account like parasitics, loading dependence between blocks and substrate coupling.

Analog synthesis can be subdivided into two parts: (1) circuit synthesis and (2) layout synthesis. A number of analog circuit synthesis systems are known from literature. These systems are mainly knowledge-based. Given a circuit specification from an IC system designer, they can produce the corresponding dimensioned circuit description. These systems include the IDAC system from CSEM [2], the OASYS system from Carnegie-Mellon University [3], the OPASYN system from the University of Berkeley [4], the Ariadne system from the University of Leuven [5] and the TOPICS system from the University of Eindhoven [6]. All these systems apply analog knowledge expressed in analytical equations, heuristic rules or predefined topologies and lead in a number of steps to appropriate dimensioned circuits.

For analog layout synthesis, tools which apply simulated annealing techniques for device or block placement include KOAN from CMU [7] and the ILAC layout system from CSEM [8]. An alternative approach based on predefined topologies is found in the OPASYN system from Berkeley [4]. Analog routing systems try to take analog constraints like parasitic coupling effects, symmetry constraints, net shielding, etc. into account. Tools which deal with these constraints are ANAGRAM from CMU [7] and the routers available in the ILAC and OPASYN system.

In this paper we will focus on our method for analog circuit and layout synthesis. In section 2, we will describe our knowledge based approach to circuit synthesis and we will demonstrate its use for the automatic generation of analog building blocks. In section 3, we will present our analog module generator and our analog routing methodology. In section 4 we will briefly describe a number of future research topics. Finally, we will present our conclusions.

2. Circuit Synthesis Automation

The number of analog circuit synthesis systems is increasing. Systems such as IDAC, OASYS and OPASYN are very well known. All these systems require that knowledge in terms of a set of rules, formulas and algorithms is embedded in the system. While the functionality of these systems is the same for the end-user, two slightly different methodologies in formalising the knowledge can be seen. The methodology used in OASYS is to describe the knowledge in a strict hierarchical fashion. Via selection and translation steps, the appropriate topology and the dimensions of all the components are derived. On the other hand, systems such as IDAC and OPASYN are not fully based on this strict hierarchical decomposition, especially for modules such as op-amps, comparators etc. These systems are based on the belief that there are too many electrical interactions between the sub-parts of these modules in order to split them up hierarchically. For larger modules, however, they do apply selection/translation principles in order to keep the knowledge well organised.

Some universities are working on the next generation of analog synthesis systems. These include the ASTRX+OBLX project at Carnegie Mellon

University and the ARIADNE project at the University of Leuven. Both of these systems strive to remove the need for the very time consuming knowledge acquisition activity. CMU is working on "equation free" synthesis. Starting from a netlist (fixed topology), a set of device models and a set of linear performance targets, simulated annealing is used to derive the device dimensions which meet the user-supplied circuit specification. On the other hand, in the ARIADNE system one tries to solve the same problem by deriving the design equations automatically using symbolic manipulation techniques. So far, both approaches have been successfully applied to relatively small circuits. These approaches, however, are unlikely to solve the problem of knowledge acquisition for modules larger than opamps, comparators, etc.

2.1. Philips' approach to automated circuit synthesis

For many of our consumer ICs, an application specific solution has to be used. Dedicated building blocks are needed to meet the required specification. Building a library with these functions makes no sense since too many instances would be necessary in order to satisfy all the required specification ranges. On the other hand, we have to reduce the total design turn-around time of complex mixed analog/digital circuits in order to remain competitive. These requirements led to the development of a tool called MIDAS[1] [9]. MIDAS is an open, knowledge based and technology independent design tool. When dedicated design knowledge is added to the system, it can be used for automatic circuit synthesis and layout generation.

MIDAS possesses extensive knowledge representation facilities. It is strictly hierarchical, breaking down analog modules into sub-blocks that may be reused throughout the hierarchy. Fig.1 is an example of the knowledge representation hierarchy used for synthesising a current-output digital to analog (D/A) convertor. The hierarchical structure of the knowledge representation promotes reusability and modularity.

To generate a circuit using MIDAS, the system designer provides a set of bounded circuit and layout specifications. MIDAS traverses the hierarchy carrying out multiple selection and translation steps, similar to the OASYS approach. Selections are made only from the limited set of topologies appropriate to that level of the hierarchy in the knowledge base. On completion of a selection, MIDAS translates the input specifications of

---

[1]MIDAS is commercially available from Silicon and Software Systems, Dublin, Ireland. It was developed according to specifications supplied by Philips' Consumer Electronics (IC Laboratory).

that block into specifications for the sub-blocks of the chosen topology. This translation is achieved by the use of both the expert's heuristic knowledge and precise transistor-level equations, as used in circuit simulators. After the generation of each sub-block, its specification values are passed back to its parent block. This mechanism allows multiple optimisation/iteration steps. On completion of the circuit synthesis, the system produces the corresponding sized circuit schematics, layouts, simulation netlists and data sheets.

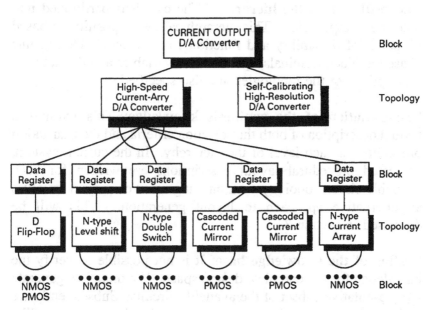

Figure 1: Hierarchical decomposition of current-output D/A convertor.

2.2. Knowledge engineering process

Analog design is commonly perceived to be one of the most knowledge-intensive IC design activities. The techniques needed to build good analog circuits are not formalised, but rely solely on the experience of individual designers. A tool such as MIDAS enables the knowledge engineer to encapsulate existing analog IC design knowledge in a methodical, stepwise manner.

The process of creating a knowledge base for a circuit topology is much more time consuming than designing one single circuit and is therefore only economical for frequently used analog modules. Only well-proven and completely engineered circuit topologies are chosen. In addition this can only be done if expert design knowledge is available. The knowledge engineering process can be divided into the following phases.

- **Knowledge acquisition:** All the relevant and available knowledge is first gathered. This knowledge is based on literature surveys and interviews with experts for the acquisition of heuristic and analytical knowledge. This knowledge can be described as a set of rules, formulas and algorithms.

- **Encapsulation of the hierarchy:** The circuit is partitioned into blocks and topologies. This hierarchical decomposition is based on potential reusability and functional consistency. During this phase the block terminals, specification variables and the network descriptions of the topologies are also specified.

- **Encapsulation of the synthesis knowledge:** This requires a textual description of both the selection criteria and the translation procedures at each level of the hierarchy. In the current system, dedicated procedural layout descriptions are written for each available circuit topology. An ongoing research project involves an alternative approach to layout generation. This will be discussed in section 3.

- **Testing of the knowledge base:** It is not possible to verify the knowledge for the complete design space. We therefore generate a representative subset of the available circuits. Subsequently we verify and test these circuits using standard full-custom tools like layout extractors, LVS and of course simulation tools. This verification phase is very time consuming but is necessary in order to get confidence in the quality of knowledge in the system. Of course, a final verification of each circuit generated by a system designer still remains necessary.

2.3. Examples

In this section, we will describe some of the knowledge of our bit-stream analog to digital (A/D) convertor which has been integrated in the MIDAS knowledge base. Firstly, we will focus on a strategy to select the proper architecture for the bit stream A/D convertor. Secondly, we will consider one of its constituent blocks, viz. a 1-bit switch capacitor D/A convertor.

2.3.1. Strategy for architecture selection

The architecture of our bit stream A/D convertor is based on a combination of two convertor types, (1) a Flash convertor and (2) a Sigma-Delta convertor. The big advantage of this combined architecture is that we can now vary two important parameters, the order of the sigma-delta convertor and the number of bits from the flash convertor [10]. This enables us to cover a broad range of specifications involving both audio and video applications. Some of the knowledge which has been encapsulated to derive the order and the number of bits of the combined architecture will now be described.

1. The quantization noise can be estimated on the basis of theoretical studies [11]. On the other hand, the maximum allowed input signal depends on the number of bits and the order of the sigma-delta convertor. These experimental data are shown in Table 1. Together with the information about the oversampling factor (fs/fb) and the requested dynamic range, the possible candidates (order and bits) can now be derived.

	1 bit	2 bits	3 bits	4 bits
1st order	-1	-0.32	-0.136	-0.063
2nd order	-3	-0.89	-0.37	-0.171
3rd order	-3	-0.89	-0.37	-0.171

Table 1: Maximum input signal (0 dB = full scale) for A/D convertors as a function of the number of bits and order of the sigma-delta convertor

2. The set of candidates will be reduced by linearity constraints and "threshold effects" considerations [12].

   • When a linearity larger than 60 dB is requested, multi bit switch capacitor solutions will not be considered since these will require infeasible matching constraints.

   • There are further restrictions on the minimum input signals due to threshold effects. Smaller input signals can not be distinguished from the noise level, because they do not disturb the idling pattern of the sigma-delta convertor. These threshold effects will influence the performance of the convertor and have to be taken care of. We use the following theoretical formulas for the minimum input signals:

1st order: $sin(\pi\, fi\, /\, fs\,)$;
2nd order: $3\, sin^2(\, \pi\, fi\, /\, fs\,)$;
3rd order: $69\, sin^3(\, \pi\, fi\, /\, fs\,)$.

3. Because of the many assumptions in the theoretical studies, which might not be valid for all cases (e.g. for low oversampling), a final functional simulation will be done for safety reasons. The quantization noise in the neighbourhood of the maximum input signal will be derived from the simulation by means of a Fast Fourier Transform. If the requested dynamic range can not be reached, another candidate will be taken.

4. The difference between the available noise power and the simulated quantization noise power will be allocated to analog noise. If during the dimensioning of the lower level modules, certain noise levels are exceeded, noise budgets will be rearranged and if this will not help, another topology will be considered.

Besides the electrical parameters mentioned above, we also deal with *context-specific* parameters like whether the convertor should be current or voltage driven, handle bipolar or unipolar signals, be symmetrical or non-symmetrical, etc. The top level specifications will influence the lower circuits in the knowledge hierarchy. For example, due to the high level context-specific parameters, we offer 12 different switched capacitor D/A topologies.

2.3.2. Switch capacitor D/A topologies

In this subsection we will describe some of the knowledge of switch capacitor D/A convertors which are used in the bit stream A/D convertor. The functionality of the switch capacitor D/A convertors is based on charging a capacitor during one clock phase (*ph1*) and discharging an equal amount of charge during the second clock phase (*ph2*) or vice versa. The specification parameters for these D/A convertors are listed below:

1. symmetrical or non-symmetrical implementation

2. unipolar or bipolar operation

3. V_{ss} or V_{ref} output level

4. inverting or non-inverting condition

5. sample frequency

6. bandwidth

7. V_{dd} voltage

8. V_{ref} voltage

9. capacitor value

The D/A selection knowledge is very straightforward. Depending on the values given for the first four specification parameters, the appropriate topology is determined.

The D/A translation knowledge has to convert the aforementioned specification parameters into specification parameters for all its sub-blocks, in this case *switches*. The switches must be dimensioned so that within a given time, the D/A capacitors can be charged/discharged. On the other hand, a specific scaling between the R_{on} resistance of the switches has to be maintained. Otherwise, some node voltages will go below zero volts. In that case, charge can leak away via the substrate.

Depending on (1) the terminal voltages of the switches, (2) the requested R_{on} resistance and (3) certain layout area considerations, these switches will be implemented as either a single n-MOS, a single p-MOS or as a double n-MOS p-MOS switch.

In Fig.2 an overview of the MIDAS synthesis process is given. Starting from a high level specification of a switched capacitor D/A convertor, the optimized schematic and the corresponding simulation circuits are generated.

2.4. Current Status

The hierarchical selection/translation principles have been employed to encapsulate the design knowledge of a number of operational amplifiers topologies, current-steering and current-calibrated CMOS D/A convertors, as described in [13,14]. The correctness of these designs has been verified by detailed circuit simulation and silicon implementation.

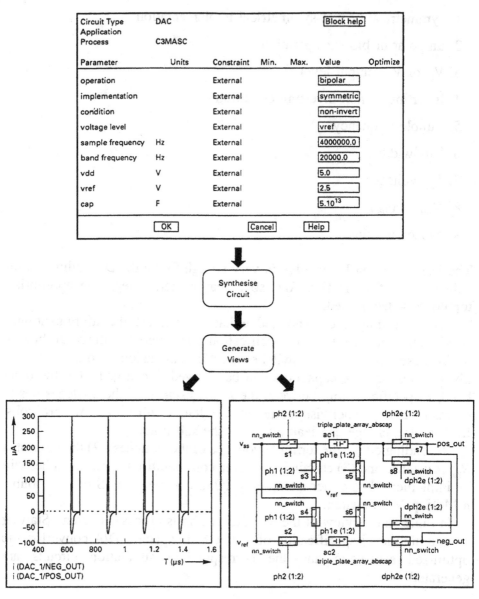

Figure 2: Switched capacitor D/A convertor

For these examples the trajectory from electrical specification down to the layout has been completely covered. The building blocks generated by MIDAS have been utilised for the design of ICs for consumer applications.

The next major goal is the automatic synthesis of more complex analog building blocks, like bit-stream A/D Convertors which are currently used in high-end consumer products. The use of such synthesis systems to produce analog circuit designs will significantly reduce the required design time.

3. Layout Design Automation

The design of analog circuit layout is a very complex task. Not only must the basic layout problems known from digital circuits be solved, but many more constraints have to be dealt with. Examples of analog constraints include matched devices, symmetry constraints, separation of noisy and sensitive nets, minimization of substrate coupling etc. These aspects heavily influence the fabricated circuit's performance. Therefore, layout design methodologies leading to solutions for digital circuits are not necessarily optimal when analog circuits are involved. Although it is recognised that the current situation is far from being ideal, it is common practice that analog circuits are manually laid out in order to meet the required performance. Since very little is yet known about optimal layout design for analog circuits, it is still seen as a work of art, performed only by experienced layout designers. With the increasing variety and complexity of required analog circuits on one hand and the lack of experienced analog designers on the other hand, layout design automation of analog circuits becomes essential to bridge the gap between demand and supply of integrated analog circuits.

In the remaining part of this section we will outline two approaches currently being investigated at our laboratories. They both aim at the automation of analog circuit layout design.

3.1. Template-based Analog Module Generation

In this section we present an analog module generator based on a "design by example" approach [15]. This approach uses a sample layout, called a "template", to graphically capture an expert's knowledge of analog device placement and routing for a given module type. An overview of the module generator is given in Fig.3.

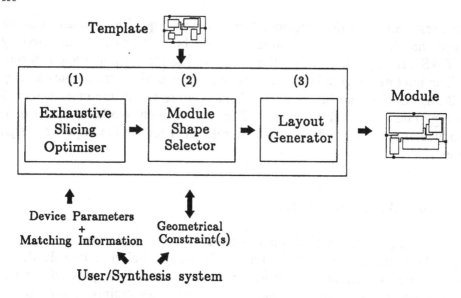

Figure 3: Overview of the analog module generation process.

Compared to procedural layout descriptions, the template approach is less time-consuming and error prone. It is also better suited for capturing geometry related expert knowledge. Furthermore, far more complex circuits can be dealt with than with procedural layout descriptions.

For a given module type, the template is created once by an expert designer. A template graphically captures an expert's knowledge of (1) device placement and orientation, (2) routing wire trajectories, material types and widths and (3) position of module terminals. A template is representative of modules of its type, since the latter have the same circuit structure but different device parameters. Template assembly proceeds quickly through the use of a layout editor with an on-line device generator and wiring tool. Note that no attention need to be paid to area minimisation at this stage.

To generate a module from a given template, the user supplies (1) the required electrical parameters for each device, (2) the sets of devices which must be matched and (3) an optional geometrical constraint on the module's shape e.g. desired aspect ratio, minimum area etc. The tool then automatically extracts all possible slicing structures from the device topology in the template. Using the area optimisation techniques described in [16,17], the optimum geometrical parameters for each device (i.e. device shape) which satisfy the user's geometrical constraint on the module's shape are then determined. Subsequently, new devices with the user-supplied electrical parameters and the determined geometrical parameters are generated. Finally, the template is transformed into the required module using compaction. During this phase, the routing in the template remains intact while the devices are replaced by the newly generated ones. During the layout generation phase, device merging, matching and alignment are supported.

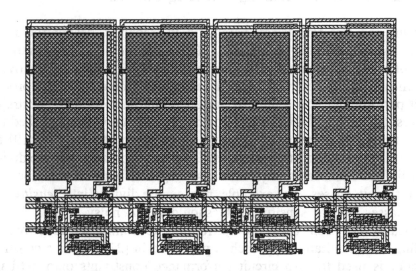

Figure 4: Generated module: four 1-bit D/A convertors connected in parallel.

The analog module generator has been extended to handle hierarchy. This means that templates may not only contain devices but also may include other templates. This allows us to generate large circuits. An example of an analog module generated using hierarchical templates is shown in figure Fig.4. This circuit represents four 1-bit D/A convertors, a sub-circuit used in bit-stream A/D convertors. Using hierarchical templates is of particular importance to us since we can use the same topology/template hierarchy for both circuit and layout synthesis. This means that the device parameters required by the layout generator for each template can now be supplied directly by each corresponding circuit topology description in the synthesis tool.

Up to now, the template based approach has been evaluated for various circuits, ranging from super-MOS circuits [18], opamps, up to multi-bit D/A convertors. These examples reveal that this technique produces good quality layout in a reasonable amount of time. This is achieved by availing of the expert designers knowledge embedded in the template and by taking analog specific features like device matching and merging into account during the layout transformation phase. Ongoing work in this area includes routing area estimation. On the basis of the wiring in the template, we must extend our floorplanning tool to take wiring area into account in order to provide the user with more accurate floorplan area estimations.

3.2. Constraint Driven Routing of Analog Circuits

During the last few years, automatic interconnect design for analog circuits has gained a lot of attention. The basic problem is to control the effect of interconnect resistance and parasitic capacitances on the overall performance of the circuit. To guide the interconnect design process, circuit performance constraints are translated into net oriented constraints.

In results published up to now, two approaches can be distinguished with respect to the formulation of these net oriented constraints. In [19] and [20], qualitative net constraints are used. In this approach, nets are divided into a limited number of disjunct classes together with relations between these different classes. For example, nets can be divided into *sensitive* and *noisy* classes, where sensitive nets should *avoid* noisy nets. This net classification is then used during routing to minimise undesired net coupling. A quantitative approach is presented in [21], where a constraint generator is used to map circuit performance constraints onto net-based parasitic constraints. With this approach, actual values for coupling capacitances are taken into account during interconnect design. In this way a direct relation between circuit performance constraints and interconnect parasitics is maintained. Two routing tools based on this approach have been reported [22,23]. Compared to net-classification, the second method provides a more accurate control of interconnect effects. This is essential to reduce the number of design iterations necessary to reach an acceptable circuit implementation. Other aspects that must be dealt with during interconnect design are variable wire widths, wire matching and wire resistance.

The layout design methodology we propose is also based on a quantitative approach to performance constraints. The input for our router consists of a netlist, a floorplan and a constraint file. The constraint file specifies a set of resistance and capacitance constraints (to substrate and to other nets) for every net. Restrictions on routing layers, wire matching and variable wire widths can also be dealt with.

The router kernel consists of a Lee-type algorithm [24]. A variant of this algorithm has been developed which deals with actual layout geometry.

Other improvements, as suggested in [25,26], are used to reduce the size of the search space. The main advantage of our algorithm is that accurate resistance and capacitance estimates can be obtained on the fly during routing. Furthermore, variable wire widths are supported in a natural way.

Since nets are sequentially routed, special attention must be paid to net ordering. An interleaved routing and rip-up/re-route strategy, as suggested in [19], is used to increase the probability that connections are found. Net ordering is also important to achieve high quality solutions. The order in which nets are be routed is partly determined by geometrical constraints and partly by electrical constraints. Nets with tight constraints are routed before nets with relatively weak constraints, for which longer detours can be accepted. Net length and symmetry account for geometrical constraints and resistance and capacitance as electrical constraints.

Results obtained thus far indicate that the router kernel provides a powerful basis for future extensions. One aspect that will be considered in the near future concerns the trade-off between interconnect resistance and capacitance. In case of conflicting resistance/capacitance constraints, wire width variations can be used to solve these conflicts.

4. Directions for Future Research

In the previous sections we have described our approach to automatic analog circuit design. In addition, an overview of existing methods has been given. In this section, we will describe our vision on future research trends. We will focus on verification, synthesis and testing tools for mixed analog/digital ICs.

4.1. IC Design Verification

In the introduction we mentioned that mixed level/mixed signal simulation tools are becoming available which simulate ideal circuit behaviour. However, the integration of analog and digital blocks on one IC requires more advanced tools. These tools must handle a wide range of effects including parasitic coupling and substrate couplings between the analog and digital blocks. Due to decreasing device dimensions, very severe demands are put on the performance of the ICs with respect to accuracy and speed, so very accurate circuit simulators are needed. Tools which can deal with thermal effects, EMC effects, and device degradation effects like hot electron degradation, electro-migration are needed. The trend in circuit simulation is going in two directions: on the one hand more accuracy is required, while on the other hand the sizes of circuits to be simulated are increasing. The ultimate goal can simply be formulated: develop tools which can simulate complete ICs with analog accuracy. Due to the large number of components on one IC, the present day circuit simulators cannot deal with these requirements. To achieve this final goal, if possible at all, much effort has to be spent on modelling subcircuits. These models must include degradation effects in such a way that they can be applied as behavioral models in mixed level simulators, instead of applying circuit simulators only.

Simulation tools are not the only means to do design verification. We have developed an in-house tool, called VERA [27], which has proven to be capable of detecting all topology errors in digital designs. By applying VERA, we can drastically reduced the verification time from several man weeks for simulation to several hours for verification. We have also applied this tool to a number of mixed analog/digital designs. The application of VERA revealed design errors which were not detected by simulations. More research, however, is needed to extend the verification principles in VERA in order to be generally applicable for analog designs.

4.2. Automated Synthesis

The approach for both circuit and layout synthesis which we have described, relies heavily on the availability of expert design knowledge. To create the knowledge base is very time consuming. The same applies for the verification of that knowledge. The creation of the module layout generators, in our case based on templates, also requires a lot of assistance from IC designers. The total effort to create knowledge-based synthesis systems can therefore only be afforded when the knowledge can be used in a production environment where many variants of well-known building blocks are needed. This situation is familiar to large consumer IC companies. Otherwise, it is presently more economic to do full custom design.

Other approaches for automating the design process have to be studied. In the literature methods based on simulated annealing are being published. Those methods ignore analog knowledge almost completely, so that excessive CPU power is used before acceptable and well understood results are achieved. Hence this approach is at the other side of the spectrum; it only uses little or no knowledge. The appropriate solution is, presumably, a combination of knowledge based and optimisation related techniques. Much research is needed to develop such automated synthesis tools which can be successfully applied in a production environment. The experiences obtained so far with digital silicon compilers have shown that the development of tools which automatically synthesise competitive layout starting from a high level specification is a major effort.

The interaction between the circuit synthesis and layout synthesis needs more attention. Feedback about parasitic effects and layout area estimations from the layout synthesis tool to the circuit synthesis tool is required to generate high-quality circuits.

Analog constrained layout synthesis needs further elaboration. Research results are being published, but in IC companies the only reliable tools which deal with the many analog constraints are the layout polygon pushers and manual routing applied by experienced IC designers.

Last but not least we want to mention that a better integration of process technology knowledge and synthesis tools is needed in order to better deal with the spread in technology.

4.3. Design for Testability

In the digital IC design community, design for testability is well accepted nowadays. Extra hardware is often added to an IC in order to make it testable. Macro test, boundary scan and built-in self test methods are well known and generally applied. Techniques to generate test patterns for a class of frequently occurring faults like stuck-at faults are known. Design for testability techniques for analog design are presently not available in such a way that they can be applied economically for mass production. Currently, dedicated tests are performed to check the functionality of analog and mixed analog digital ICs. For high speed ICs or high resolution ICs these tests require much preparation and precautions. For example, due to EMC problems the measurement of the signals on the pins of the ICs can be very problematic.

Techniques are needed to develop macro test models for analog building blocks in such a way that they are generic i.e. that they can be applied to instances of one specific class of building blocks, instead of being generated in a dedicated way for each instance. Boundary scan and built-in self test techniques for analog IC design need much more research in order to be practically applicable.

In the current literature some papers deal explicitly with testing certain details or describe algorithms which can be used in signal processing, but an overview of design for testability for mixed signal designs is lacking.

5. Conclusions

In this paper we have described our approach to automate the analog synthesis process. We have shown that a knowledge based approach is a feasible solution to automate the circuit synthesis of analog building blocks like A/D and D/A convertors. We have shown that the results of our approach are very well suited to designers who work in a production environment, where many variants of a circuit architecture are required. In this case the design time can drastically be reduced. We have applied our analog synthesis system in practice to efficiently generate a number of circuit blocks for industrial consumer ICs.

Automated analog synthesis tools have still a long way to go. We enumerated a number of future research topics which are of particular interest to us. At Philips, we have learned that successful practical analog CAD tools can only be realised when there is a strong interaction between expert analog IC designers and CAD teams.

Acknowledgements

We would like to thank our colleagues in the research analog design groups and in the analog MOS Design group within Philips' Consumer Electronics (IC Lab) for their many valuable contributions and for their fruitful cooperation.

References

[1] A.C.J. Stroucken et al., *MILES: A mixed level simulator for analog/digital design*, ICCD 88 Conf. Proc., pp.120-123.

[2] M.G.R. Degrauwe et al., *IDAC: an Interactive Design Tool for Analog CMOS Circuits*, IEEE J. of Solid-State Circ., Vol. SC-22, 1987, pp.1106-1115.

[3] R. Harjani et al., *OASYS, a Framework for Analog Circuit Synthesis*, IEEE Trans. On CAD, VOL. 8, 1989, pp.1247-1266.

[4] H.Y. Koh et al., *OPASYN, a Compiler for CMOS Operational Amplifiers*, IEEE Trans. on CAD, VOL. 9, 1990, pp.113-125.

[5] K. Swings et al., *HECTOR: A Hierarchical Topology Construction Program for Analog Circuits Based on a Declarative Approach to Circuit Modelling*, IEEE CICC 1991, Conf. Proc., pp.5.3.1-5.3.4.

[6] D.M.W. Leenaerts, *TOPICS, a Contribution to Analog Design Automation*, Ph.D. Thesis, University of Technology Eindhoven, 1992.

[7] J. Cohn et al., *KOAN/ANAGRAM II: New Tools for Device Level Analog Placement and Routing*, IEEE J. Solid State Circuits, 1991, pp.330-342.

[8] J. Rijmenants et al., *ILAC: An Automated Layout Tool for Analog CMOS Circuits*, IEEE J. Solid State Circuits, 1989, pp.417-425.

[9] Silicon & Software Systems, *MIDAS: A Knowledge Based Analog Synthesis Environment*, User Guide.

[10] P.J.A. Naus et al., *Multibit Oversampled Sigma-Delta A/D Convertors as front end for CD players*, IEEE journal of solid-state circuits, Vol.26, No.7, July 1991

[11] E.F. Stikvoort, *Some remarks on the stability and performance of the noise shaper or sigma-delta modulator*, IEEE Trans. On Communications, Vol.36, 1988, pp.1157-1162.

[12] P.J.A. Naus et al., *Low Signal-Level Distortion in Sigma-Delta Modulators*, 84th AES Convention Proc., Paris, March 1988.

[13] C. Bastiaansen et al., *A 10-bit 40 MHz 0.8 μm CMOS Current Output D/A Convertor*, IEEE J. on Solid State Circuits, Vol. 26, 1991, pp.917-921.

[14] W. Groeneveld et al., *A Self Calibrating Technique for Monolithic High Resolution D/A Convertors*, ISSCC 1989 Conf. Proc., pp.1517-1522.

[15] J.D. Conway et al., *An Automatic Layout Generator for Analog Circuits*, EDAC 1992 Conf. Proc., pp.513-519.

[16] R.H.J.M. Otten, *Efficient Floorplan Optimization*, ICCD 1983, Conf. Proc., pp.499-502.

[17] L.P.P.P. van Ginneken, *Optimal Slicing of Plane Point Placements*, EDAC 1990, Conf. Proc., pp.322-326.

[18] K. Bult et al., *The CMOS Gain-Boosting Technique*, Analog Integrated Circuits and Signal Processing, Vol.1, 1991, pp.119-135.

[19] J.M. Cohn et al., *KOAN/ANAGRAM II: New Tools for Device-Level Analog Placement and Routing*, IEEE J. of Solid-State Circ., Vol.26, 1991, pp.330-342.

[20] M. Degrauwe et al., *The Adam Analog Design Automation System*, ISCAS 1990 Conf. Proc., pp.820-822.

[21] U. Choudhury et al., *Constraint Generation for Routing Analog Circuits*, 27th DAC 1990, Conf. Proc., pp.561-566.

[22] U. Choudhury et al., *Constraint-Based Channel Routing for Analog and Mixed Analog/Digital Circuits*, ICCAD 1990, Conf. Proc, pp.198-201.

[23] E. Malavasi et al., *A Routing Methodology for Analog Integrated Circuits*, ICCAD 1990, Conf. Proc., pp.202-205.

[24] C.Y. Lee, *An Algorithm for Path Connections and Its Applications*, IRE Trans. on Electr. Comp., Vol.10, 1961, pp.346-365.

[25] F. Rubin, *The Lee Path Connection Algorithm*, IEEE Trans. on Comp., Vol.23, 1974, pp.907-914.

[26] Y. Lin et al., *Hybrid Routing*, IEEE Trans. on CAD, Vol.9, 1990, pp.151-157.

[27] T. Kostelijk et al., *Automated Verification of Library-Based IC Designs*, IEEE J. of Solid State Circuits, pp.394-403, 1991.

Biographies

Gerard F.M. Beenker received the M.Sc. degree in mathematics from the University of Technology in Eindhoven, The Netherlands, in 1980. He joined Philips Research Labs in Eindhoven in September 1980. He was with the mathematics group from 1980 to 1986 working on discrete mathematics. From 1986 to 1988 he was involved in logic synthesis. Since January 1989 he has been department head of the group CAD for VLSI Circuits. His current interests include: 1) simulation and verification, 2) analogue synthesis and 3) design management.

John D. Conway received the B.E. degree in electronic engineering from University College Galway, Ireland in 1984 and the M.Sc. degree in microelectronics from Trinity College Dublin, Ireland in 1990. Since 1984 he has worked in the CAD for VLSI Circuits group at Philips Research Laboratories, Eindhoven, The Netherlands. In this period he has worked on layout verification and automatic layout generation for both digital and analog circuits. His current interests are in layout synthesis, design management and object-oriented programming.

Guido G. Schrooten received the M.Sc. degree from the Department of Electrical Engineering, University of Leuven, Belgium, in 1982. In 1982 he joined Philips Research Laboratories, Eindhoven. In this period he has worked on switch level simulation, IC design verification, placement optimisation and analog synthesis. His current research interests include analog synthesis, optimisation and analog layout automation.

André G.J. Slenter received the M.Sc. and Ph.D. degree in electrical engineering from the University of Eindhoven, in 1985 and 1990. In 1990 he joined Philips Research Laboratories in Eindhoven. Since then he has worked on the layout design automation of analog circuits. His research interests are in CAD for VLSI and Computational Geometry.

Tools for Analog Design

C. Meixenberger, R. K. Henderson, L. Astier, M. Degrauwe

Centre Suisse d'Electronique et de Microtechnique
Neuchâtel
Switzerland

Abstract

This paper presents a synopsis of the state-of-the-art in the field of analog design automation. The advantages and the limitations of the existing approaches are reviewed. Based on these considerations a new design tool for analog circuits is proposed. Its developement is driven by the EDA framework technology and key user's requirements such as "zero-programming" interfaces. The tool is fully integrated in the Mentor Graphics FALCON Framework and is intended to suit modern mixed-mode design methodologies.

1. Introduction

The realization of microsystems has been made possible by the fast evolution of VLSI technology, the maturity of modern (Bi)CMOS processes and the progress in micropatterning techniques. In a wider sense, a microsystem contains, on the same chip, sensors / actuators together with signal processing. In a narrower sense they contain analog and real time digital processing as well as data processing on the same mixed-signal ASIC.

The movement of whole mixed-signal systems onto a single chip is driven by customers calling for more integrated functionality. Their specifications imply greater speed and density on one hand, while on the other, necessitate reduced power dissipation, lower chip counts and less cost. A number of mixed-signal markets such as telecommunication and signal processing are already experiencing rapid growth. New mixed signal opportunities are emerging in domains such as optical communication and numerical television. Further vast potential is in neural networks, voice recognition, artificial vision and computer interfaces.

The success of mixed-signal systems has been confirmed by market research. Integrated Circuit Engineering Corporation estimates the total mixed mode IC market to be 320 million dollar in 1990, and predict a growth of 30% from 1990 to 1995. In parallel, the proportion of all ICs containing analog circuits has reached 30% and is expected to double in the next decade.

Mixed-signal system design requires a long development cycle, which refines the system into sub-systems through a top-down methodology. This decomposition is based on a Hardware Description Language (HDL) and rigorous design validation. Time-to-market constraints are complicated by the risk that the design may not work the first time. These constraints promote a great demand for a variety of tools to assist and speed up the design process.

In most catalogues the IC designer will find a section on "Analog Chip Design Tools" and imagines that he will find the ultimate electronic design automation (EDA) package: a silicon compiler able to automatically generate a dense layout starting from the circuit described in his favorite analog behavioral language. Although this is worthy goal for EDA design tools, this ideal solution will not be fully achieved in the immediate future.

With the exception of the commercialized silicon compilers IDAC and MIDAS (see table 1 in section 3), the market of analog design tools is widely dominated by simulators. This is due to three major factors. Firstly, the EDA market is principally driven by digital design. Secondly, high non-recurring engineering expenses for ASIC approval have led simulation to become mandatory for successful design. Thirdly, analog synthesis is commonly perceived to be one of the most knowledge-intensive IC design tasks. The techniques needed to build and size good analog circuits are not formalized and rely solely on the experience of individual designers. The most important difficulties are:

- Direct relations between components and the overall circuit which implies a less consistent and accepted hierarchy.

- Tighter coupling between the circuit, the layout and the technology.

- Large number of schematics for each class of circuits.

- Large set of diverse specifications.

- Loose form of abstraction.

Unfortunately universities do not provide sufficient instruction in systematic analog system design. This results in a chronic lack of analog designers and strengthens the need for EDA tool allowing designers to document their know-how and to make it re-usable.

The next section is an overview of the mainstream of EDA market. It is centered on simulation and framework concepts which are the key parts of mixed-mode design automation. Section three presents the history of analog silicon compilers and compares different approaches. The limitations of the available tools are highlighted. Section four details the target users and key requirements concerning design tools. Degree of flexibility, quality of interfaces and speed are reviewed. Section five examines the trends in analog synthesis. The emergence of a new generation of analog system design tools is presented. A conclusion ends this paper and summarizes the topics presented.

2. EDA Mainstream

Fields of activity such as layout and verification, design entry, simulation and test are supported by numerous EDA tools. Among these, simulators and frameworks are essential topics to be considered in the realization of analog system design tools.

Isolated simulators are being replaced by emerging simulation environments. Their purpose is to integrate many different analog and digital algorithms to offer a single multi-technology simulation platform (e.g. LSIM from Mentor Graphics). A complete survey on mixed analog-digital simulation is provided in [1]. This symbiosis is also supported by the efforts invested in the definition of an analog extension to VHDL. However the available simulation environments have still far to go and present many weaknesses. The analog designer must switch between different behavioral syntaxes (e.g. MAST from Analogy, FAS from Anacad). He must find an optimal way to partition his mixed-signal system between several available algorithms. For each component of his system he has to determine the appropriate level of modelling for an efficient system simulation.

Another area of improvement for simulators is their interfaces. Many simulators already offer easy to use human interfaces but with limited flexibility (e.g. Accusim from Mentor Graphics, HSPICE from Metasoftware, ELDO from Anacad and SABER from Analogy). Their objectives are to support a correct-by-construction editing of simulation files, the numerical and graphical post-processing of simulation results and the automation of different analysis under various conditions. However, none of them are flexible enough or present a sufficiently satisfactory solution for analog design automation. Some interfaces allow parameter sweeping for multiple sequential runs but prevent modifications of the electrical schematic. Thus it is not possible, for example, automatically to extract a gain-bandwidth product or an input offset voltage. The user has to adapt by hand the netlist between analyses. Other interfaces provide processing facilities restricted to results available during simulation. Consequently, it is, for instance, not possible to extract a voltage swing. Generally, these interfaces are attached to a parent simulator and do not allow the user to extend they capabilities with his own simulation menus or to re-use them within different tools.

Although macromodelling is becoming a key activity in system design, little significant progress has been made in model back-annotation. All available techniques are based on a numerical optimiser leading to low modelling turn-around time.

The management of complex system design flows combined with time-to-market constraints and product quality needs new engineering methodologies. Since the 1990's, concurrent engineering has continued to pick up steam. This promising method is an efficient and systematic approach to system design. It is intended to encourage the designer starting from the system specifications, to consider all components of the system life cycle, from conception to disposal, including quality, cost, schedule and user requirements. It focuses on the interaction of all engineering activities. Compared to the traditional design methodology, it results in a transfer of effort from the verification and validation phase to conception, with more "what-if" analysis during detailed design. A 30% to 40% time saving is estimated. This method works if all necessary tools are integrated in a common framework accessible by multiple users in a network. Such a framework is called a concurrent design environment. The ultimate framework will allow the designer to choose from a large selection of tools and build a user-configurable environment tailored to his design methodology. However, many frameworks present this ambition, but there is a poor database sharing and a lack standard software interfaces between them. One of the most advanced commercialized framework is FALCON from Mentor Graphics. It implies six significant changes in design methodology:

1. In-house tools can be encapsulated into the framework due to database access routines and proprietary language for the software.

2. The frameworks offer a package of integrated tools covering a wide range of engineering and management activities.

3. The frameworks manage and communicate data between the integrated tools through a library of callable subroutines.

4. Their human interfaces can be tailored to the end user.

5. The frameworks facilitate the use and the propagation of software standards.

6. The frameworks accommodate different design flows.

The development of dedicated tools for analog synthesis is a difficult and expensive process which can barely be supported due to it's relative small market compared to the digital field. Meanwhile the accession of EDA frameworks leverages the development cost of design tools. The frameworks offer reliable libraries of subroutines which reduce the software development time and increase its overall quality. Therefore it is essential to built the analog synthesis tool into a framework for electronic design. This is a general conclusion for most design software which is increasingly finding that isolated tools cannot survive. The price to pay for tool encapsulation is the need for a high degree of competence in modern programming techniques and a familiarity with the architecture of the software framework.

3. Existing approaches

Analog synthesis tools have awaited the development of reliable analog simulators, numerical algorithms and faster computers in the 1970's. The increased demand for full-custom integrated analog circuits drove the development of the first analog circuit compilers in the mid 1980's. The way was lead by the first version of IDAC [2], a tool which was one of the first to demonstrate in practice the synthesis path from specifications to silicon.

Key features to the success of IDAC were its library of schematics, its speed based on hard-coded design algorithms and inbuilt simulation. The advance of IDAC on similar systems was achieved at the expense of making compromises on the openness and flexibility of the software. The general trend towards more user-friendly, interfacable and customizable design tools has led to a host of synthesis programs which aim for more complete automation of the design process. Some approaches take purely numerical strategies, employing powerful global optimization engines while others make use of rule-based algorithmic methods. The motivation for synthesis tools is clearly to make better use of the facilities and to deal more in mixed analog/digital systems.

At the start of the 1990's, more than twenty analog synthesis tools were presented. Significantly, only a few of these have made it to commercial availability, the majority being the products of universities, suggesting that certain important issues have still to be resolved. The different competing approaches to analog design automation and their difficulties will now be reviewed.

A successful CAD tool should aim to accumulate the knowledge of its users and to adapt easily to new technologies, circuit topologies and performance specifications. This leads directly to the key ideas of openness and re-usability. The following approaches have been taken to capture design knowledge

1. *Algorithmic encapsulation of a designer's knowledge in a programming language form* [3]. The so-called *design plan* is a sequence of instructions or calls to subroutines written in a high level programming language to determine the dimensions of elementary devices [4], [5]. This approach suffers from the difficulty of capturing such knowledge. The designer has to think as a programmer, which demands a high level of computer skills. Developing a design plan is error prone since their are few aids to ensure correctness and the user must know the program architecture well. One problem is that the designer's knowledge may not be formalized enough to be expressible in the form of a design plan. The major advantage of this technique is a high degree of control by the designer and its ability to gain speed by hardwired shortcuts. The time to development of a design plan would be typically one month for an engineer.

2. *Rule based knowledge entry.* Introduce into a knowledge base a set of "if-then" rules about a block and allow a task co-ordinator to select rules appropriate to the stage of the design. This allows heuristic knowledge to be expressed easily but suffers from a relatively slow knowledge-entry (up to one year for a full-custom design) and computation time. This approach has also been criticized for taking too much control from the designer.

3. *Equation based synthesis.* The designer enters equations describing the relations of design variables in his circuit. An equation manager manipulates the system of equation to determine unknowns or to request more information. Advantages are that this is a natural expression of analog design relationships. A design can be viewed in many different directions, either from specifications to design variables or the inverse or combinations of the two. This is a useful aid to design exploration since both analysis and design can be done within the one framework. On the negative side, the equation solution system must be sophisticated. Since not all designs can be broken down into linear or inversable nonlinear equations, numerical methods must sometimes be employed with inevitable convergence and speed problems.

Approaches 2 and 3 have an important drawback in that they have no interface with an external simulator. They rely entirely on the internal device model and circuit equations entered by the user for the precision of characteristics of the dimensioned circuits.

All the approaches above are based, to a certain extent, on having available equations describing the behavior of the circuit in terms of the design variables. The task of extracting useful equations of circuit performance measures is traditionally performed by hand analysis. This process is knowledge intensive and laborious and hence various attempts have been made at automation by *symbolic simulation* [6]. The most successful systems are able to extract from linearized circuit equations an arbitrary transfer function in the frequency domain e.g. gain or PSRR. In general, symbolic analysis suffers from the following dilemma: if the equation sought is simple and contains only a few dominant terms, it can probably be written down by inspection. But if it is complex and contains many terms, it no longer offers any real insight. A major criticism of current symbolic analyses is that they work only in the small-signal domain and neglect many important characteristics. Nor is it the case that these tools obviate the need for analysis skills, since a great deal of interpretation or manipulation of the equations is necessary before they can be usefully employed in a synthesis strategy. For instance, the computation of the gain-bandwidth product, the poles and zeros of a two stages amplifier depends on the style of compensation (pole spliting or pole loading), which implies additionnal design expertise to the straightforward circuit analysis. Thus, symbolic analysis is still a long way from providing knowledge which is directly compatible with automatic synthesis methods.

A critical question which has hindered the development of a successful commercial analog synthesis package is that of the transistor model. There is no consensus of which transistor model is best and much depends on the circuit application and the speed and accuracy desired. As a result, many models exist proprietary to a given design house or simulator. This makes it exceedingly difficult to have a synthesis tool which adapts to all needs. Either the tool works with a commonly accepted simplified model but neglects certain effects desired by a particular designer and is too inaccurate, or a specific MOS model is inbuilt which needs parameters to be customized for each proprietary model. Another possibility is all-numeric optimization based on the proprietary simulator. This methods suffers from long run times and the difficulty of interfacing to different simulator file syntaxes.

Table 1 gives a comparison of the most important present-day analog synthesis packages. It can be seen that very few tools can claim to be complete in that they lack some important functionality. Some specific criticisms are:

- Few tools design both MOS and bipolar circuits.

- Worst-case design is rarely performed.

- Some tools lack a link to an external simulator for validation of synthesized circuits.

- Certain tools have limited standard libraries of circuit topologies leading to mistrust by designers.

- The introduction of know-how in the systems is tedious.

- During the set-up of design know-how there is little interactivity or feedback.

- Hierarchy is limited to single topology selection in some systems.

- Not all tools have layout synthesis.

In summary, despite mushrooming of research design tools, there is still a difficulty in providing an acceptable commercial solution. Various problems are still to be faced concerning openness, integration with frameworks and limited functionality.

Name	Developed by	Layout synthesis	External simulator	Only OTAs	Optimisation	Commercialy available	Symbolic analyser	Hierarchy	Open Interface	(Bi)CMOS	Worst-case design	Sizing Method
ACACIA	Carnegie Melon University	X	X	X	X			X	X			Knowledge-based synthesis plus expert system
OPASYN	University of Berkeley	X	X	X	X			X				Heuristic circuit selection plus steepest descent optimisation based on analytic models
MIDAS	Silicon and Software System	X				X		X	X			Knowledge based synthesis via compilable language.
BLADES	AT&T		X			X	X				X	Knowledge based synthesis based on an expert system
ARIADNE	Katholieke Universitei Leuven	X	X		X		X	X	X	X	X	Simulated annealing optimisation based on analytic models drawn from symbolic analysis
SEAS	Twente University			X	X		X					Knowledge based synthesis plus combinatorial optimisation sizing
CNET	CNET	X	X	X	X		X					Structure recognition plus optimisation. Hard-coded design equations.
OASE	University of Dortmund		X	X			X	X			X	Knowledge based synthesis plus rule-based sizing with simplified device equations.
CHIPAIDE	Imperial College London	X	X	X	X		X			X	X	Hard-coded design equations plus quadratic programming optimisation.
STAIC	University of Waterloo	X			X			X	X	X	X	Object-oriented program knowledge entry. Classical optimisation.
An-Com	General Electric	X		X				X	X			Knowledge-based synthesis.
ADAM	CSEM	X	X		X	X	X		X	X	X	Hard-coded design plans based on analytic circuit knowledge. Optimisation by simulated-annealing and steepest descent.

Table 1: Comparison between different analog synthesis tools.

4. Target Users and Key Requirements

Analog designers are highly skilled engineers who are most accustomed to exercising their creativity on a sheet of paper and are often resistant to change. This partially explains the relatively poor acceptance of the present analog sizing tools based on hard-coded design rules. The new generation of synthesis tools must provide high flexibility and be easily applied. One method is to use a spreadsheet. It allows sizing strategies to be entered in a computer executable form without a special skill in software.

To facilitate the set-up of a sizing strategy, the design tool should offer a library of design primitives. These are model manipulators (e.g. calculation of the W/L gate ratio for a given drain current and a given transconductance), device calculators (e.g. design of an MOS transistor for a given W/L ratio under random offset, 1/f noise and gain conditions) and structure calculators (e.g. design of a bipolar current mirror for a given output current precision). These design primitives are based on simplified models equations. In this way, the stored sizing strategies are almost independant of the technology. In return, the circuit characteristics must be verified through the designers simulator. According to the simulation results, a control machanism (embedded feedback loops or optimizers) automatically defines new targets for the sizing strategy.

The system designer is mostly active during the circuit conception. He needs a lot of automation to evaluate different architectural trade-offs. The lowest level of detail he is concerned with is the building block level. A critical building block is only re-designed from scratch if all workarounds at the system level have been exhausted. Hence, it is essential to offer libraries of fixed standard building blocks, popular schematics sizeable for a wide range of specifications and reliable macromodels of different complexity.

Macromodelling and circuit characterization are important activities in mixed-signal system design, firstly for system specification in top-down design methodology and secondly for a bottom-up system verification. This implies that the design tool monitors the simulators, extracts the simulation results and process them. The necessary sequence of operations should be stored in a documentation plan associated with the circuits. The design tool should offer a pool of simulation primitives to assist the setup of that plan.

Speed improvement of simulators is still not sufficient on its own: HSPICE for instance takes around seven seconds to compute the gain-bandwidth product and the phase margin of a twenty device amplifier. A synthesis tool loops around the simulator between ten and twenty times before the target amplifier specifications are met. In this way, the sizing process will spend an average of two minutes to design an amplifier under basic AC conditions. To keep tool efficiency for large scale systems, a task manager becomes necessary to automatically decompose the overall design process in independant sub-tasks, which can be devided among the different processors across a computer network.

A successful system design implies an extensive overhead in simulation effort. In this field there is a place for design aids which prevent excessive simulation in order to improve development quality and time. A system verification tool can check the electrical compatibility between system components before simulation and select the most appropriate level of description for an efficient system simulation.

As shown in Tables 2 and 3, system design know-how is less mature than that of building blocks. Four levels of maturity can be defined. The first level is *feeling* and represents the heuristic design rules. The second level is *equation*. The third level, partial *strategy*, describes how equations are processed. The ultimate level, *algorithm*, incorporates general methods for circuit design [5]. Design methods for mixed-signal systems are complex and mostly based on heuristics, feeling and intuition. Its formaliztion needs many investigations and represents an active area of research for universities. However, until a well developed methodology is available, the gap could be bridged by "fuzzy techniques". Such an approach, suggested by fuzzy logic is illustrated in Fig.1.

	Feeling	Equation	Strategy	Algorithm
Simulation			X	
Synthesis			X	
Layout		X		
Worst-case design				X
Floorplaning		X		
Testing			X	

Table 2: Building block design knowledge maturity.

	Feeling	Equation	Strategy	Algorithm
Simulation			X	
Synthesis	X ———→			
Layout		X ———→		
Worst-case design	X ——→			
Floorplaning	X			
Testing			X	

Table 3: Mixed signal system design knowledge maturity.

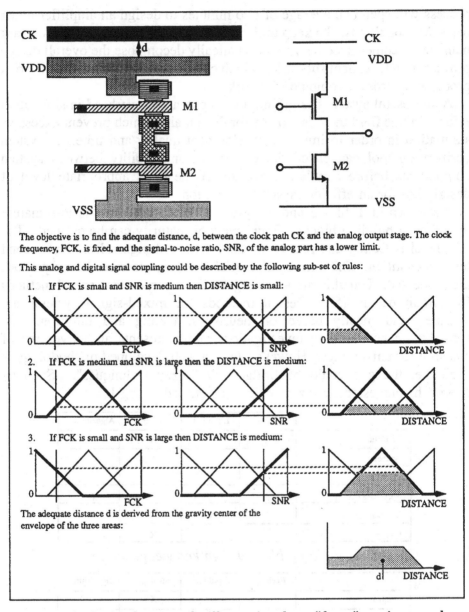

The objective is to find the adequate distance, d, between the clock path CK and the analog output stage. The clock frequency, FCK, is fixed, and the signal-to-noise ratio, SNR, of the analog part has a lower limit.

This analog and digital signal coupling could be described by the following sub-set of rules:

1. If FCK is small and SNR is medium then DISTANCE is small:

2. If FCK is medium and SNR is large then the DISTANCE is medium:

3. If FCK is small and SNR is large then DISTANCE is medium:

The adequate distance d is derived from the gravity center of the envelope of the three areas:

Figure 1: A simple example illustrating how "fuzzy" engine can be used by design tools to handle heuristic rules.

In summary, the key requirements for a design tool suitable to both analog and system designers are:

- "Zero programming" interface for knowledge entry.

- Knowledge portability on different models and technologies.

- Toolbox of design and simulation primitives.

- Library of standard circuits and datasheets, schematics and macromodels, sizing and documentation plans.

- Support for hierarchical top-down and bottom-up design method.

- Design aids for system verification.

5. Advanced Analog Design Tool

In this section, the concept behind a new analog design tool, CADI ILAC, is presented. The starting point for the tool is integration with an EDA framework, namely Mentor Graphic's FALCON Framework.

5.1 Integration with EDA Frameworks

Mentor Graphic's FALCON Framework is a state-of-the-art EDA framework written with the philosophy of openness. Thus design data can be shared and concurently accessed from simulators, schematic capture tools, layout tools and circuit synthesis tools. Customization of the tools (or "userware") is done using user-defined software written in a proprietary language called AMPLE. The four key tools used for the new concept are: the schematic entry tools Design Architect (DA) and Design Viewpoint Editor (DVE), the spreadsheet Decision Support System (DSS) and the layout tool IC-Station.

DA contains a sophisticated design database capable of storing both schematic and VHDL models. The ideas of VHDL have been adopted in its construction allowing the user to change easily between equivalent behavioral or schematic models. After using DA to edit independant sheets, a design hierarchy is built with DVE. This hierarchy is obtained by the choice of a model (schematic or behavioral) for each cell instance included in this design. During this phase the design is configured ready so as to be passed to a downstream tool, such as a simulator or a layout editor.

Another important FALCON service is an inter-process communication aid tool (IPC) which makes it easy for two applications to exchange data or commands through a socket. This is used to join the tools DA and DSS together. Fig. 2 illustrates the integration of the new design tool CADI ILAC into FALCON.

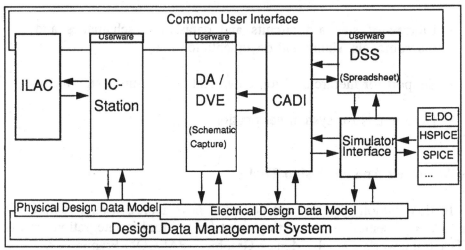

Figure 2: View of the design tool CADI ILAC integrated into FALCON.

5.2 A Spreadsheet as a Knowledge Entry Mechanism

The problem of offering a flexible environment for entry of analog design knowledge, which does not require a designer to be a software specialist is particularly pertinent. Current systems err too much on the side of sophisticated artificial intelligence approaches or towards a full-blown computer language, both of which require programming expertise. A compromise is the *spreadsheet* which is a numerical calculation environment already familiar to many non-programmers. It has advantages of flexibility, openness and "zero programming". Indeed it's an interactive programming environment with advanced debugging and testing capabilities. Spreadsheets are ideal vehicles for simple programs without complicated data structures or algorithms, such as is likely to be the case for most analog design.

The DSS spreadsheet system from Mentor Graphics provides more capabilities than most conventional spreadsheets in that C code can be called, design databases can be accessed and graphical interfaces can be created. The spreadsheet can contain simple subroutines or macros with loop and if structures as well as data tables. These spreadsheet macros, once developed and verified, can be translated into C and called as design primitives by higher level routines. In this way, there is a cycle of knowledge re-use as circuit specific knowledge is moved from development in the spreadsheet to wider availability as a tested library routine.

Analog designers should be encouraged to take a structured approach to the design of their analog circuit, rather than be forced into it by the organization of the system. This can be done by demonstrating re-usability with the provision of libraries of circuit design primitives and by showing examples of structured design plans which have wider applicability. There is still very little formalization in analog design, and so the system must allow free expression of an individual designers 'style'.

5.3 User Modes

Different users need different modes of access to the design tool. This principle is embedded in the tool by recognizing the following modes:

1. A *foundry client designer* is a user of a proprietary library. As a customer of the foundry design aid services, he has only access to the execution of the library design plans. He does not have read or write privileges. The only control he has over a cell synthesis is the definition of the specifications.

2. A *system designer* is regarded as someone who is a user of the design tool knowledge library. He has read and execution privileges for the design plans. The possibility remains for the system designer to be a novice or an expert but it is assumed that he will simply want to use the knowledge in the system without wishing to change or add to it. He may still wish to have more or less detail on which operations are being performed and on guiding those operations. This is the role of a flexible system of defaults, help, checking and options. The novice is a user who will make most use of the built-in defaults, help and auto-selection mechanisms of the system. The expert will be more likely to override these defaults.

3. A *knowledge engineer* is a user who wishes to extend the knowledge base of the program by combining existing design primitives. This person receives read, write and execution access to the design plans. There are various possibilities to enter knowledge.

a) Structural knowledge: a new circuit architecture or a variant of an existing architecture is to be introduced. This will be done with a schematic editor.

b) Design knowledge: a sequence of instructions to design a particular *circuit variant*. This will be written in terms of a library of pre-defined model primitives or circuit design tools. This may equally well be the design of a macromodel from high level specifications.

c) Analysis knowledge: a sequence of instructions to extract a particular performance measure from a simulator.

d) Circuit topology selection knowledge: a performance measure to be used to compare the relative merits of a set of functionally equivalent circuit topologies plus a sequence of instructions.

Access to the knowledge introduced by the knowledge engineer may be controlled.

4. A *model expert*: is someone who wishes to develop new design and simulation primitives. Such a user is likely to want to write more complex or widely used algorithms which are no longer suited to development in a spreadsheet environment and needs access to the C/C++ code level. The changes made by the model expert will be globally available to other users of the system. All rights are given to the model expert: read, write, execution and compilation of design plans. Only specially authorized people/ourselves are allowed here.

5.4 Hierarchical Design

The user of the system must be encouraged to design circuits in a structured manner by providing support for hierarchical entry of design knowledge. This means that both the storage of circuit schematics and associated design knowledge is held in a hierarchical form. Users are encouraged (though not compelled) to observe the hierarchy for maximum design knowledge re-use. Schematics are entered using DA.

The storage of a set of equivalent schematic, macro or behavioral models is a feature of the most modern design entry tools. The synthesis of the

385

latter two can be accomplished in a unified manner from the spreadsheet enabling both top-down and bottom-up design strategies. A library of common schematics, macromodels and symbols is already available for the user.

Design plan hierarchy is supported by the ability of one design plan to call other designplans(normally for sub-components). The different specifications translation mechanisms are reprensented in Fig.3. other designplans(normally for sub-components). The different specifications translation mechanisms are reprensented in Fig.3. The specifications decomposition combining these mechanisms is illustrated in Fig.4 for an A/D converter. A screenshot of the hierarchical systhesis of a cascaded filter is illustrated in Fig.5.

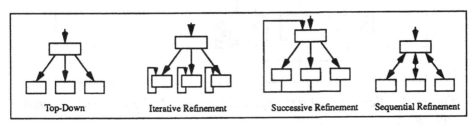

Figure 3: Four major specifications translation mechanisms.

5.5 Smart Simulator Interface

The difficulty of exchanging data with simulators has been mentioned as a impediment to portable analog designs. The reason is the diversity of simulator input/output formats and functionalities. The smart simulator interface aims to provide a toolbox of simulation primitives to protect the user from knowing the syntax of a particular simulator. Simulation files can then be exchanged and translated between simulators.

The simulator input and output syntaxes are described in an input file which is used to set up a parser. The simulation files can then be read and a database of simulation run data created. The user accesses the simulation database via a toolbox of functions from the spreadsheet DSS. The spreadsheet is a natural environment to do post-processing of simulation results because it deals easily with data in tabular form and can prepare graphs or do calculations on columns of figures easily.

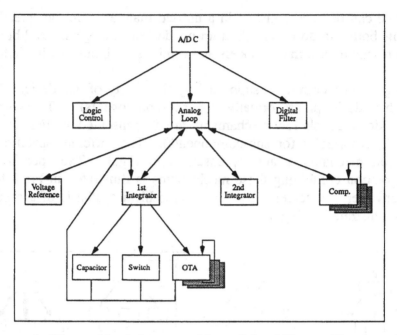

Figure 4: Hierarchical decomposition of the A/D specifications into sub-blocks.

5.6 System Verification

The degree of flexibility offered by the new tool to build up designs in a hierarchical manner from a wide range of library blocks is accompanied by a certain risk of human error. It is a simple matter to take a library block and place it mistakenly in a system for which it was not originally designed. Elimination of such errors can be done by means of a verification tool which checks electrical specifications of a block against expected performances of its environment.

The user is able to describe rules through flexible interactive query menus. This way, he can configure a set of rules to the needs of a specific block. A check is then run using these rules every time a new block is instantiated or specifications are changed. The errors and warnings found can be report to the user either in an interactive or batch mode. In addition, the tool proposes a sufficient description level (gate model or macromodel) for each block to minimize the system simulation time.

387

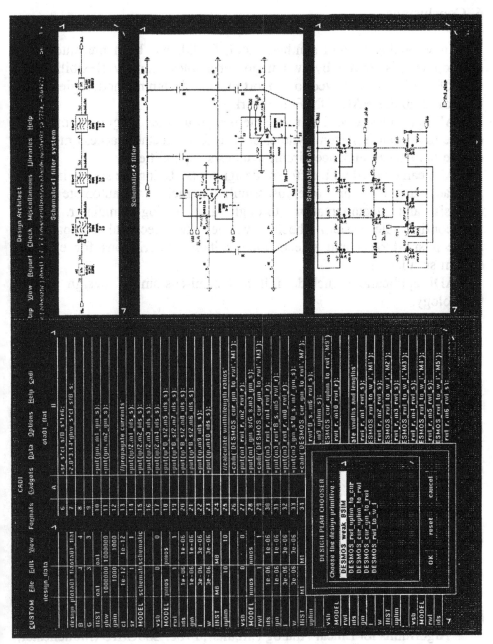

Figure 5: Screenshot example of a cascaded filter synthesis.

6. Conclusions

A new analog system synthesis tool, CADI, has been presented. Its development is driven by two major principles, namely flexibility and ease-of-use. It is conceived to support modern design methodologies in the Mentor Graphics FALCON Framework.

CADI provides a pool of re-usable design utilities ranging from single device to complete system synthesis. It offers selectable access modes to enable the user to enhance circuit synthesis capabilities.

CADI can be used with different design tools. Using the smart simulator interface, the user can store his simulation recipes to automate circuit analysis and documentation through any analog simulator. The user-configurable system checker verifies the electrical compatibility between sub-blocks and selects their suitable descriptions for efficient system simulations.

CADI applications include full and semi-custom IC design in any technology.

References

[1] Hans W. Klein, *Mixed Analog-Digital Simulation in the 1990's*, Technology Information Publishing 1991.

[2] M. Degrauwe et al., "IDAC: An Interactive Design Tool for Analog CMOS Circuits", *IEEE J. Solid-State Circuits*, vol. SC-22, no.6, December 1987, pp.1106-1116.

[3] J. Jongsma, C. Meixenberger, B. Goffart, J. Litsios, M. Pierre, S. Seda, G. Di Domenico, P. Deck, L. Menevaut and M. Degrauwe, "An Open Design Tool for Analog Circuits", *Proc. ISCAS*, Singapore, 1991, pp.2000-2003.

[4] C. Meixenberger, B. Goffart, M. Pierre, and M. Degrauwe, "Sizing Algorithms for Linear Analog Circuits", *Proc. ESSCIRC* (Manchester, England), Sept. 21-23, 1988, pp.190-193.

[5] R.K. Henderson, C. Meixenberger, B. Goffart, J. Jongsma, M. Pierre and M. Degrauwe, "Sizing of Analog Circuits for Small-Signal Gains", *Proc. EDAC* 1992.

[6] S. Seda, M. Degrauwe, and W. Fichtner, "A Symbolic Analysis Tool for Analog Circuit Design Automation", *Proc. ICCAD* (Santa Clara, CA), Nov. 1988, pp.448-491.

Biographies

Christian Meixenberger was born in Bern, Switzerland, on October 11, 1960. He received the engineering degree in electronic physics from the University of Neuchâtel in 1986. After graduation, he joined the Centre Suisse d'Electronique et de Microtechnique S.A. (CSEM) in Neuchâtel, Switzerland, where he is deputy head of the design automation section.

Luc Astier was born in Paris, France on March 13, 1966. He received the diplome d'ingenieur in electrical engineering from the Institut Superieur d'Electronique du Nord (ISEN), Lille, France, in 1988. From 1989 to 1992 he worked at Silicon Compiler Systems/Mentor Graphics Co., San Jose, CA, in the area of analog and mixed-signal synthesis and simulation front-end. Presently, he is with the Swiss Centre of Electronics and Microtechnics (CSEM), Neuchâtel, Switzerland, involved in analog design and layout synthesis.

Robert K. Henderson received the B.S. (Hons.) degree in electronics and electrical engineering with computing science and the Ph.D. degree in electrical engineering in 1986 and 1990, respectively, both from the University of Glasgow, Scotland.
From 1989 to 1990 he was engaged at the University of Glasgow as a research assistant. He is currently a Research Engineer at the Centre Suisse d'Electronique et de Microtechnique S.A., Neuchâtel, Switzerland. During his doctoral work he co-authored a compiler for analog filters called PANDDA. His research interest are in the areas of analog signal processing, numerical algorithms, and computer-aided circuit and system design.

Strategies and Routines in Analog Design

E.H. Nordholt

Delft University of Technology
Catena Microelectronics B.V.
Delft
The Netherlands

Abstract

This paper addresses fundamental aspects of analog design automation. Rather than following the present trends of basing the development of design automation tools on the existing expert knowledge of circuit topologies, proven circuit implementations and application examples, it emphasizes the need for the development of strategies and routines for generating specific solutions for design problems, starting from a definition of the required behaviour of a basic signal-processing function. The proposed design approach differs from the traditional approach in that it explicitly takes the signal information as a starting point for setting up performance requirements. As a prerequisite for true design automation, it thereby attempts to explicitly formulate all steps required to find adequate solutions, formulated in terms of information processing fidelity. It isolates the various performance aspects, tries to find optimization strategies for each of them individually, and subsequently, tries to find strategies to minimize their interaction. It concentrates first on three essential limitations of all physical systems: speed, noise and power, the three elements of Shannon's formula for channel capacity. By attaching behavioral models to active devices with respect to each of these aspects in combination with the above mentioned strategies, one can quickly decide on the fundamental feasibility of a function, given a certain implementation technology. The strategies refer to circuit topology generation as well as to modification of the behaviour by applying error-reduction techniques in several subsequent hierarchical steps. During this process, other performance requirements such as those with respect to offsets, accuracy, non-linearity, etc., are taken into account. The routines refer to calculation methods needed for quick analysis and to the tools for the final verification.

1. Introduction

Present analog design practice is based on a traditional approach of a highly heuristic nature and is therefore unsuitable as a basis for design automation. The approach lacks rigour in making various design steps explicit. Finding smart solutions for design problems seems to be the prerogative of clever and experienced magicians who are able to communicate their tricks to their apprentices only in terms of paradigms. Looking at most textbooks on analog electronics and studying the design practice of professional electronic designers, one striking common feature is their paradigmatic approach. Rather than systematically using general methods to solve electronic design problems, they intuitively or routinely select circuit topologies and analyze their behaviour with respect to certain behavioural aspects usually not in relation to a specific environment or signal type. Attempts to synthesize electronic circuits are scarce and usually lack practical relevance. The most common ones relate to active filter design. They emphasize the network-theoretical aspects rather than the information-processing aspects.

Since electronic circuits derive their very right to exist from their excellent suitability for information processing, it is mandatory to take the information to be processed as the starting point for a synthesis approach. In this paper we will try to explicitly define a design sequence based on this view. It is believed to provide a much better basis for design automation than the present one. This approach is characterized as follows.

1. An inventory is made of information processing and reference functions needed to build electronic systems.

2. Subsequently, their possible imperfections with respect to the ideal function in relation to the information characteristics are listed, so that we can set-up behavioural models of these functions.

3. The next step is to find the mechanisms in active and passive devices that can be used to realize the most basic approximation of the function. An analysis of these mechanisms is needed to model their physical behaviour in terms relevant from the viewpoint of information processing.

4. Then, error-reduction techniques (compensation, errorfeedforward, negative feedback and isolation) are formulated and modelled from a synthesis point of view and their capabilities of improving performance aspects are investigated.

5. One or more of these techniques are applied to improve the basic function realisation to the required performance level.

6. If we don't succeed in realizing this performance level, even if we use the optimum circuit topology, parameters and technology, we have to go back to the system level and change the system architecture in order to modify the information so that it becomes less susceptive to imperfections.

For the latter purpose we can use a different category of error-reduction techniques, which includes modulation and demodulation, sampling, quantisation and coding. These will not be addressed in this paper. We will use the amplification function as a vehicle, and restrict the discussion to some aspects of the error-reduction techniques. A similar approach can be followed for all other functions.

2. Ideal information processing functions; modelling

A list of basic processing and reference functions from which a selection can be made for the implementation of an electronic system is given in Table 1. Some of the choices are more or less arbitrary. However, these functions are believed to fit into the present electronic design practice sufficiently well.

Before we discuss implementation strategies, we need definitions of these functions and their ideal behaviour, together with the types of imperfections we can expect to occur in their practical approximations. These imperfections relate to the types of signal information to be processed and to the environment in which the function has to operate.

394

They are therefore usually application specific. Setting up a specification of a function usually poses one of the most difficult (interpretation) problems of the design sequence.

<u>Information processing</u>   <u>Associated references</u>

1. Addition
2. Distribution
3. Amplification
4. Filtering   Frequency
5. Quantization   DC level
6. Memorization   Time
7. Switching   Time, level
8. Distortion

Table 1: Information processing functions with associated reference functions

Taking the amplification function as an example, we can define an ideal time-invariant amplifier as a linear two-port with an available power gain larger than unity, where one or more of its transmission parameters have a constant value different from zero. The transmission parameter representation of a two-port is given as

$$V_i = AV_o + BI_o,$$
$$I_i = CV_o + DI_o.$$

The input quantities are thus expressed in the output quantities.

The ideal amplifier types with only one parameter different from zero are the well-known transactors or controlled sources. They have input and output impedances which are either a short circuit or an open circuit. The other amplifier types, where more than one parameters differ from zero may have non-zero finite input and/or output impedances. With the complete set of ideal amplifier types, we can meet any termination requirement for both the source and the load. The termination requirements follow from the properties of the signal source and the load. An illustration is a piezo-electric transducer, which should be terminated by a zero impedance in order to eliminate the influence of the non-linear source capacitance. A filter or a transmission line should be terminated by an accurate and linear resistance.

We can summarize the requirements on various ideal amplifier types as shown in Fig.1. The voltage-current *(v-i)* relations as well as the input-output *(i-o)* relations can be plotted as straight lines through the origin in their respective planes. As a result, also the relations between source and load quantities (the *s-l* relations) can be made accurate and linear.

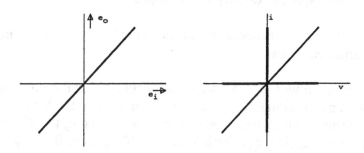

Figure 1:   *i-o relation of an ideal amplifier(a)*
v-i relations of the input and output ports(b)
1. short circuit 2. open circuit   3. finite resistance

3. Definitions of imperfections; specifications

As any physical system, an amplifier will have a limited information-handling capacity related to the essential limitations *noise, speed and power.* However, the fidelity of information transfer is limited as well by the non-perfect predictability of the parameters of the devices with which the amplifier is built. This non-perfect predictability prevents that adjustments or compensation techniques can fully eliminate the imperfections.

In order to describe the behaviour of a real amplifier, a list of quantifiable data (a specification) is needed. For the purpose of easy evaluation, normally, most of this data will preferably be related to deterministic test signals. Since the characteristics of information signals can only be described in terms of probability and averages, the specification cannot describe the amplifier performance completely for real information signals. However, the use of deterministic test signals becomes justifiable if these signals relate well to important characteristics of the information.

In view of the earlier mentioned essential limitations, the most relevant signal characteristics are:

1. the probability density function of signal levels (voltages or currents).

2. the power spectral density of the signal.

3. the amount of random noise power already present in the information signal.

In general, we will have to design an amplifier for the information as it is produced by the signal source and required to drive the load. This means that we normally don't have the opportunity of coding the signal in order to exchange dynamic range for bandwidth. Therefore, from the viewpoint of the required fidelity of information processing both the signal levels and the rates of change should be handled by the amplifier adequately, thereby allowing a limited amount of errors, acceptable or just not noticeable by the observer. Moreover, we require the amplifier to contribute not more noise than acceptable or relevant in the perspective of characteristic 3.

Adequate handling of the signal levels and the rates of change means that we require a unique correspondence between the source quantity and the load quantity. Without applying error-reduction techniques to the basic amplification mechanism this unique correspondence criterion can usually not be met due to the earlier mentioned parameter dispersion.

In order to characterize the behaviour of a real amplifier we can anticipate on the properties of the active devices, realizing that their behaviour can be described by non-linear dynamic differential equations with spreading and varying parameters and that they inherently generate noise. As a first requirement for linearity, the i-o relation should approximate a straight line through the origin. Departures from this linear instantaneous behaviour can be described in terms of

- operating point
- offset and offset drift
- small-signal relations
- accuracy
- clipping points
- (differential) non-linearity
- differential gain

Departures from the small-signal instantaneous behaviour due to speed limitations are described in terms of

- magnitude and argument plots
- poles and zeros
- (im)pulse response

Departures from the linear instantaneous or dynamic behaviour are described in terms of

- differential gain and phase
- harmonic distortion at specified level and frequency
- intermodulation distortion at specified levels and frequencies
- intercept points
- slewing
- power bandwidth
- transient intermodulation distortion
- settling time
- overdrive recovery

The noise behaviour can be described in terms of

- signal-to-noise ratio
- noise figure (only if the source immittance has a real part)

Useful characterizations, combining the power, distortion and noise aspects are the:

- dynamic range
- intermodulation or distortion free dynamic range

measured in a given bandwidth and at certain frequencies.

The specifications of all these behavioural aspects allow us to build a behavioural model of an amplifier and to relate this to the properties of the active devices. For the latter purpose, however, we need to develop a design strategy that allows us to minimize the interaction between the behavioural aspects. This design strategy heavily relies on the availability and applicability of error-reduction techniques with respect to the basic amplification mechanism of the active devices. In this section we didn't mention the imperfections related to the power supply. They are very much determined by the biasing methods and can be combated by error reduction techniques as well, including isolation. We don't deal with these in this paper.

4. The amplification mechanism; device and stage modelling

In order to realise a real amplifier as an appropriate approximation of an ideal amplifier, we have to use the amplification mechanism available in active devices and realize that the amplification of signal power is only possible by supplying (DC) power to the device. Under control of the input signal this (DC) power is partly converted into signal power available at the output. Non-deterministic power should be sufficiently suppressed at the output in order not to unacceptably mutilate the signal information.

Most important for amplification purposes are the almost invariably used three-terminal semi-conductor devices such as the bipolar transistor and the various types of field-effect transistors. We can model these devices as non-linear two-ports where two terminals are interconnected. For the bipolar transistor and the various types of field-effect devices, we consider the common-emitter (CE) and common-source (CS) configurations, respectively, as the basic amplification stages. The reason for this choice is that these configurations appear to have the largest available power gain. More arguments for this choice will be given later. Without loosing generality, we will restrict our considerations to bipolar transistors.

The modelling issue is very important from a viewpoint of design strategy. As long as we don't have a complete schematic of the amplifier, it makes no sense to use the complete model equations. The models are needed for synthesis rather than for analysis purposes. The following approach is suggested.

1. For characterizing the low-frequency behaviour, a reduced set of model equations is used. For the bipolar transistor we use, for example, the Ebers-Moll model, together with a forward Early voltage. This model is sufficiently accurate to allow us to relate the amplifier stage behaviour to the device behaviour in terms of the first set of specification items.

2. From this model, we can derive a small-signal model, where we can add the small-signal capacitances and thus obtain for the bipolar transistor the hybrid-π model, the parameters of which can be expressed in those of the large-signal model and the operating

quantities. To this model we also add the (stationary) noise sources expressed in the operating currents and the bulk resistances. This model allows us to make calculations and estimations with respect to gain, impedances, poles and zeros, impulse response, equivalent noise sources, etc. closely related to the physical device parameters.

Besides these two device models, however, we need models that allow us to investigate the effectiveness and the consequences of error-reduction techniques in terms of modification of the behaviour.

3. For investigating the influence of error-reduction techniques on the non-linear behaviour, we use a generalized non-linear s-l relations of the form

$$e_l = \hat{e}_l(e_s),$$

where e_l and e_s are the source and load quantities, respectively.

4. For investigating the effects on the small-signal behaviour, we use the transmission-parameter representation as introduced in section 2.

It will be shown that these models, together with some related behavioral models allow us to set up a design strategy. The final verification of the amplifier structure will have to be performed with the complete model equations. Notice, however, that the noise performance cannot be verified by means of the common types of simulators since they allow only a stationary representation of the noise. The assumption of the noise being stationary is usually incorrect in real amplifiers, not to mention oscillators, mixers, comparators, etc.

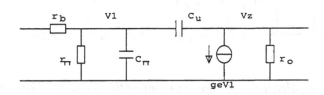

Figure 2: Hybrid π model of the bipolar transistor

From the device models, various conclusions can be drawn. The large-signal model indicates the region where the device has to be operated in order to retrieve all signal information at its output. The small-signal model provides the means for estimating the pole and zero locations and the various gain parameters with a given source and load in relation to the operating point. It may be illustrative to show how the original model can easily be adapted to a model better suited to estimate the small-signal dynamic behaviour. For this purpose, the hybrid-π model is given in Fig.2. If the source and the load are connected to this model, the gain quantities (voltage gain, transadmittance, transimpedance and current gain) can be calculated routinely by using the MNA method. A modification of the model, however, provides us with much more insight. The steps involved in obtaining this modified model are shown in figures 3 to 5. The first step replaces the capacitor C_μ using the substitution theorem by two current sources, equal to $sC_\mu \cdot V_1$ and $sC_\mu \cdot V_o$, respectively.

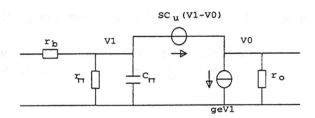

Figure 3: Application of the substitution theorem to replace C_μ by a current source

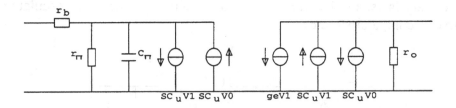

Figure 4: Application of the current-split theorem. The current flowing into the nodes do not change

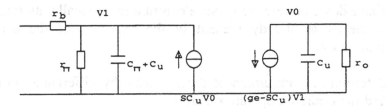

Figure 5: Application of the substitution theorem to replace two current sources by C_μ at both the input and the output

The second step uses the current splitting theorem in order to obtain the representation of Fig.4. Two of the current sources can now be replaced by capacitors by using once more the substitution theorem as shown in Fig.5. This modified model allows us to immediately find the right half-plane zero and to more easily calculate the poles and zeros in the impedance and transfer functions.

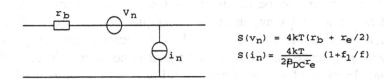

$$S(v_n) = 4kT(r_b + r_e/2)$$
$$S(i_n) = \frac{4kT}{2\beta_{DC}r_e}(1+f_1/f)$$

Figure 6: Low frequency noise model for the bipolar transistor

The small signal model also allows us to set up a very simple behavioural noise model for the device in terms of equivalent input noise sources. This low-frequency noise model is shown in Fig.6 for a bipolar transistor. The parameters of the spectra of these sources can be given in terms of operating point quantities and/or in terms of small-signal parameters. In combination with a source impedance specification, the model gives us all the information we need for the optimization of the signal-to-noise ratio. The steps involved in this optimization are simple.

1. A Norton or Thevenin transformation gives the total equivalent noise in series or in parallel with the source.

2. Find a device where the r_b noise contribution is small with respect to the noise already present in the real part of the source immittance.

3. Determine the minimum of the spectrum by differentiation and find the optimum operating current.

Though admittedly this optimization is performed for a CE stage which has very limited practical relevance from the viewpoint of information processing, it can be shown that it has a general relevance for all amplifier configurations including those with negative feedback. The inclusion of high-frequency noise influences is only slightly more involved.

5. Behaviour modifications and topology generation by applying compensation techniques to CE and CS stages

The CE and CS stages form poor approximations of the ideal amplifier. If an improvement is needed, error-reduction techniques will have to be used. The first one to be discussed is the application of compensation techniques. The first compensation technique that can be applied has a multiplicative character. The network that compensates for predictable errors with respect to non-linearity or speed is connected in cascade with the stage. In view of the poor predictability due to parameter dispersion, this method is hardly useful in the case of a CE- or CS stage. The second method has an additive character. It is based on the synthesis of an odd s-l function, where the original s-l relation contains both even and odd terms. The odd function contains no offset term and no even-order terms and therefore provides an improved approximation of the ideal amplifier characteristic. We will illustrate this method by some examples.

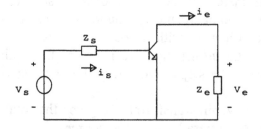

Figure 7: CE stage between source and load

Fig.7 shows for this purpose a CE stage inserted between a signal source and a load. The source-load relation in terms of source and load quantities (voltages or currents) can then be written as

$$e_{l1} = \hat{e}_l(e_s) .$$

An odd function can be synthesized if we have a second source-load relation given by

$$e_{l2} = \hat{e}_l(-e_s) ,$$

or

$$e_{l2} = -\hat{e}_l(-e_s) .$$

In the first case we have to subtract the load quantities in the second case we have to add them to obtain

$$e_l = \hat{e}_l(e_s) - \hat{e}_l(-e_s)$$

We find in this way eight configurations. Four of them use two identical stages and are shown in Fig.8. The other four use two complementary stages and are shown in Fig.9.

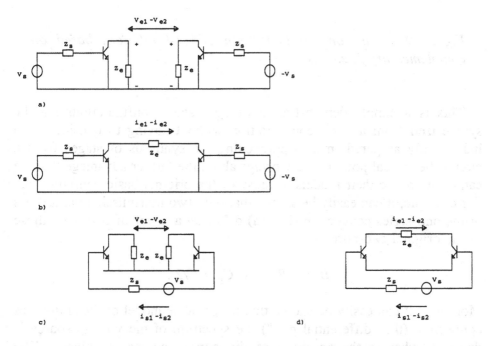

Figure 8: Four configurations based on identical stages with odd s-l relations

404

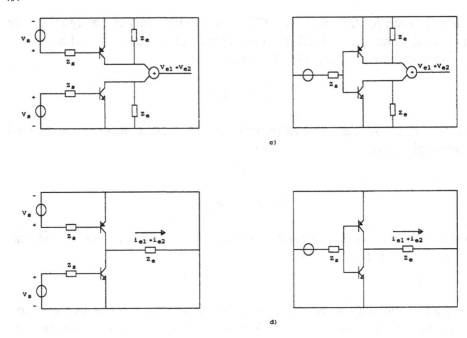

Figure 9: Four configurations with odd s-l relations based on complementary functions

This is a simple demonstration of how stage configurations can be synthesized from a basic configuration without having to consider each individually as paradigms. Approaching the synthesis of stages from a circuit-theoretical point of view brings about the further advantage that we can easily relate their models to those of the original basic configuration. For example, it can easily be shown that only two transmission parameters of the anti-series connection (Fig.8d) differ by a factor of two from those of the original two-port:

$$B_s = 2B \quad and \quad C_s = C/2 .$$

Moreover, it can easily be shown that in the noise model of the anti-series connection (the "differential pair") the spectrum of the voltage source is doubled, whereas the spectrum of the current source is halved. The behavioral model of the differential pair is thus simply related to that of the CE stage. The relations between the models of the other stage combinations and the CE stage are of a similar simplicity.

6. Behaviour modification and topology generation by the application of negative feedback.

Same as with compensation techniques, from an analysis point of view it is not necessary to make the presence of feedback explicit. From a synthesis point of view, however it appears to be very useful since it allows us to generate an abundance of configurations which provide improved approximations to the ideal amplifier functions. Feedback is then considered as a technique for exchanging power gain versus information-processing quality. To illustrate this, we will first introduce the nullor concept as an infinite-gain two-port. This concept allows us to separate the first step in feedback amplifier design from a second step which concerns the implementation of a nullor approximation with real active devices.

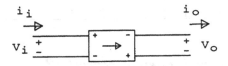

Figure 10: Symbol for the nullor. The arrow indicates the direction of the transmission. the + and - signs indicate the mode (inverting or non-inverting

The nullor has transmission parameters which are all equal to zero. The so-called transfer parameters, being the reciprocals of the transmission parameters are all infinite. Both inverting and non-inverting *i-o* relations are available by interchanging the terminals at one of the ports. The nullor concept is useful, since we can consider a basic amplifier stage or a cascade of amplifier stages as an approximation of the nullor as will be illustrated in the next section. The symbol we will use for the nullor is shown in Fig.10. The nullor as such is not a useful element to model an actual amplifier. The nullor with an external network, however, can make the output quantities finite when the input quantity if finite. As an illustration, we consider the configuration of Fig.11, where the input and the output are interrelated by means of an impedance network. Knowing that v_i and i_i equal zero with a finite output voltage, we can write by inspection

$$v_l = v_s \left(\frac{Z_1 + Z_2}{Z_1} \right).$$

The voltage across the load obviously doesn't depend on the load and therefore, this configuration is completely equivalent to the voltage-controlled voltage source, which is characterized by a finite transmission parameter A, whereas all other transmission parameters equal zero.

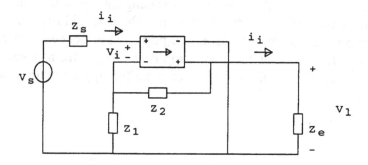

Figure 11: Nullor with external network (Z_1, Z_2 inserted between source and load

From a design point of view we can label the network with Z_1 and Z_2 as a negative feedback network. The adjective negative is used since we have an inverting transmission around the loop formed by the nullor and the external network. The result of the feedback is that we can fix a transmission parameter at a value completely determined by the external passive network. We can do the same for the other transmission parameters. For this purpose we have to follow the following strategy. In order to make the voltage across the load independent of the load impedance, we have to sense the load voltage at the input of the feedback network. This is normally called output shunt feedback. Similarly, sensing the current through the load, labelled as output series feedback, will yield an infinite output impedance. At the input we realize an infinite input impedance by inserting the output voltage of the feedback network in series with the input (input series feedback, or input voltage comparison).

Similarly, the input impedance becomes zero if we feed the output current of the feedback network into the same input node as the current from the signal source (input shunt feedback). By using only one feedback loop, which means that we sense one of the output quantities and take either the output voltage or the current of the feedback network in series or in parallel with the corresponding source quantity, we can give one transmission parameter an accurate finite value, completely determined by the (passive) feedback network. All other transmission parameters retain their zero values. By choosing the proper single-loop configuration, we can realize the equivalents of all transactor types.

We can also sense the combination of the output voltage and the output current. Similarly, we can feed back both a voltage and a current derived from the output quantity(-ies) to the input. In this way we are able to define two, three or all four transmission parameters by means of multi-loop feedback networks.

We can choose various types of negative feedback networks. The most versatile among these and the first to be discussed are the non-energetic types of networks, either short circuits, open circuits or transformers and gyrators. These elements have in common that they have no memory and no power dissipation. The gyrator and the transformer, having galvanically isolated ports can be used both in an inverting and in a non-inverting mode. We can therefore implement all 15 types of amplifiers by using one or more of these elements in the feedback network. Fig.12 gives an example where all the transmission parameters are fixed at values determined solely by the parameters of these elements.

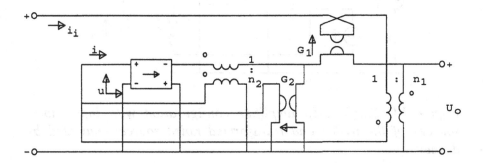

Figure 12: Amplifier with the maximum number of feedback loops. All transfer parameters are determined by the feedback-network transfer function

Assuming now that the nullor has two imperfections, we can study the influence of such a feedback network on the related performance aspects. The first imperfection we assume is that the nullor generates noise as represented by the model of Fig.13. The second is that it is linear in a restricted range of output signals so it can handle a limited signal power at the output. The non-energeticness of the feedback elements implies that they give the amplifier optimum properties with respect to the noise performance and the signal-handling capability. In a single-loop amplifier as shown in Fig.14, we can easily show that the equivalent input noise sources of the amplifier are equal to those of the nullor. The same holds for the power handling capability.

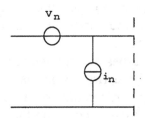

Figure 13: Noise model of the nullor

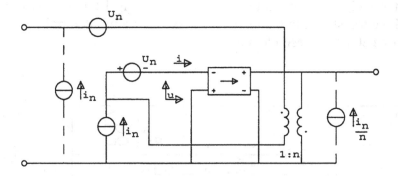

Figure 14: Single loop amplifier configuration with input noise sources of the nullor and transformed noise sources indicated by dashes

Though the feedback configurations with transformers and gyrators may have a very limited practical relevance, they are of great significance from a design-theory point of view since they allow us to find all the possible amplifier configurations and show that non-energeticness of the feedback network is a condition for optimally exchanging power gain versus information transfer quality in wide-band amplifiers. The configurations found by applying non-energetic feedback with shorts and opens are the voltage follower and the current follower shown in figures 15a and 15b, respectively. It will be immediately clear that these configurations fix the parameters A and D at unity, respectively, without deteriorating the noise and power performance. They have non-inverting i-o relations due to the fact that an inversion in the feedback network is not possible.

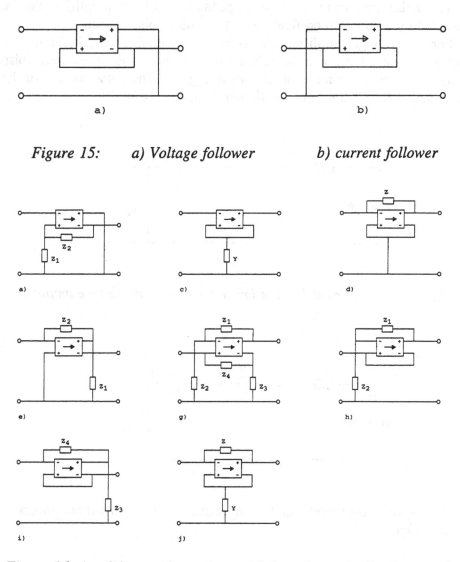

Figure 15:   *a) Voltage follower*   *b) current follower*

Figure 16: Amplifier configurations with impedance feedback networks

The second class of feedback networks includes impedance networks either reactive or resistive or a combination. Since also these networks do not allow to make an inversion, the number of possible configurations is now restricted. All possibilities where the source, the load and the configuration have a common terminal are shown in Fig.16. Now, even when the feedback network comprises reactive elements exclusively and is therefore lossless, the noise and power performance are affected. We refer to literature [1] for illustrations and only formulate a conclusion which provides us with a simple model.

The noise performance of an impedance feedback amplifier can be modeled by a simple modification of the nullor noise model.

For the voltage amplifier, the sum of the admittances, including its thermal noise source can be taken in series with the equivalent noise voltage source of the nullor as shown in Fig.17. The noise models of the other single-loop amplifiers are shown in figures 18 to 20.

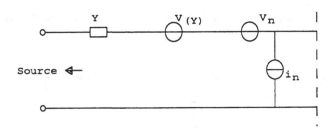

Figure 17: Noise model of the impedance-feedback voltage amplifier

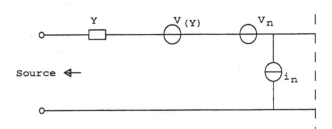

Figure 18: Noise model of the impedance-feedback transadmittance amplifier

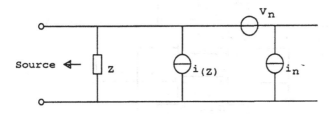

Figure 19: Noise model of the impedance-feedback tranimpedance amplifier

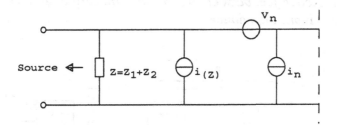

Figure 20: Noise model of the impedance-feedback current amplifier

The power performance model is equally simple. For the voltage amplifier, the sum of the impedances is in parallel with the load. Similar models are found for the other amplifier types.

In order to implement the other amplifier types which are not available when we use only impedance feedback networks, we can consider the use of active feedback networks which can be either inverting or non-inverting. These active networks are required to be linear for making approximations to the ideal amplifier. This means that we have to use a feedback configuration in the feedback network. An illustration of this technique, realising an amplifier with a finite input resistance and a zero output resistance is shown in Fig.21.

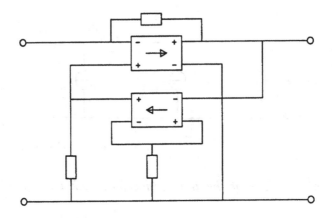

Figure 21: Active-feedback configuration with zero output impedance and a finite input impedance

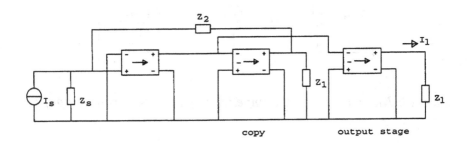

Figure 22: Example of an indirect-feedback inverting current amplifier

Finally we will discuss the application of indirect feedback, which can be regarded as a combination of negative feedback and compensation. This type of feedback can be used when the actual load quantity cannot be sensed or when the feedback quantity cannot be compared directly with the corresponding source quantity. The general idea of indirect sensing is illustrated in Fig.22. An identical copy of the non-linear two-port that drives the load is used within the feedback loop in order to fix the current in that stage by means of the feedback network. If the two-port is an inverting type, characterized by its transmission parameters we will obtain a current gain given by

$$A_{i\infty} = -\left(1 + Z_2/Z_1\right)\left(\frac{AZ'+B}{AZ_l+B}\right)$$

where

$$Z' = \frac{Z_1 Z_2}{Z_1 + Z_2}$$

From this expression it is seen that only two of the parameters of the two-port are relevant in the *i-o* relation. By using a two-port in which these transmission parameters are sufficiently small, or where $Z'=Z_l$, we can realize an accurate *i-o* relation. The indirect feedback scheme can be used for example to make an inverting current amplifier or a non-inverting transadmittance, which is not possible with a passive impedance feedback network. The indirect feedback configuration can be used as a feedback network around the nullor as well in order to realize an active feedback configuration.

To all these negative feedback amplifier types we can apply the odd function synthesis methods from the previous sections. The number of useful topologies thus further increases.

7. Negative feedback amplifier modelling; topology generation of the active part

In the previous section we considered feedback configurations where nullor properties were assumed for the active part and were able to formulate criteria for the synthesis of the feedback network. The use of the nullor concept allows us furthermore to find simple behavioural models for noise and power-handling. The approximation of nullor properties by using suitable topologies of active devices is the next step in the design. For this purpose we need a model that allows us to evaluate the success of the approximation. A model closely connected to the nullor approximation is the so-called asymptotic gain model, first formulated by Rosenstark and elaborated for input and output impedance modelling in [1]. The model is formulated in terms of direct transmission from source to load, loopgain and asymptotic gain. The latter being the source-load relation when the active part is assumed to have nullor properties. The model for the source-load relation can be written as follows

$$A_f = \frac{E_l}{E_s} = A_{f\infty} \cdot \frac{-A\beta}{1-A\beta} + \frac{\rho}{1-A\beta} \, ,$$

where E_l and E_s are the load and source quantities, respectively, $A_{f\infty}$ is the asymptotic gain, ρ is the direct transmission when the loopgain $A\beta = 0$. The asymptotic gain is realized when the active part has nullor properties.

A and β are strictly defined. The reference variable A is the relation between one of the input quantities of the active part and an arbitrarily chosen controlled source in that active part. The feedback factor β is the relation between that controlled source, however now considered as being independent, and the same input quantity. The nullor representation of the active part then coincides with the assumption that the loopgain approaches infinity. In some cases, the reference variable A can't be chosen so that the loopgain goes to infinity along with A. Then some internal feedback is present within the active part that prevents both input port quantities from becoming zero simultaneously. In that case, one can either correct in the analysis the value of $A_{f\infty}$, or modify the internal topology of the active part so that a better approximation of the nullor can be obtained.

In order to generate topologies for the active part, we can start with the most simple structure, the CE stage. The CE stage can be identified with the nullor symbol as done in Fig.23. Since one terminal of input and output are now in common, we have a restricted number of configurations with passive feedback networks. Not considering the transformer and the gyrator, all possible configurations are shown in Fig.24.

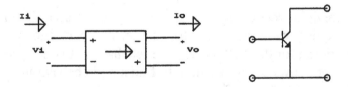

Figure 23: CE stage identified with the nullor symbol

Two of these configurations use non-energetic feedback with opens and shorts. They are recognized as the CC stage and the CB stage, respectively. Obviously, they are the one stage realizations of the voltage follower and the current follower. The transmission parameters A and D of these stages have values close to unity. The other parameters have nearly the same value as those of the CE stage. Due to the non-energeticness of the feedback networks, they have nearly the same equivalent input noise sources and signal-handling capabilities as the CE stage in the same operating point. Notice that these conclusions can be drawn immediately on the basis of the conceptual approach of applying negative feedback as an error reduction technique. For the configurations using impedances in their feedback networks, we can draw conclusions about their behaviour equally quickly.

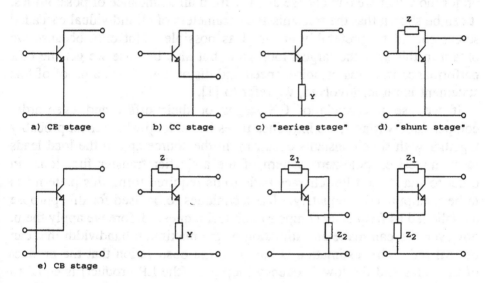

Figure 24: Feedback configurations based on the CE stage

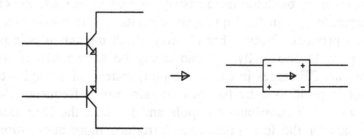

Figure 25: Anti-series connection of two CE stages identified with the nullor symbol

A second step in the approximation of the nullor is the use of an active two-port with input and output ports which can be floating relative to each other. The most basic realization is the anti-series connection of two CE stages as shown in Fig.25. We now find the same possibilities for the passive-feedback configurations as we found with the nullor. The behaviour of these configurations can be evaluated as easy as the single-stage configurations. In order to make a better nullor approximation, we have to decrease the transmission parameters of the active part. We can do so by cascading stages. The question is what kind of stages we should prefer, now that we can choose already from an abundance of possibilities. It can be shown that the transmission parameters of all individual cascaded stages should be preferably as small as possible. Not only because we obtain in this way the largest loop gain, but also because we get the best performance in terms of noise, speed and distortion. Since a proof of this statement is rather involved, we refer to [1].

If we use a cascade of CE stages or their offset and even-order compensated variants, we introduce poles in the loop which most probably together with time constants occurring in the source and in the load leads to an undesired pole-zero pattern of the amplifier transfer function. In order to shape the filter characteristic to its required form, this pattern has to be manipulated adequately. The techniques to be used for this purpose are labelled as frequency compensation techniques. Before we apply them, however, we can make an estimation of the maximum bandwidth that can be realized. This estimation is based on the observation that the product of the poles and the low-frequency loop gain (the LP product) is equal to the product of the poles of the negative-feedback amplifier. So, if we realize a filter characteristic where we would manage to locate all the poles in Butterworth positions, the amplifier would have a bandwidth equal to the n-th root of the LP product, where n is the number of poles. The LP product can usually be determined easily. In most situations, one can leave out the capacitor C_μ in the equivalent circuits of the transistors without altering this product. Notice that C_μ may affect the actual pole positions in the loop quite drastically. It can easily be shown that if we use a cascade of two CE stages in the active part, instead of a single stage, the LP product is multiplied by the current-gain transit frequency f_T. Each cascaded stage will contribute one pole and increase the LP product. As the source and/or the load impedance introduce more attenuation, along with lower-frequency poles in the loop transfer function, the number of stages to realize the required bandwidth will become larger. As a matter of course there is a limit to the number of poles that we can successfully manipulate into the required positions. Normally, two poles don't pose any problem and three poles can be handled with some difficulties. The frequency compensation techniques that can be applied can be arranged in a preference order. The use of phantom-zero techniques has the highest preference since it minimally affects the noise and distortion behaviour and may even improve it. Local feedback techniques within the active part of the amplifier are second in preference. These can be used for pole splitting. Finally, if these two techniques cannot be applied one can consider the use of shunt impedances in the active part. These can be used for pole splitting as well or for changing the pole positions. The latter two techniques have much more impact either on the noise performance or on the distortion performance or on both.

In some situations, one may prefer the use of other stages than CE stages in the active part for reasons which are related to the ease of manipulating pole positions or to biasing problems. The non-energetic feedback Strategies and Rstages (CC and CB) then have the first preference since these introduce no losses in their exchange of power gain versus information transfer quality.

A special place in the selection of stage configurations in the active part is reserved for indirect feedback stages, in particular for the current mirror. Though we will not deal with biasing methods in this paper, we will just shortly mention the importance of the current mirror, being a special case of the indirect feedback configurations. The most simple version uses a CE stage together with a copy of this stage around which feedback is applied. The collector current of this copy is completely returned to the input, thereby making it very nearly equal to the current from the source. With sufficiently low values of the parameters A and B of the CE stage, the collector current in the output stage is very close to the one in the copied stage. The limited amount of feedback in the copied stage gives rise to some influence of the parameters C and D as well. The configuration provides an inverting unity current gain. It is therefore useful in the synthesis of odd functions, where the output quantities have to be subtracted in order to get rid of offsets. From a power point of view this can be done very efficiently, because we need only a small voltage drop across the devices to keep them in their active region. From a noise point of view, however, the indirect feedback is such an inferior technique that it should be avoided at noise sensitive places in the active part of the amplifier.

Although there is much more to say about the synthesis of the active part, in particular in relation to biasing methods, we restrict ourselves to the very general strategies outlined in this section.

8. Conclusions.

Though certainly not all performance aspects and synthesis methods have been discussed in this paper, it has been shown to be possible to define explicit strategies and routines for the design of amplifiers. Such an explicitness is believed to be a prerequisite not only for real design automation, but also for educating future generations of analog designers. Though the discussion in this paper has been restricted to time-invariant amplifiers, the approach can equally well be followed for all other types of information-processing functions. On the basis of this approach, the program Ampdes has been developed in the Electronics Research Laboratory of the Delft University of Technology [2,3]. Although this program is not yet as mature as one should wish, it is able to automatically generate amplifier structures for specific applications, optimized with respect to noise, bandwidth and signal handling.

That the above described approach is not just an academic curiosity is demonstrated by the fact that many application specific IC's have been designed on the basis of this approach, in particular in the area of high-performance circuits for radio communication.

References

[1] Design of high-performance negative-feedback amplifiers, E.H. Nordholt, Amsterdam, Elsevier, 1983.

[2] A Design strategy for the synthesis of high-performance Amplifiers, J. Stoffels. Submitted to IEEE

[3] Ampdes: A program for the synthesis of high-performance amplifiers, J. Stoffels and C. v. Reeuwijk, EDAC '92

Biography

Prof.Dr.Ir. Ernst H. Nordholt (1940) graduated in electrical engineering at Delft University of Technology in 1969. In 1964 he became an employee of the Electronics Teaching Laboratory. He was in charge of that laboratory from 1970 to 1977. In 1977 he joined the Electronics Research Laboratory of the same department in order to prepare his Ph.D. thesis on the design of high-performance negative-feedback amplifiers (Elsevier, 1983). He was awarded a doctoral degree in 1980. For several years he was involved in research, lecturing, research management and fund raising in the area of analog electronic systems. He is author of numerous scientific papers and he holds several patents. In 1984 he became a part-time senior consultant of the IC design house Sagantec in Eindhoven, The Netherlands. In 1985 he was one of the founders of Catena Microelectronics B.V. in Delft, The Netherlands, where he now supervises a research program on design-driven process-technology development and design automation. Prof. Nordholt is the general manager of Catena Microelectronics B.V. Moreover, he is one of the authors and instructors of the DESC seminars and courses.

Open Analog Synthesis System based on Declarative Models

Georges Gielen*, Koen Swings and Willy Sansen

Katholieke Universiteit Leuven
Electrical Engineering Department, ESAT-MICAS
*Senior research assistant National Fund of Scientific Research
Belgium

Abstract

The flexible, open and extendable analog design system ARIADNE is presented. The approach is based on a clear separation between the declaration of the knowledge about the behavior of an analog circuit (described in declarative equation-based models) and the procedures (general computational techniques such as constraint propagation and optimization) that utilise the knowledge to synthesize a circuit for a particular application. The use of declarative models together with symbolic simulation also provides an easy interface for the designer to include new design knowledge and circuit schematics into the design system himself. The ARIADNE system can be used by system designers, experienced and unexperienced designers in both automatic and interactive mode. The different tools in the system are described in detail and the key ideas are illustrated for some practical examples.

1. Introduction

Advances in VLSI technology nowadays allow the realization of complex integrated electronic circuits and systems. Application-specific integrated circuits (ASICs) are moving towards the integration of complete systems on a single chip, including both digital and analog parts. Besides the specific problems associated with integrating both digital and analog

422

circuitry on one substrate, the analog portion of the chip is usually much more time-consuming and error-prone to design than the relatively larger digital portion. The design of the analog circuits is therefore often the bottleneck in bringing new high-performance mixed-signal solutions fast to the market. There are two main reasons for this [1]. One is the highly complex and knowledge-intensive nature of analog design. The quality (and often the functionality) of the resulting design strongly relies on the insight and expertise of the analog designer. The other reason is the current lack of computer-aided design tools that are adapted to the peculiar nature of the analog design task. This situation highly contrasts with the digital design case, which is inherently more structured and which is already supported by large numbers of mature design and synthesis tools. All this explains the growing interest and increasing industrial demand observed nowadays for the realization of computer tools that support or eventually automate the design of analog and mixed-signal integrated circuits and systems.

In recent years, interesting work has been done in the field of analog synthesis [2-7]. Approaches in this direction are mainly based on a compilation of the design experience of experienced designers into design procedures, design plans, decision trees or design rules. The main problems with these systems are the hard-coded mixture of knowledge declaration and context-dependent knowledge exploitation, and the difficulty with which expert knowledge is extracted and encoded into procedures, plans or rules. This seriously limits the performance flexibility of the design systems and the interaction capabilities with experienced designers. It also makes these systems difficult to extend with new circuits and design knowledge, and effectively restricts them to the design capabilities initially included by the tool developer.

This paper presents an alternative approach for a flexible, open and extendable analog design system, called ARIADNE [8,16], which is a further evolution of the ASAIC strategy presented in [1]. The approach is based on a clear separation between the declaration of the knowledge about the behavior of an analog circuit (described in declarative equation-based models) and the procedures (general computational techniques such as constraint propagation and optimization) that utilise the knowledge to synthesize a circuit for a particular application. The use of declarative models together with symbolic simulation also provides an easy interface for the designer to include new design knowledge and circuit schematics himself. According to this ARIADNE approach, tools have been implemented which can be used in interactive mode by both inexperienced and experienced designers for analysis and design, and which can be integrated into a design system for the automated synthesis of analog integrated circuits.

This paper is organized as follows. Section 2 describes the basic requirements for an analog design system and evaluates existing approaches. Section 3 then presents a general strategy for the synthesis of analog and mixed-signal integrated circuits and systems. Section 4 describes the ARIADNE approach as one implementation of this general strategy, based on declarative models and general computational techniques. The different subtools are discussed in more detail and illustrated with some examples. Final conclusions are provided in section 5.

2. Basic requirements for an analog design system

The general objectives of analog CAD tools can be summarized as:

1) reduce the total design time and cost

2) allow the designer to produce first-time error-free designs

3) allow the designer to produce high-quality optimum designs

4) enhance the extraction, formalization and transfer of analog circuit design knowledge

5) reduce the training time for novice, inexperienced designers.

These objectives already indicate that design knowledge plays a crucial role in analog design automation, in the same way as in manual design. This design knowledge consists of both formal/analytic and intuitive/heuristic knowledge. When confronted with a new design, good designers often rely on this analytic and intuitive knowledge, which is based on their large experiences and accumulated over many years. Yet intuition is difficult to teach or capture into tools. An important issue in analog circuit design tools is therefore the formalization of the design knowledge into a description form that is both understandable by the human designer and executable on the computer.

Current approaches in the direction of analog synthesis mainly focus on the application of AI-based knowledge description forms [2-7]. In OASYS [2], synthesis is performed in an alternating sequence of design style (topology) selection and specification translation down the hierarchy.

Topologies are selected from a set of hierarchically structured alternatives stored in the database. A planning mechanism is then used to translate higher-level specifications to specifications for the subblocks of the lower level. OPASYN [3] uses a non-hierarchical topology selection process that is based on decision trees and that is followed by an optimization-based sizing process based on manually derived analytic models for the circuit behavior. The BLADES [4] and CAMP [5] approaches are similar in that they use if-then rules to encode expert knowledge concerning topology choices and sizing. A separate rule-based inference engine is then used to solve the design problem. In OAC [6], topologies are selected from a library of already designed transistor-level schematics. Knowledge-based modification plans are used to modify the design parameters to obtain a rough device sizing. These sizes are then fine-tuned in an optimization loop which also includes procedural layout generation. In IDAC [7], knowledge and its manipulation are hard-coded into a circuit-specific formal procedure which performs the sizing of the circuit chosen from a fixed set of flat transistor schematics.

The main advantage of these systems is that they are usually very fast in designing circuits available in the system's database, if the specifications are within the target performance range of the encoded knowledge. This allows the designer to explore trade-offs by interactively synthesizing multiple alternatives. These design systems however also have some serious disadvantages:

- The hard-coded mixture of knowledge declaration and context-dependent knowledge exploitation, such as in the design plans of OASYS [2] or the formal procedures of IDAC [7], limits the performance flexibility of the design systems. They are difficult to apply in a design context which differs from the context originally envisioned by the tool developers. For example, a design plan or a formal procedure for a certain topology targeted towards minimum power consumption cannot be used without modifications for minimizing the noise or optimizing the speed of the same topology.

- The design systems are difficult to extend with new circuit topologies and design knowledge. It is very hard for expert designers to encode their knowledge into general procedures, plans or rules. This effectively restricts these systems to the design capabilities and topologies initially included by the system developer, which is in large contrast with the large spectrum and range of performance criteria and the continuous evolution in circuit design solutions characterizing analog design.

- The interaction capabilities with experienced designers are often limited: important design decisions are hard-coded in the design plans or procedures, and are taken without the designer having any control. This has a negative effect on the willingness of designers to accept and use the system.

The ARIADNE approach [8,16] presented in section 4 tries to overcome the above disadvantages, by keeping a clear distinction between the declaration of circuit knowledge and the use of that knowledge in a particular application context. The basic format in which knowledge is represented are equations, generalized into the concept of constraints. This closely matches the natural way of reasoning of expert designers. The equations are then manipulated upon with powerful mathematical routines during synthesis. The use of declarative models together with symbolic simulation also provides an easy interface for the designer to include new design knowledge and circuit schematics himself, making ARIADNE an open and extendable system.

Similar extensions to create an open analog design system are presently being implemented for IDAC [7] and OASYS [2] as well. The knowledge which can be introduced in the ADAM system (containing IDAC as a subtool) [9] for example, is knowledge concerning the topology and functionality of a customer circuit, complemented with knowledge about its device sizing sequence (application-specific design plans). A symbolic simulator can assist the designer with the first type of knowledge [9]. The design plans however have to be encoded by the designer himself, which is a very tough job. ADAM does not contain any general knowledge manipulation routines which can manipulate the same declarative knowledge for different application targets as in ARIADNE [8,16].

The basic requirements for an analog computer design system can now be summarized as:

- it must be very flexible with respect to the attainable performances and optimization targets, in order to be useful in a broad range of application domains

- it must provide different implementation styles to the designer, ranging from analog/mixed arrays over standard and parameterized cells to custom design. The actual implementation style can then be selected based on the application and other technical and economical motivations.

- it must be independent of technology as much as possible, and at least allow easy process updates

- it must be an open system, extendable and accessible by the designer, to keep track of the continuous evolution of analog design solutions. A closed system will soon be outdated, especially for high-performance applications where the specifications keep on pushing the limits. This means that it must be possible for designers to easily include new topologies and new design knowledge into the system, or modify the existing knowledge.

- finally, there are two different kinds of users with different expectations:

  * *ASIC/system designers*, unfamiliar with analog circuit design details, want the system to automatically design the analog circuits and modules they need at the ASIC/system level in a very short time. This automatic mode, which relies on the built-in knowledge, can be used for designs with not too aggressive specifications.

  * *expert circuit designers* perceive the CAD system more as a design assistant freeing them from the laborious routine tasks and allowing them to concentrate fully on the creative tasks (such as designing circuits for aggressive specifications or exploring new circuit schematics). These expert designers will always want to take the ultimate design decisions themselves relying on their own expertise. They expect the CAD system to be interactive, open, extendable, and most of all easy to use.

The above requirements are of course hard to unite in one and the same system.

In the following section, a general strategy for the synthesis of analog and mixed-signal integrated circuits and systems is presented.

3. Design strategy for analog and mixed-signal systems

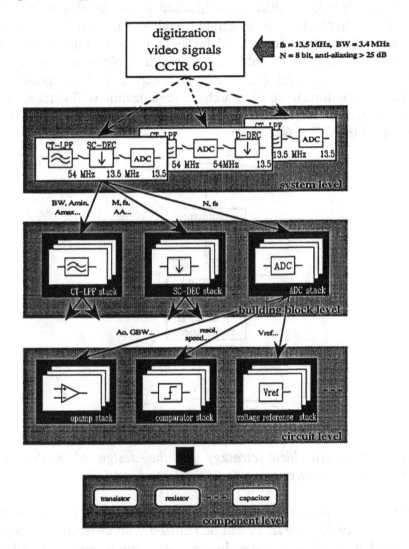

Figure 1: Hierarchical design organization for an analog-digital interface system for video signals.

Mixed-signal systems contain many different parts, both digital and analog, which perform many different functions and contain many thousands of transistors. In order to handle this complexity, a design environment automating or assisting the design of mixed-signal systems has to be organized hierarchically. The overall design task has to be broken up in more manageable subtasks, which are again broken up in smaller subtasks, etc. This results in a hierarchy of different levels of design abstraction, where the entities being handled at each level have a different functional abstraction. For integrated analog and mixed analog-digital systems, the following hierarchical levels could be distinguished: system level (modem chip, ISDN interface...), building block level (filter, data converter...), circuit level (comparator, opamp...), and component level (transistor, capacitor...). This hierarchical organization is illustrated in Fig.1 for the design of an analog-digital interface system for video signals according to the CCIR 601 norm.

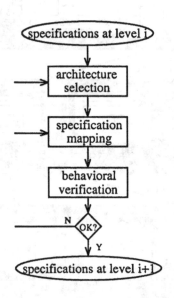

Figure 2: Hierarchical strategy for the design of analog and mixed-signal systems: going from level i to level i+1.

An additional reason for the complexity of mixed-signal design is that each function in this hierarchy can be realized with different architectures, also called topologies at lower levels. An architecture or topology is defined as an interconnection of blocks from the next lower level. This means that a system architecture is defined in terms of functional building blocks (low-pass filter, decimator and A/D converter for example in Fig.1), and a building block architecture in terms of circuits, etc. Each architecture has its own performance space and implementation cost (power, area), and is more or less appropriate depending on the specifications of the particular application. Fig.1 for example shows 3 possible architectures for the video signal A/D interface. Similarly, different architectures exist for data converters, as well as for amplifiers, etc.

The design in this hierarchical environment then consists of an alternating sequence of architecture selection and specification translation. At each level in the hierarchy, a synthesis routine for the block under design is executed which always consists of the same three basic steps also shown schematically in Fig.2:

1) *architecture/topology selection*

First, an architecture is selected based on the current specifications for the block and the technology data. For example, for the video signal interface of Fig.1, the first architecture could be selected from among the three alternatives. When the design system is used in automatic mode, the architecture is chosen from among predefined alternatives stored in the system's database. The selection can be based on a technology-independent encoding of expert designers' selection heuristics for example into rules, on algorithmic techniques such as interval analysis which provide boundaries for the performance range of each topology, or a combination of such techniques. The selection of a particular architecture may of course lead to different types of building blocks (e.g. switched-capacitor or digital decimators), even different specifications for the same block (e.g. the ADC operating at 13.5 *MHz* or 54 *MHz*), and also highly affects the resource consumption (area, power) of the final design. At the highest level, this step also includes the partitioning between analog and digital.

2) *specification mapping*

Next, the specifications of the block under design are translated or mapped into specifications for the subblocks within the selected architecture. For example, at the top level in Fig.1, specifications for the A/D interface are mapped into specifications for the CT-LPF, the SC-DEC and the ADC if the first architecture has been selected. At the lowest level, this mapping is equivalent to device sizing. The mapping can be based on mathematical procedures, heuristics and/or optimization techniques. At the same time, geometrical information (e.g. floorplanning information) can be passed to the lower level to direct the layout generation process later on.

3) *behavioral verification*

Before going down in the hierarchy and in order to reduce the number of design iterations, the translation of the specifications can be verified by means of mixed-signal behavioral simulation. The architecture is simulated in terms of the composing circuit blocks and their specifications as determined in the previous mapping step. If the design is satisfactory, the appropriate synthesis routines are called to design the subblocks in the architecture (for example the ADC in Fig.1). If not, the mapping of the specifications can be adapted or the topology changed.

This process of architecture selection, specification translation and behavioral verification is repeated down the hierarchy until a level is reached which allows a physical implementation (component level for custom layouts, higher levels when using standard cells for instance). The hierarchy is then traversed bottom-up, at each level generating the block layouts by assembling the subblock layouts. This results in the layout of the overall system at the top level. A final verification can then be performed by simulating the whole system with an appropriate electrical, mixed-mode or behavioral simulator, this time using the exact parameter values extracted from the layout and including parasitic effects and statistical variations.

Such a hierarchical performance-driven top-down design strategy allows for the largest designer flexibility if the synthesis routines at the different levels operate independently from one another. The same data converter synthesizer for instance can then access a wide variety of lower-level design tools such as standard cell libraries, parameterized circuit module generators (for capacitor arrays for instance) and circuit-level compilers (for opamps and comparators for instance). At the same time, the same lower-level tools can be shared by all higher-level tools.

Practical analog design systems of course differ in the way how the hierarchy is conceived and how the different tasks are actually performed. Such a hierarchical, performance-driven analog design system is for example under development at UC Berkeley. It is however a closed system consisting thus far of the OPASYN (certain opamp types) [3] and CADICS (cyclic A/D converters) [10] programs. In the next section, the ARIADNE approach is described as another implementation of this general strategy, based on declarative models and general computational techniques to establish openness and flexibility [8,16].

4. ARIADNE: an open analog synthesis system based on declarative models

ARIADNE is an open and flexible synthesis system for analog functional modules such as operational amplifiers, comparators, buffers, filters, data converters, etc. [1,8,16]. The system's flow diagram is shown in Fig.3. ARIADNE consists of a synthesis part and a modelling part for the inclusion of new circuit schematics. The synthesis part covers the whole design path from topology generation over optimal circuit sizing down to layout.

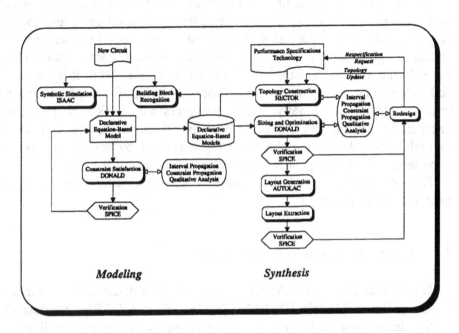

Figure 3: Diagram of the ARIADNE open analog design system.

The key idea behind the synthesis methodology is to clearly separate the knowledge (both analytic and heuristic) about analog behavior from the procedures that actually carry out the synthesis. Knowledge about a circuit is generally described by equations, which are grouped in so-called *declarative equation-based models* (DEBMs) [8,11]. These DEBMs simply declare the knowledge without indicating any way to use them. In this way, DEBMs can easily be extended with new circuit design knowledge and they make it easy for designers to include new schematics. The synthesis procedures on the other hand are general computational techniques such as constraint propagation, interval propagation and optimization. They can be applied to arbitrary DEBMs, resulting in a general approach that can easily be applied to new design descriptions and to any application-dependent design objectives.

The main purpose of the modelling methodology is to ease the process of constructing efficient DEBMs for new circuit topologies. The modelling methodology consists of symbolic simulation, building block recognition, constraint propagation, interval propagation and qualitative analysis. The availability of software tools for modelling turns the ARIADNE system into an open design system.

The ARIADNE synthesis and modelling system is intended to be used by system designers, inexperienced designers and expert designers. System designers will mainly use the system in automatic mode in order to rapidly obtain designs that satisfy the specifications. Inexperienced designers will also use the system in automatic mode but will learn from it by watching and interacting with the decision process, asking for explanations and testing alternatives. Finally, expert designers will work with the system interactively, taking the important decisions themselves, and will be able to update the design capabilities of the system by modifying existing knowledge or including new knowledge and schematics [1,8,16].

The different parts of the synthesis and modelling frame are now explained in more detail.

4.1. Declarative equation-based models

The central part of the ARIADNE system is a database of Declarative Equation-Based Models (DEBMs) [8,11]. Such a DEBM consists of a netlist description of the circuit, a set of equations (both equalities and inequalities, linear and nonlinear, implicit and explicit) describing the circuit behavior and additional circuit knowledge, and intervals on variables in the model. Consider for example the CMOS two-stage Miller-compensated opamp of Fig.4. The figure also shows a set of analytic equations describing the behavior of this amplifier. For the sake of clarity, only a subset of the complete analytic model of this amplifier is included in the figures here. The corresponding DEBM of this amplifier is then shown in Fig.5.

The equations in the DEBM are a direct translation of the analytic model equations of the Miller opamp of Fig.4. The important conceptual difference is that the equations in the DEBM are actually definitions of constraints between the design variables. They declare mathematical relationships. They do not specify a specific direction of computation and can therefore be solved for any of their variables. They also do not have to be ordered into a specific sequence.

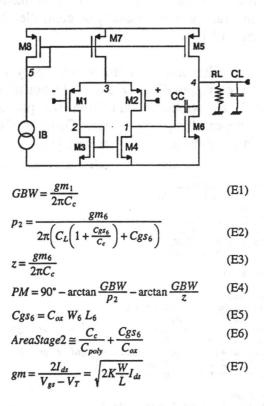

$$GBW = \frac{gm_1}{2\pi C_c} \tag{E1}$$

$$p_2 = \frac{gm_6}{2\pi\left(C_L\left(1 + \frac{C_{gs_6}}{C_c}\right) + C_{gs_6}\right)} \tag{E2}$$

$$z = \frac{gm_6}{2\pi C_c} \tag{E3}$$

$$PM = 90° - \arctan\frac{GBW}{p_2} - \arctan\frac{GBW}{z} \tag{E4}$$

$$Cgs_6 = C_{ox}\, W_6\, L_6 \tag{E5}$$

$$AreaStage2 \cong \frac{C_c}{C_{poly}} + \frac{Cgs_6}{C_{ox}} \tag{E6}$$

$$gm = \frac{2I_{ds}}{V_{gs} - V_T} = \sqrt{2K\frac{W}{L}I_{ds}} \tag{E7}$$

Figure 4: CMOS two-stage Miller-compensated opamp and example set of analytic equations describing its behavior.

The variable definitions contain upperbounds and lowerbounds that result from an in-depth interval analysis. They indicate the boundaries of the feasibility space of the circuit. In other words, the interval characterization defines a multidimensional cube in the design space which contains the feasibility space as a subset [8,16].

The main advantages of the DEBMs are:

- circuit schematics can be modeled by simply pasting together DEBMs of their constituting parts, allowing for large flexibility in topology construction [11].

- circuit knowledge is declared, not programmed. The actual solution order is determined at run time when synthesizing a circuit for a particular application. The same DEBM can thus be applied for different application areas and optimization targets, and the solution strategy can eventually be controlled by an expert designer himself [12,13].

- including a new schematic into the system only requires the creation of the corresponding DEBM. Powerful algorithmic techniques such as symbolic simulation are provided within ARIADNE to support this process. Also, DEBMs of constituting parts already in the database can be reused [1].

```
(def-topology P_MILLER_OPA
   :of-functionblock OTA
   :network
       (M1 2 6 3 3 PMOS)
       (M2 1 7 3 3 PMOS)
       (M3 1 1 8 8 NMOS)
       (M4 2 1 8 8 NMOS)
       (CL 4 0)
       (RL 4 0)
       ...
   :equations
       (gbw = gm.m1 / (2 pi cc))
       (z = gm6 / (2 pi cc))
       (pm = pi / 2 - atan (gbw / p2) - atan (gbw / z))
       (cgs6 = cox w6 16)
       (areastage2 = (cc / cpoly + cgs6 / cox))
       (gm6 = 2 ids6 / (vgs6 - vtn)
       (gm6 = sqrt (2 kn ids6 w6 / 16))
       ...
   :variables
       ibias  :lb 1.0e-7 :ub 1.0e-3
       w.m1   :lb 3.0e-6 :ub 1.0e-3
       ...
)
```

Figure 5: Declarative equation-based model (DEBM) of the Miller opamp of Fig.4, including a netlist, an analytic description and an interval characterization.

4.2. Topology generation

The input to the ARIADNE system (see Fig.3) is a description of the required function of the block to be designed, the performance specifications and technology specifications. The first step is then the generation of a topology which can achieve these specifications. This is the task of the program HECTOR [11]. It has to build both an unsized circuit schematic and the corresponding DEBM for that schematic.

Topology generation in HECTOR is implemented as an expert system, following a hierarchical, library-based refinement process. Functionality is abstracted into a *function block*. Each function block can be implemented as one or more alternatives, called *topologies*. A topology is defined as an interconnection of function blocks of the next lower level in the hierarchy. At each level, HECTOR uses rules to select the best topology from a predefined set of alternatives stored in ARIADNE's database. In this process, information can be used resulting from the comparison of the performance specifications with the interval characterizations of the alternative topologies. By means of interval propagation, the specifications for the selected topology can then be translated into specifications for each of the lower-level blocks. These translated specifications are then again used to select a topology for each of the subblocks.

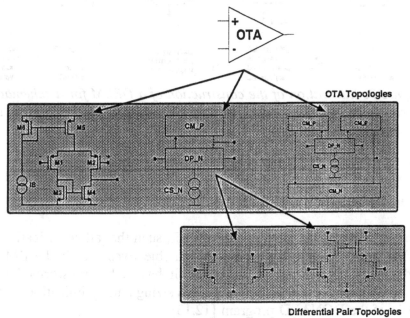

Figure 6: Illustration of the hierarchical refinement process for building a schematic during topology generation.

This refinement process is repeated down the hierarchy until an unsized transistor schematic is obtained at the lowest level. The DEBM for this final schematic is then constructed by simply pasting together the DEBMs of all the constituting topologies in the circuit hierarchy. Note that the circuits are constructed hierarchically in ARIADNE, but that after the DEBMs have been pasted together low-level circuits such as amplifiers and comparators are synthesized as flat circuits. This is the only way to really optimize the performance of these circuits because of the very strong interaction between the different subcircuits. For higher-level modules where the subcircuits are less interacting, the synthesis is performed hierarchically as well according to the strategy described in section 3.

The expert-system-based hierarchical refinement process for topology generation has been prototyped on a LISP machine using the expert system development tool KEE. The refinement process for building the schematic and the pasting of the DEBMs are illustrated in Fig.6 and Fig.7.

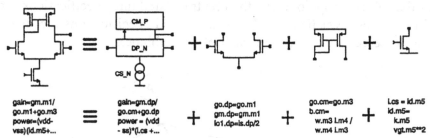

Figure 7: Illustration of the construction of a DEBM for a schematic during topology generation, by pasting together the DEBMs of its constituting subblocks.

4.3. Design equation manipulation

The next step is the sizing of the circuit, such that all specifications are satisfied, possibly optimizing some design objectives. As the DEBM is an unordered declarative set of equations, it has to be transformed into a solution plan which is appropriate for this sizing and optimization. This is the role of the DONALD program [12,13].

First, in DONALD, a possible set of independent variables is constructed [1]. This set can be determined automatically by the program or interactively by the user. Note that any variable, even a specification, can be used as independent variable, not only transistor widths, lengths and currents. The number of independent variables is equal to the degrees of freedom in the design. Next, DONALD finds out how all the other, dependent variables are to be computed out of the independent ones by means of the analytic equations in the DEBM. This results in a computational path, which essentially is a sequential list of calls to numerical routines to calculate all the dependent variables given values for the independent variables [12]. These values can be provided interactively by a designer, or can be determined automatically by an optimization program which executes the computational path within each iteration.

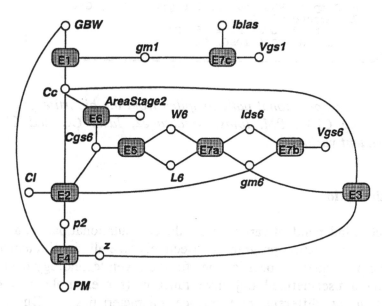

Figure 8: Bipartite graph representation of the DEBM of Fig.5.

The DONALD program has been implemented in LISP on a workstation environment. Internally in the program, the analytic equations of the DEBM are considered as constraints which are represented in a bipartite graph, in which the nodes either represent equations or variables [12]. Constraint propagation is used to construct a possible set of independent variables and the corresponding computational path [12]. The bipartite graph for the example Miller opamp of Fig.4 and Fig.5 is shown in Fig.8. The corresponding computational path for a certain set of independent variables is given in Fig.9. Sensitivity information is also provided in DONALD to guide a designer in his interactive design process or to guide an optimization program.

```
1. GBW, PM, IBIAS, AREASTAGE2, IDS6, CC and CL
   specified by USER
2. Compute GM1  out of E1  given CC, GBW
3. Compute VGS1 out of E7C given IBIAS, GM1
4. Compute CGS6 out of E6  given AREASTAGE2, CC
5. Compute Z, P2, GM6
     out of  CLUSTER1 = E2, E3, E4
     given   PM, GBW, CC, CGS6, CL
6. Compute VGS6 out of E7B given IDS6, GM6
7. Compute W6, L6
     out of  CLUSTER2 = E5, E7A
     given   CGS6,IDS6, GM6
```

Figure 9: Computational path compiled from the bipartite graph of Fig.8 using GBW, PM, Ibias, AreaStage2, Ids6, CC and Cl as independent variables.

4.4. Optimization

The set of independent variables and the computational path can then be used by an optimization program to effectively size all circuit elements, to satisfy the performance specifications for the given technology data while optimizing a user-defined objective function (for example a weighted combination of different performance characteristics).   During this optimization, the independent variables are varied according to the optimization algorithm. Within ARIADNE, multiple optimization programs can be used. One possibility, which has already been implemented in VAX PASCAL, is the OPTIMAN program [14].   OPTIMAN is based on simulated annealing. It is a general and global optimization program which can size any analog circuit described by a DEBM with a high quality but at the expense of larger CPU times [1,14].

4.5. Verification and redesign

The design is then verified in the verification routine [1]. Because of the approximate nature of the equation-based models and the heuristic nature of any design rules used, a detailed verification must be performed on the sized schematic.   This verification consists of several tests, such as controlling the operation region of each transistor, and an accurate performance check by means of a numerical simulator such as SPICE.

If the design is accepted, the circuit schematic and the device sizes are passed to the layout program. If some specifications are violated, the redesign system is entered. Based on the failure information, DONALD uses the constraint network to trace the characteristics and if possible the parts of the circuit that are responsible for the failure [8,16]. Consequently, a backtrace is carried out to tighten some internal specifications, to modify or reselect a new topology or the user is queried to trade off some input specifications. The redesign in this way is based on mathematical evidence of the failure information and not on experience compiled into rules.

4.6. Layout generation

The layout generator then lays out the analog module, starting from the circuit schematic and size information and taking into account analog requirements such as minimum area, symmetry, matching, balancing, reduction of parasitic capacitances, minimization of capacitive couplings and crosstalk, etc [1]. This layout step consists of component-level placement and routing, eventually followed by analog compaction. Both the placer and router generate layouts automatically, but will additionally be part of an interactive layout editor which enables the designer to control the behavior of the placer and the router and to fine-tune the resulting layout.

A final extraction and verification process is then performed. This will either deliver a fully functional layout or will enter the redesign system for an update of the design (sizes and/or topology) or a re-specification request. If the design is accepted, a complete behavioral model is generated to simulate the module at the system level.

4.7. Analytic modelling

To enable the fast inclusion of new circuit ideas into the synthesis system's database, a separate modelling frame has been developed. Its design flow is also illustrated in Fig.3. The cornerstone in this modelling process is the symbolic simulator ISAAC [1,15] which automatically generates symbolic expressions for the AC characteristics (including harmonic distortion) of both time-continuous and switched-capacitor analog circuits. If desired, the expressions can be simplified heuristically in ISAAC up to a user-defined error based on relative size information of the different elements in the circuit [15]. ISAAC has been implemented in COMMON LISP on a workstation environment.

The symbolic simulator is used in combination with other powerful techniques, such as building block recognition, which help the design system in extracting as much information as possible about the new topology to simplify the modelling effort. The main idea behind the building block recognizer is to recognize in a user-specified schematic those building blocks that have already been described by DEBMs in the database. Those DEBMs can then be pasted as part of the global DEBM of the schematic. Another important use of building block recognition is to generalize a user-specified schematic in such a way that similar schematics can be constructed automatically. For instance, a user-specified opamp can be generalized into an interconnection of current mirrors, a differential pair, a current source, etc. By using implementations for these modules that are different from the ones used in the original schematic, different transistor schematics can be constructed that all exhibit the same functionality as the original schematic but with different performance ranges. The building block recognizer has been prototyped on a LISP machine using the expert system development shell KEE.

The equations of ISAAC, the netlist of the user-specified schematic and the DEBMs resulting from recognized subblocks are all assembled into a first DEBM for the circuit. This model can be further extended by the user, for example to include expressions for characteristics which cannot be obtained with a symbolic simulator (e.g. the slew rate) or to add additional design heuristics (e.g. a rule to reduce the offset in a circuit) [1]. An interval propagation routine can then be applied to characterize the feasibility space of the circuit. The consistency and completeness of the model can be checked with DONALD, and the accuracy of the model can be verified with a SPICE-like simulator. If necessary, the user can update the model.

4.8. Design example

To illustrate the ARIADNE approach, the results of an example design of the Miller opamp of Fig.4 are listed in Table 1. The design is based on a more elaborate model than the one indicated in Fig.4. Results obtained from SPICE simulations are also included in Table 1 and show an excellent agreement with the ARIADNE data. As the layout generation program is still under development, no comparison with measurement results can be provided for the time being.

Variable	Units	Spec	DONALD	SPICE
Unity Gain Freq.	MHz	> 1	1	0.958
Gain	dB	> 80	80	80.18
Phase Margin	Degrees	> 60	63.1	61.3
Power Consumption	mW	MINIMIZE	0.1323	0.1324
Active Area	μm^2	-	6,110	-
Slew Rate +	V/μs	-	1.408	1.58
Slew Rate -	V/μs	-	1.408	1.23
Input Range	V	-	-2.3/0.9	
Output Range	V	> +/- 2	-2.3/2.0	
Load Capacitance	pF	10	10	10
Load Resistance	kO	100	100	100
Supply Voltage	V	+/- 2.5	+/- 2.5	+/- 2.5
W.M1	μm	-	3.6	3.6
W.M3	μm	-	3	3
W.M5	μm	-	59.2	59.2
W.M6	μm	-	121.6	121.6
L.M3	μm	-	12.25	12.25
Ibias	μA	-	0.5	0.5
Cc	pF	-	0.355	0.355
V1	V	-	-1.400	-1.400
V3	V	-	1.100	1.100
V4	V	-	0	3.49E-3
First Pole	Hertz	-	-112.2	-98.7
Second Pole	Hertz	-	-2.02E6	-1.76E6
Gm.M1	μA/V	-	2.450	2.450
Gm.M6	μA/V	-	254.7	254.7
Go.M4	μA/V	-	4.750E-3	4.653E-3
Go.M6	μA/V	-	4.839E-1	4.620E-1
Ids.M1	μA	-	0.2500	0.2500
Ids.M6	μA	-	25.47	25.47

Table 1: Results of a design of the opamp of Fig.4 and comparison with SPICE simulations.

5. Conclusions

We have presented the ARIADNE system which is a flexible, open and extendable design system for the analysis and synthesis of analog integrated circuits. The approach is based on a clear separation between the declaration of the knowledge about the behavior of an analog circuit (described in declarative equation-based models) and the procedures (general computational techniques such as constraint propagation and optimization) that utilise the knowledge to synthesize a circuit for a particular application. The use of declarative models together with symbolic simulation also provides an easy interface for the designer to include new design knowledge and circuit schematics into the design system himself. The ARIADNE system can be used by system designers, experienced and unexperienced designers in both automatic and interactive mode.

The design flow and the different tools in the system have been described in detail and the key ideas have been illustrated for some practical examples. The cornerstones of the system are the symbolic simulator ISAAC for the modelling of new circuits and the design equation manipulator DONALD which uses constraint propagation to derive the computational path needed for synthesis. The other tools are the topology generator HECTOR, the optimization program OPTIMAN, the verification and redesign system and the layout generator. A prototype version of the ARIADNE system is presently being developed.

Acknowledgements

The authors are grateful to Piet Wambacq, Stéphane Donnay and Hans De Doncker for their contributions to the ARIADNE system, and to the Belgian National Fund of Scientific Research and Philips Research Laboratories (The Netherlands) for their support.

References

[1] G. Gielen and W. Sansen, *Symbolic analysis for automated design of analog integrated circuits*, Kluwer Academic Publishers, 1991.

[2] R. Harjani, R.A. Rutenbar, and L.R. Carley, "OASYS: a framework for analog circuit synthesis", *IEEE Transactions on Computer-Aided Design*, vol.8, no.12, December 1989, pp.1247-1266.

[3] H.Y. Koh, C.H. Séquin, and P.R. Gray, "OPASYN: a compiler for CMOS operational amplifiers", *IEEE Transactions on Computer-Aided Design*, vol.9, no.2, February 1990, pp.113-125.

[4] F.M. El-Turky and E.E. Perry, "BLADES: an artificial intelligence approach to analog circuit design", *IEEE Transactions on Computer-Aided Design*, vol.8, no.6, June 1989, pp.680-692.

[5]   A.H. Fung, B.W. Lee, and B.J. Sheu, "Self-reconstructing technique for expert system-based analog IC designs", *IEEE Transactions on Computer-Aided Design*, vol.36, no.2, 1989, pp.318-322.

[6]   H. Onodera, H. Kanbara, and K. Tamaru, "Operational-amplifier compilation with performance optimization", *IEEE J. Solid-State Circuits*, vol. SC-25, no.2, April 1990, pp.466-473.

[7]   M. Degrauwe et al., "IDAC: an interactive design tool for analog CMOS circuits", *IEEE J. Solid-State Circuits*, vol. SC-22, no.6, December 1987, pp.1106-1116.

[8]   K. Swings, W. Sansen, "Topology selection and circuit sizing within the analog circuit design system ARIADNE", *Proc. Summer Course on Systematic Analogue Design*, Leuven, Belgium, 1990.

[9]   J. Jongsma et al., "An open design tool for analog circuits", *Proc. ISCAS*, 1991, pp.2000-2003.

[10] G. Jusuf, P.R. Gray, and A.L. Sangiovanni-Vincentelli, "CADICS - Cyclic analog-to-digital converter synthesis", *Proc. ICCAD*, 1990, pp.286-289.

[11] K. Swings, S. Donnay, and W. Sansen, "HECTOR: a hierarchical topology-construction program for analog circuits based on a declarative approach to circuit modelling", *Proc. CICC*, 1991, pp.5.3.1-5.3.4.

[12] K. Swings, G. Gielen, and W. Sansen, "An intelligent analog IC design system based on manipulation of design equations", *Proc. CICC*, 1990, pp.8.6.1-8.6.4.

[13] K. Swings and W. Sansen, "DONALD: a workbench for interactive design space exploration and sizing of analog circuits", *Proc. EDAC*, 1991, pp.475-479.

[14] G. Gielen, H. Walscharts, and W. Sansen, "Analog circuit design optimization based on symbolic simulation and simulated annealing", *IEEE J. Solid-State Circuits*, vol. SC-25, no.3, June 1990, pp.707-713.

[15] G. Gielen, H. Walscharts, and W. Sansen, "ISAAC: a symbolic simulator for analog integrated circuits", *IEEE J. Solid-State Circuits*, vol. SC-24, no.6, December 1989, pp.1587-1597.

[16] K. Swings, W. Sansen, "Ariadne : A constraint-based approach to computer aided synthesis and modeling of analog ICs", to be published in the Special Issue on Analog CAD of the Journal of Analog Integrated Circuits and Signal Processing, 1992.

Biographies

Georges Gielen was born in Heist-op-den-Berg, Belgium on August 25, 1963. He received his masters degree in Electrical Engineering from the Katholieke Universiteit Leuven in 1986. In the same year he was appointed by the National Fund of Scientific Research (Belgium) as a Research Assistant at the ESAT laboratory of the K.U. Leuven. He obtained his Ph.D. degree on symbolic analysis as part of an analog design system in 1990. His thesis has been published as a book by Kluwer Academic Publishers in 1991. After this, he has taken up a postdoctoral position at U.C. Berkeley for one year, where he has been active in research in analog design automation. In October 1991, he was appointed by the National Fund of Scientific Research (Belgium) as a Senior Research Assistant at the ESAT laboratory of the K.U. Leuven. His Research is in design of analog integrated circuits and in analog design synthesis, optimisation and layout. He has participated in several conferences such as ISSCC, CICC, EDAC, ISCAS and ICCAD, and has several publications in the IEEE JSSC. He organized and participated in the K.U. Leuven Summer Course on Systematic Analogue Design, June 6-8, 1990, as well as the international course on "Knowledge-based techniques for automated analogue design", February 11-14, 1992.

Ir. Koen Swings was born in Deurne, Belgium, on October 2, 1964. He received his masters degree in Electrical Engineering in 1987 from the Katholieke Universiteit Leuven with a thesis on the architectural synthesis of a DSP algorithm for adaptive interpolation of audio signals. He then joined the ESAT-MICAS laboratory of the KUL where he is currently pursuing a Ph.D. degree with the development of CAD tools for analog integrated circuit design. His research interest include analog circuit synthesis, mathematical software and artificial intelligence.

Willy Sansen was born in Poperinge, Belgium, on May 16, 1943. He received the masters degree in Electrical Engineering from the Katholieke Universiteit Leuven in 1967 and the Ph.D. degree in Electronics from the University of California, Berkeley in 1972.

In 1968 he was employed as a research assistant at the K.U. Leuven. In 1971 he was employed as a teaching fellow at the U.C. Berkeley. In 1972 he was appointed by the National Fund of Scientific Research (Belgium) as a Research Associate at the ESAT laboratory of the K.U. Leuven, where he has been a full professor since 1981. During the period 1984-1990 he was the head of the Electrical Engineering Department.

In 1978 he spent the winter quarter at Stanford University as a visiting professor. In 1981 he was a visiting professor at the Federal Technical University Lausanne and in 1985 at the University of Pennsylvania.

Prof. Sansen is a member of several editorial committees of journals such as the IEEE Journal of Solid-State Circuits, Sensors and Actuators, High Speed Electronics, etc. He serves regularly on the programme committees of Conferences such as ISSCC, ESSCIRC, ASICTT, EUROSENSORS, TRANSDUCERS, EDAC, etc. He has been involved in numerous analog integrated circuit designs for telecom, medical applications and sensors, and also in design automation. He has been supervisor of 25 Ph.D. theses in the same field. He has authored and coauthored four books and 300 papers in international journals and conference proceedings.

Index